WITHDRAWN

```
QL 698.95 .B585 1993
Birds as monitors of
                       190486
```

DATE DUE

DEC 16 1994	
MAR 2 6 1996	
MAR 1 8 1999	

BRODART — Cat. No. 23-221

Birds as Monitors of Environmental Change

Birds as Monitors of Environmental Change

Edited by

R.W. Furness

Department of Zoology
Glasgow University
Glasgow, UK

and

J.J.D. Greenwood

British Trust for Ornithology
Thetford
Norfolk, UK

CHAPMAN & HALL

London · Glasgow · Weinheim · New York · Tokyo · Melbourne · Madras

Published by Chapman & Hall, 2-6 Boundary Row, London SE1 8HN, UK

Chapman & Hall, 2-6 Boundary Row, London SE1 8HN, UK

Blackie Academic & Professional, Wester Cleddens Road, Bishopbriggs, Glasgow G64 2NZ, UK

Chapman & Hall GmbH, Pappelallee 3, 69469 Weinheim, Germany

Chapman & Hall Inc., One Penn Plaza, 41st Floor, New York, NY10119, USA

Chapman & Hall Japan, Thomson Publishing Japan, Hirakawacho Nemoto Building, 6F, 1-7-11 Hirakawa-cho, Chiyoda-ku, Tokyo 102, Japan

Chapman & Hall Australia, Thomas Nelson Australia, 102 Dodds Street, South Melbourne, Victoria 3205, Australia

Chapman & Hall India, R. Seshadri, 32 Second Main Road, CIT East, Madras 600 035, India

First edition 1993
Reprinted 1994

© 1993 Chapman & Hall

Typeset in 10/12pt Palatino by Mews Photosetting, Beckenham, Kent
Printed in Great Britain by T.J. Press (Padstow) Ltd, Padstow, Cornwall

ISBN   0 412 40230 0

Apart from any fair dealing for the purposes of research or private study, or criticism or review, as permitted under the UK Copyright Designs and Patents Act, 1988, this publication may not be reproduced, stored, or transmitted, in any form or by any means, without the prior permission in writing of the publishers, or in the case of reprographic reproduction only in accordance with the terms of the licences issued by the Copyright Licensing Agency in the UK, or in accordance with the terms of licences issued by the appropriate Reproduction Rights Organization outside the UK. Enquiries concerning reproduction outside the terms stated here should be sent to the publishers at the London address printed on this page.

The publisher makes no representation, express or implied, with regard to the accuracy of the information contained in this book and cannot accept any legal responsibility or liability for any errors or omissions that may be made.

A Catalogue record for this book is available from the British Library

Library of Congress Cataloging-in-Publication Data available

∞ Printed on permanent acid-free text paper, manufactured in accordance with the proposed ANSI/NISO Z 39.48-199X and ANSI Z 39.48-1984

Contents

List of contributors		ix
1	**Can birds be used to monitor the environment?**	**1**
	R.W. Furness, J.J.D. Greenwood and P.J. Jarvis	
	1.1 Introduction	1
	1.2 Monitoring environmental change	5
	1.3 Advantages and disadvantages of birds as monitors	27
	1.4 Subjects considered in detail in this book	29
	1.5 Other uses of birds as monitors	29
	References	35
2	**Environmental changes**	**42**
	P.J. Jarvis	
	2.1 Introduction	42
	2.2 The nature and scale of environmental changes	43
	2.3 Climatic changes	46
	2.4 Changes in economic activities	54
	2.5 Habitat alteration and fragmentation	63
	2.6 Introduced species	68
	2.7 Pollution	70
	2.8 Conclusions	76
	References	77
3	**Birds as monitors of pollutants**	**86**
	R.W. Furness	
	3.1 Introduction	86
	3.2 Pesticides	89
	3.3 PCBs	100
	3.4 Heavy metals	102

	3.5 Plastic	120
	3.6 Nutrients	122
	3.7 Radionuclides	125
	3.8 Acidification	127
	3.9 Oil	128
	3.10 Air pollution	129
	3.11 National pollution monitoring schemes using birds	131
	References	131
4	**Birds as monitors of radionuclide contamination**	**144**
	I. Lehr Brisbin, Jr	
	4.1 Introduction	144
	4.2 Characteristics of environmentally important radioactive contaminants	146
	4.3 Evaluating birds as monitors of radionuclide contamination	152
	4.4 Selecting a monitor to describe environmental radionuclide contamination	169
	4.5 Conclusions – can birds be used as effective monitors of environmental contamination with radionuclides?	172
	Acknowledgements	174
	References	175
5	**Birds as indicators of changes in water quality**	**179**
	S.J. Ormerod and S.J. Tyler	
	5.1 Introduction	179
	5.2 Pollution and other sources of change in lakes and rivers	182
	5.3 Monitoring aims	185
	5.4 Birds as indicators of contaminants in freshwaters	187
	5.5 Birds as ecological indicators of water quality	198
	5.6 Current perspectives on water quality assessment in Britain	209
	5.7 Synthesis: achievements, problems and possibilities	210
	References	212
6	**Birds as indicators of change in marine prey stocks**	**217**
	W.A. Montevecchi	
	6.1 Introduction	217
	6.2 Avian roles in marine food webs	220
	6.3 Monitoring seabirds	225
	6.4 Population response to food supply	229
	6.5 Physiological response to nutrition	234
	6.6 Behavioural response to prey availability	239

	6.7 Seabirds as fisheries indicators	244
	6.8 Conclusions	250
	6.9 Research directions and recommendations	251
	Acknowledgements	253
	References	253
7	**Integrated population monitoring: detecting the effects of diverse changes**	**267**
	J.J.D. Greenwood, S.R. Baillie, H.P.Q. Crick,	
	J.H. Marchant and W.J. Peach	
	7.1 The scope of this chapter	267
	7.2 Representativeness of study areas	272
	7.3 Monitoring distribution	276
	7.4 Surveillance of abundance	276
	7.5 Measuring changes in abundance	287
	7.6 The surveillance of reproduction	294
	7.7 The surveillance of survival	301
	7.8 Immigration and emigration	314
	7.9 Integrated population monitoring	315
	References	328
	Index	343

Contributors

S.R. Baillie
British Trust for Ornithology,
The Nunnery,
Nunnery Place,
Thetford, Norfolk IP24 2PU, UK

I.L. Brisbin, Jr.
Savannah River Ecology Laboratory,
P.O. Drawer E.,
Aiken, South Carolina 29802, USA

H.P.Q. Crick
British Trust for Ornithology,
The Nunnery,
Nunnery Place,
Thetford, Norfolk IP24 2PU, UK

R.W. Furness
Applied Ornithology Unit,
Zoology Department,
University of Glasgow,
Glasgow G12 8QQ, UK

J.J.D. Greenwood
British Trust for Ornithology,
The Nunnery,
Nunnery Place,
Thetford, Norfolk IP24 2PU, UK

P.J. Jarvis
School of Geography,
University of Birmingham,
Edgbaston,
Birmingham B15 2TT, UK

J.H. Marchant
British Trust for Ornithology,
The Nunnery,
Nunnery Place,
Thetford, Norfolk IP24 2PU, UK

W.A. Montevecchi
Ocean Sciences Centre,
Memorial University of Newfoundland,
St John's,
Newfoundland, AC1 5S7, Canada

S. Ormerod
UWCC Catchment Research Group,
School of Pure and Applied Biology,
University of Wales College of Cardiff,
Cathays Park,
Cardiff CF1 3TL, UK

W.J. Peach
British Trust for Ornithology,
The Nunnery,
Nunnery Place,
Thetford, Norfolk IP24 2PU, UK

S.J. Tyler
The Royal Society for the Protection of Birds,
Brynn Aderyn,
The Bank,
Newtown, Powys SY16 2AB, UK

1
Can birds be used to monitor the environment?

R.W. Furness, J.J.D. Greenwood and P.J. Jarvis

1.1 INTRODUCTION

The need for environmental monitoring has never been greater. Burgeoning human populations, the greater demands they make on resources, and technological developments all result in massive and continuing increase in the impact of people on their environments. We have passed through a period in which environmental monitoring has had little academic respectability. This has been to some extent justified, for many monitoring programmes have had poorly-defined objectives and poorly-designed methodology, producing few or no useful results. Monitoring has also suffered because it relies largely on correlative analyses rather than on the manipulative experiments that have been more fashionable among ecologists.

Unfortunately, adequate experiments in ecology may demand unrealistic resources. As a result, many experiments have been conducted with inadequate controls or replication, rendering their results almost worthless. Furthermore, there are important and scientifically valid questions which, because they are essentially about unmanipulated nature, are scarcely amenable to an experimental approach – such as the question of what factors are mainly responsible for the variation in numbers of bird populations from year to year. Experiments can be used to address some such questions but only after observational and correlational studies have established the basic facts. Indeed, Pimm (1991) has argued that, because most ecological

Birds as Monitors of Environmental Change. Edited by R.W. Furness and J.J.D. Greenwood. Published in 1993 by Chapman & Hall, London. ISBN 0 412 40230 0.

studies are short-term, based on small areas, and involve few species, they are bound to fail to address the important questions in applied ecology, which are about processes that may take many years, over wide areas, and involve many species. Simple descriptive correlations may also be of greater practical value than the analysis of detailed models if these latter are too complex (e.g. McGowan, 1990). The appreciation that short-term and experimental approaches have their limits has led to increased acceptability of long-term studies (e.g. Dunnet, 1991a; Godfray and Hassell, 1992; Hassell, 1989; Likens, 1989; Wooller et al., 1992) and some increased interest in monitoring coupled with refinement of both the theoretical basis and the techniques of monitoring programmes (e.g. Goldsmith, 1991; Spellerberg, 1991).

But there is still a long way to go. Thus, a report of a workshop held 'to bring together a small group of young practising ecologists to discuss priorities in ecological research' (Hassell, 1989) raises monitoring under only one of the three broad topic areas discussed (pollution), though the other topics concerned management of populations and of communities and habitats, where we would argue that monitoring is essential.

In practical terms, though development projects in many parts of the world are often preceded by environmental-impact assessments, monitoring of the effects of the developments is rare. Perhaps even rarer is the proper monitoring of the potentially huge environmental consequences of changes in land management or land use at national, continental or even global levels brought about by new technologies or by new policies. The failure to establish proper monitoring would be rational only if we knew enough about the environment to make confident predictions about the effects of such changes. The fact is that we cannot. (Indeed, even for much better understood systems, the reliability of predictions often turns out to be strikingly less than it appears at the time they are made (Shlyakter and Kammen, 1992).)

Worse than the failure to implement new monitoring schemes is the discontinuation of existing ones, for the accumulation of data is an important element in the interpretation of monitoring results, and the data become more and more valuable as a monitoring scheme continues. Yet the number of stations producing long-term, precipitation data suitable for monitoring climatic change declined from 4700 in 1970 to 700 in 1990 (and from 2500 to 900 for temperature) (Atkinson, 1992) and 40% of long-term marine monitoring projects were terminated in the late 1980s (Duarté et al., 1992). Even the Continuous Plankton Recorder Survey, with 60 years of proven worth (Aebischer et al., 1990; MAFF Fisheries Laboratory, n.d.; Colebrook, 1986; Colebrook et al., 1991; Dickson et al., 1988), has come under threat. Organizations specifically set up to provide the administrative and financial stability to enable long-term, broadly-based research programmes to continue have been responsible for some of these short-sighted decisions (Jeffers, 1989).

Those responsible for bird-monitoring programmes in many countries would probably echo Koskimies's (1992) assertion that funding is poor and declining. He states: 'In principle, Finland runs the most versatile bird monitoring scheme in the world, but in practice lack of funding prevents us from reaching the proper level of activity in several projects.' Our aim in this book is to show the growing power and refinement of environmental monitoring schemes, especially those based on bird studies. We hope that this will aid both the establishment of new schemes and the strengthening of existing ones, as matters of priority in environmental science.

The idea that birds can be used to monitor the environment is not new. Folk lore and natural history observations, some dating back to ancient times, suggest that some aspects of bird behaviour can be used to predict changes in the weather. Aristotle, in his *Historia Animalium* of 342 BC, describes how the behaviour of cranes *Grus grus* shows the weather to come: 'they will fly to a great distance and high up in the air, to command an extensive view; if they see clouds and signs of bad weather they fly down again and remain still' (Thompson, 1910). According to Inwards (1869) the early arrival of cranes in autumn signals that there is going to be a particularly severe winter. The same prediction is made by early autumn arrival of the fieldfare *Turdus pilaris* in its winter range (Swann, 1913). In a similar vein, early northwards movements of greylag geese *Anser anser* in spring are considered to indicate that a period of settled spring weather is coming (Swann, 1913).

Nowadays we use satellites, radar and other advanced technologies to forecast the weather and little attention is given to the old folk beliefs. However, another centuries-old tradition is still in use to this day in some areas: the watching of seabirds to give clues to mariners as to their location with respect to land or shallow banks and to provide a means of locating schools of fish. The presence of seabirds was used as an important indication of approaching landfall or fishing banks in eastern Canada around 1700 (Montevecchi and Tuck, 1987). The *English Pilot*, published in 1706, contained crude drawings (Figure 1.1) of great auks *Alca impennis* which, it stated, were so restricted in distribution that they were reliable indicators of the Grand Banks for fishermen, where the auks could be found in numerous small flocks. To this day, some fishermen make use of seabirds as indicators of the location of shoals of fish. For example, fishermen seeking tuna around the Azores will go to areas where terns are fishing for small fish, for

Can birds be used to monitor the environment?

Figure 1.1 Drawings of great auks in *The English Pilot*. These birds were recommended as indicators of the location of the Grand Banks area as they were found there in numerous small flocks but were rarely seen at sea over deeper water. (After Montevecchi and Tuck, 1987.)

tuna feed on the same fish and are often found in the same places as the terns (Batty, 1989). Unfortunately the information that birds can provide is often neglected. Batty (1989) has argued, for example, that the Peruvian anchovy fishing industry would not have collapsed in 1972 if notice had been taken of the effects of overfishing on the local seabird populations.

Recently, the potential use of birds as monitors of environmental change has been reviewed in three publications. Two (Morrison, 1986; Temple and Wiens, 1989) concluded that bird population densities were often unsatisfactory as monitors of environmental change, largely because the causal links between the numbers of birds and environmental changes were not clear. We agree with the conclusion of Morrison that habitat change is better measured directly rather than through studies of bird communities. However, this is not a reason to disregard the many opportunities to use birds for other forms of environmental monitoring. Indeed, Morrison makes clear the value of birds in pollution monitoring and Temple and Wiens make clear their value for revealing long-term effects. Selected aspects of the successful use of birds as monitors were also discussed by authors in the second section of a book edited by Diamond and Filion (1987). We intend that this book will give a wider appraisal of birds as monitors.

In this chapter we will consider the nature of monitoring, the ways in which it can be done, and the extent to which studies of birds may help. We will begin to develop the argument that bird studies have a valuable role to play, as one of several independent ways of monitoring some aspects of environmental change and, in a few cases, the best or the only available monitor.

1.2 MONITORING ENVIRONMENTAL CHANGE

1.2.1 Definition of monitoring

Though some still equate monitoring with surveillance (repeated survey using standardized methods) most modern authorities would go further than this (Baillie, 1990, 1991; Goldsmith, 1991; Greenwood and Baillie, 1991; Hellawell, 1991; Hinds, 1984; Koskimies, 1992; Koskimies and Väisänen, 1991; Pienkowski, 1990, 1991; Spellerberg, 1991; Usher, 1991; Verner, 1986). Current opinion is that proper monitoring consists of surveillance plus:

1. Assessment of any changes against some standard or target.
2. The gathering of data in such a way that the reasons for the departures from the standard may be illuminated.
3. Clear understanding of the objectives of the programme.

The standard against which changes are assessed may be some norm or natural level. Such can sometimes be established through the study of control areas, free from the effects of the activity whose results are being monitored. Often, however, such effects may potentially be so widespread that such controls are impossible and the norm has then to be established from historical records: this is an important reason why long-term continuity is valuable in monitoring programmes. Alternatively, monitoring may be relative to the target of some management action, such as the recovery of some species following the control of a detrimental pollutant or the holding of a pollutant below some predetermined level. If the monitoring is to be useful, mechanisms must exist whereby appropriate action is taken if the standard or target is not attained. Such action may consist of measures to achieve better compliance with the standard or, if the reasons for the standard not being attained are not known, research to elucidate them.

Component 2 above is important because if the data have been gathered with interpretation in mind, such interpretation may indicate the appropriate action to take if the desired standard is not being attained. It is, of course, particularly important to be able to distinguish natural causes from those resulting from human activity. At the most basic level, we need sufficient information on natural fluctuations in the variable being measured to be able to discount changes within the normal range. False alarms will otherwise be frequent.

Definition of objectives, the third point above, might seem so obvious as not to be worth stating, were it not for the number of monitoring schemes established on the grounds that 'we need to know what's going on'. Such an approach is rarely crowned with success. In designing monitoring strategies for examining environmental change it is important to identify precise objectives and to select key indicators that can be measured in an appropriate manner and that can cut through 'environmental noise' to provide a clear indication of what, if any, environmental changes are taking place. Having determined what key indicators are appropriate and feasible, a monitoring prescription can be devised to accommodate method and frequency of sampling. The sensitivity and precision inevitably vary according to the environmental system being considered, the aims of the monitoring programme and the criteria being used (Spellerberg, 1991). Of course, if objectives are too rigidly defined, we may not only miss new problems but, even worse, merely set up schemes to monitor yesterday's problems. To avoid this, a scheme to use studies of birds to monitor the effects of acid deposition on aquatic communities, for example, could also, with little expansion, be used to monitor the aquatic ecosystem more widely, in a way that might give pointers at least to new and unexpected problems.

There are obviously strong links between censuses, long-term population studies, surveillance and monitoring programmes. Indeed, it is possible for the descriptive or analytical population studies to evolve into monitoring schemes as reasons for monitoring in relation to newly perceived hazards become evident. Such an evolution is particularly to be encouraged because it is essential that a wide knowledge of the biology of animals used in monitoring should be available in order to guide the design of a monitoring procedure appropriate for the species and problem, and also to permit sensible interpretation of any trends determined by the monitoring work. It can be seen that environmental monitoring schemes can use plants and animals, including birds, to identify trends in the nature and concentration of pollutants, and in the nature and consequences of a range of environmental changes. Similarly, in monitoring bird populations, one can examine the extent to which environmental changes contribute to changes in bird behaviour, reproduction, distribution and status. Chapter 2 considers in greater detail the nature of some these recent environmental changes.

Holdgate (1979) distinguishes between target-monitoring programmes which measure actual or potential targets (anything that may be liable to show change in distribution or performance) and factor-monitoring programmes which measure anything that might cause changes in the environment or in living targets. Target monitoring may involve physical systems (air, water and terrestrial systems) or biological ones (ecosystems, communities, species, populations, or individuals). Factor monitoring may involve physical attributes of the environment (such as temperature, radiation or water availability), chemical variables (such as oxidants, hydrocarbons or sulphur dioxide), or biological variables liable to affect other organisms (such as pathogens, herbivores, predators or decomposers).

Monitoring can be undertaken through remote sensing or by direct survey or sampling of features of the ground, water or air. The key indicators can be abiotic, biotic or a combination of the two. A review of techniques is given by Clarke (1986), though he concentrates on the factors covered by the Global Environmental Monitoring System (GEMS) of the United Nations Environment Programme (UNEP) which are largely physical, chemical and land-use variables. This book is concerned with the use of bird data to provide direct measurements (such as levels of contaminants in the tissues of top predators), or to provide proxy measurements (as by using seabird breeding success to indicate fish stocks), or to provide representative indicators of environmental impacts (as by using dippers *Cinclus cinclus* as representatives of stream communities).

Given the objectives of the monitoring programme and the methods to hand, one needs to consider whether the objectives can be achieved given the resources available or, more optimistically, what are the resources required to achieve the objectives? If the needs for adequate replication of measurements, for enough information to cope with natural variability, for the right information to allow the results to be adequately interpreted, and for resources to sustain the entire programme in the long-term are not assessed in advance, the scheme may be an expensive failure (Hinds, 1984). Formal analysis is desirable, aimed at optimizing the design of the scheme by minimizing the cost for a given power or by maximizing power for a given cost (Millard and Lettenmaier, 1986).

1.2.2 Remote sensing

It might be imagined that environmental changes can nowadays all be monitored by satellites, so that studies of birds may be of very limited use; in fact the use of satellites and other remote sensing technologies have considerable limitations.

While it is true that 'the ability of remote sensing to map and inventory terrestrial vegetation on a global scale is a key to the study of the biosphere' (Estes and Cosentino, 1989), remote sensing is by no means confined to the land or to plant communities; nor, indeed need the scale be global. Remote sensing, a term coined in the early 1960s, involves a range of techniques and scales, from satellites as earth observing systems to low-level aerial photography and radar. Valuable information can be provided on marine and freshwater ecosystems, on individual environmental components such as temperature, and on animal groups, including birds. Whether viewed in terms of vegetation types in themselves, as habitats for birds, or in other contexts, remote sensing data always need ground survey for refining and validating the initial survey information, interpretation and classification.

In 1972 the first Earth Resources Technology Satellite ERTS-1 (later renamed Landsat-1) was launched in the USA. It used multispectral scanners which digitally recorded electromagnetic energy in selected wavelength bands. Progress was made with Earth resources sensors systems, which collect thermal infrared and radar data. In 1980 the advanced multispectral scanner, Thematic Mapper (TM), was launched on Landsat-4 and in 1981 a satellite carrying the first Advanced Very High Resolution Radiometer was launched. SPOT, a broad-band multispectral High Resolution Visible system developed by France, was launched in 1986.

Satellite imagery has been used with considerable success in monitoring natural vegetation, for example in examining the rate and extent of deforestation in the tropics (Malingreau et al., 1989; Sader et al., 1990). It has been valuable in assessing productivity and health in agricultural land, forests and rangeland: the near-infrared spectral region (0.7–1.1 μm) is sensitive both to total plant biomass (through low absorbance) and healthy green vegetation in particular (characterized by high reflectance, 45–50%) (Clarke, 1986; Zonneveld, 1988).

An accurate assessment of changes in vegetation and of habitat fragmentation or disappearance through a series of satellite images obviously provides valuable proxy information in examining changes in the distribution and status of bird species. Veitch (1990), for example, used Landsat-5 TM data to predict the distribution and abundance of habitats likely to serve as breeding grounds for moorland birds in the Grampian Region, Scotland. Avery and Haines-Young (1990) had earlier shown that the distribution of dunlin *Calidris alpina* in northern Scotland was well correlated with components of satellite imagery and had used the correlation to estimate the number of dunlin lost through afforestation of moorland from satellite images made before the afforestation took place.

How precise can satellite-based surveys be? The key lies in the pixel (picture element), which represents the areal extent of the instantaneous field of view of a scanning remote sensor system and thus determines the resolution of the system being used. When two or more types of vegetation are present within a given pixel, the probability of correct classification (identification) using standard statistical pattern recognition procedures can be seriously decreased. Because Landsat spectral signature pixel resolutions are 80 m for the multispectral scanner (MSS) and 30 m for the more advanced Thematic Mapper, impressive though these resolutions are, it is often difficult to find pure pixels in heterogeneous habitat types such as agricultural areas and many kinds of natural vegetation (Estes and Cosentino, 1989). Thus Frank (1988) was unable to distinguish between forest vegetation types in the Colorado Rockies using Landsat TM, and in a Landsat MSS-based vegetation map of part of the Arctic National Wildlife Refuge overall map accuracy was estimated to be only 37% (Felix and Binney, 1989). Estes and Cosentino (1989), however, report that Landsat MSS can generally distinguish between hardwood, softwood, grassland and water with an accuracy of more than 70%; if general classification levels (such as forest and grassland) are sufficient, 90% accuracies are obtainable. Bird species diversity in 2 × 2 km squares in eastern England correlates well with habitat diversity measured from land cover types in photographs provided by Landsat where each pixel represents a 25 × 25 m square (S. Gates, D.W. Gibbons, R.M. Fuller and D.A. Hill, in prep.).

Satellite images can provide effective sequential pictures over the year, provided cloud cover does not mask the images too much. Aerial photography, by contrast, is generally too expensive to be repeated at short time intervals. For monitoring certain aspects of the environment, however, low-level survey can be a useful alternative to satellite images (Clarke, 1986; Budd, 1991) and analysis can be specifically directed towards bird life. Baines (1988), for instance, showed how colour infrared aerial photographs at a scale of 1:2500 could not only be used in establishing detail about wildlife habitat in an urban area but might also be analysed to predict breeding populations of woodland birds. Robertson *et al.* (1990) also used habitat analysis derived from infrared aerial photographs to infer population changes of farmland birds in South Sweden.

Bird counts and the establishment of population change using aircraft surveys are well established: for example studies by Takekawa *et al.* (1990) on guillemots *Uria aalge* in Central California, using oblique photographs of cliffs, and by Garnett (1987), flying at 30–100 m altitude, censusing waders along the north-eastern Australian coast. Sightability of birds is a key problem. Hone and Short (1988) compared densities of emus *Dromaius novaehollandiae* estimated from aerial survey over open shrubland with results from a drive count; aerial survey estimates were only 60% and 47% of true density in 1985 and 1986, respectively.

Aircraft can also be used to monitor animal movements. In this way Duncan and Gaston (1990) tracked radio-tagged ancient murrelets *Synthliboramphus antiquus* flying in family groups away from their colony on the Queen Charlotte Islands, British Columbia.

1.2.3 Geographic information systems

Information from remote sensing can be directed towards different

10   Can birds be used to monitor the environment?

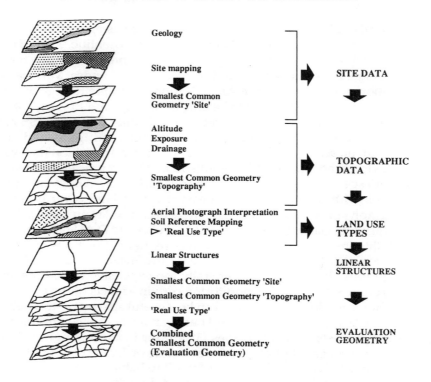

Figure 1.2 Levels of information suitable for inclusion in a typical GIS (Geographic Information System). In addition to these habitat layers the distributions and abundances of birds and other organisms can be superimposed. Any combination of variables can be used, depending on the data available and the aims of the study. The GIS can then be used to examine combinations of layers as required. (After Haslett, 1990.)

goals and interpreted in a number of ways. Various kinds of geographic information systems (GIS) have been more and more favoured as computerized systems have become increasingly available and refined. GISs store, analyse and display spatially referenced data. Raster-based systems record spatial information as points in a regular network of grid cells, while vector-based systems use patterns of points, lines and areas to represent the data (Haslett, 1990). Cartographic information is obtained from existing maps, aerial photographs or field surveys, and converted into digital form. Any number of such digital maps, each providing different kinds of information, environmental or otherwise, can be superimposed upon one another, permitting quantitative analyses using statistics that characterize patterns in the spatial distribution of the data (Figure 1.2).

Applications of GISs in ecological research include interpretation of environmental patterns in space and time (environmental change), and distributions of organisms, as well as links between the two. Thus, Miller *et al.* (1989) were able to characterize relations between environmental characteristics (essentially soil and vegetation data derived from published maps) and distribution patterns of rare bird species in eastern Tanzania using a GIS.

1.2.4 Non-biological monitoring

Environmental change can be monitored biologically or non-biologically, directly in the field or, using field samples, in the laboratory. Non-biological monitoring may involve periodic measurement of physical components of the environment such as temperature (for example, maximum, minimum, seasonality, and cumulative degree-days); precipitation (type, amount and seasonality); water availability (free water, groundwater, seasonality, etc.); and light (quality, daylength, seasonality, etc.). Chemical analyses include measurements of salinity, nutrients and heavy metal pollutants. Water, air and soil compartments may be examined. Proxy information about past conditions may be derived from analyses of glacier ice, sediments and peat cores (which may also involve biological material such as pollen, diatom frustules and plant and animal macrofossils).

It is important to distinguish between continuous recordings and periodic sampling (whether the time interval is at the scale of seconds or years). Data from continuous or high-frequency recordings are more useful (though the sheer quantity of data may need to be reduced by time-series smoothing, which reduces the amount of 'noise' and draws out salient and significant long-term features) but are often costly to obtain. Less frequent sampling may, however, be misleading. For example, intermittent river-pollution events may be missed. Worse, they may have a regular, pulsed frequency that matches sampling frequency, so that the level of pollution measured will be high if sampling takes place immediately after the pollution pulses, but will suggest little if any pollution if undertaken just before the pollution pulses.

The possibility of significant events falling between sampling occasions is a disadvantage of non-biological monitoring that can be overcome by using biological monitoring, since organisms will generally integrate the effects of environmental conditions over a period of time. Another disadvantage of non-biological monitoring is its specificity. In contrast, the reactions of living organisms may reveal quite unexpected and novel changes in the environment.

1.2.5 Biological monitoring

Biomonitors are usually selected to complement physical monitoring, but may in some instances provide the only available means of monitoring. For example, while it might be possible to carry out research trawls to monitor the abundance of some marine fish of no direct commercial importance, such work is expensive and so is generally not done. However, it might prove practical to use the diet composition of some seabird to provide crude information on the abundance of the fish. Such possibilities are discussed in Chapter 6.

For a biomonitor to be useful, it must respond in a sensitive way to changes in the variable for which it is a proxy measure. The response must be predictable and easy to measure, and should also provide a high signal-to-noise ratio, so the response is clearly distinguishable from background variations due to other factors. It is best if the causal mechanism behind the response is understood. The speed of response is also important. Biomonitoring breeding numbers of puffins *Fratercula arctica* as a measure of a fish stock, for example, would probably be of no practical use if the effect of a collapse in fish stock were to be breeding failure but not reduced adult survival, since a response in terms of breeding numbers of puffins would be delayed by the five to ten years of deferred maturity. It would be more appropriate in this case to monitor some aspect of puffin biology that responded strongly and immediately to a fall in fish stock; perhaps breeding success, meal size or feeding frequency would be a more appropriate variable to consider. In this context, Croxall *et al.* (1988) provide a careful, largely empirical, analysis of the relative merits of different species and aspects of numbers, breeding and feeding biology of marine mammals and seabirds in the southern oceans as possible monitors of krill *Euphausia superba* stocks for when commercial exploitation of krill develops. This is an excellent example of the thoughtful selection of biomonitors. A first step in this direction can be made by use of ecological theory to predict the best parameters for biomonitoring, as done by Furness and Ainley (1984) and Cairns (1987). However, it is clear that theoretical arguments cannot replace empirical assessment of the natural variations in parameters and the strength of signal that can be detected in any biomonitoring programme. Indeed, theoretical ecology is not a sufficient tool to permit accurate prediction of the functioning of complex ecosystems, so that relationships predicted by theory may not actually be found in the real world. For example, Cairns (1987) predicted that seabird chick growth would respond to reduced food supply before any effect on breeding success was evident, but the reverse was found to be the case in a study of great skuas *Catharacta skua* in Shetland, apparently because adults attempted to compensate for food shortage by spending more time foraging and so left the chicks unguarded and vulnerable to predation (Hamer et al., 1991).

Samiullah (1990) has distinguished between biological monitoring for environmental and ecological reasons. Environmental biological monitoring is concerned with determining changes in physiological, anatomical and numerical state due to environmental stress, for example by correlating levels of chemicals in environmental media with concentrations found in living tissue. In an ecological context, biological monitoring may estimate absolute numbers of individuals and establish species composition and variation in community structure. In practice, these categories overlap. The importance of biological monitoring lies in the use of organisms to establish the integrated, collective impact of environmental stresses upon plants and animals.

In the specific context of pollution studies, Samiullah (1990) used the term 'biological monitoring' either as 'the measurement, usually repeated, of concentrations of environmental contaminants in free-living organisms, or as the measurement, either singly or in combination, of such changing genetic, biochemical, physiological and ecological parameters as have been demonstrated by research to be influenced by measured contaminant concentrations.' In looking at the relationship between birds and environmental change, the use of birds as biological monitors must extend beyond the monitoring of pollution to examine other aspects of stress (for example, climatic change) or altered ecological functioning (for example, responses to introduced species, or to habitat fragmentation and alteration), but here we will consider biomonitoring of pollution first.

Measurement of pollution stress is often undertaken in terms of lethal concentrations or of functional processes such as scope for growth or biochemical oxygen demand. Lethal concentrations are standardized as the concentration of a pollutant or pollutants that will kill a certain proportion of a population within a particular time. The 96-hour LC_{50}, for example, is commonly used in aquatic acute-toxicity studies, where the concentration that kills 50% of the population (LC_{50}) after 96 hours is established. (The proportion need not be 50%, nor the time four days.) Such an index is conveniently established but often has little ecological or environmental significance. It gives no indication of the morbidity of the remaining half of the population; yet very often sublethal effects of contaminants, which may have deleterious consequences on food handling, digestion and reproduction, for instance, are more important at a population and community level than are the deaths of some individuals. Furthermore, the species used in the toxicological studies are often chosen for their convenience rather than their ecological relevance. Also, although an increasing number of studies go beyond single species and single pollutant experiments, they rarely produce results that are more than a pale reflection of the community dynamics, the fluctuating levels of contaminants, and the additive, synergistic or reducing effects of contaminant mixtures found in nature. This is not always appreciated by the technical managers and process chemists who may be responsible for environmental monitoring at individual sites, whose level of ecological understanding may be no better than that of the chemist who believed that 'Half a life-cycle is better than no life-cycle at all' (Trett, 1992).

Even though it measures the often sublethal consequences of chronic pollution, similar criticisms can be directed at the median effect concentration (EC_{50}), which is that concentration estimated to cause a particular response in 50% of a test population. A related test involves identification of threshold limits, which represent the combination of concentration and duration of exposure required to cause initial perceptible damage. A rapidly growing area of ecotoxicological research, and potentially of monitoring in the future, is the replacement of these gross assays by analysis of toxic effects at a more subtle biochemical-marker level (Peakall, 1992).

One link between physiological and population processes, with implications for ecotoxicological bioassay of environmental impact, is the notion of scope for growth (Calow and Sibly, 1990; Gray, 1980; Warren and Davis, 1967). This is the difference between energy intake and total metabolic losses, equivalent to production rate in the standard energy budget. It specifies the power available for various activities. Environmental factors that cause scope for growth to become zero or negative must ultimately be lethal.

Such indices of environmental stress are mainly used during the stage of hazard-assessment of potentially toxic discharges, rather than during monitoring, though they are also used for bioassay in the monitoring of pollutant levels. They are also useful in interpreting the apparent effects of discharges to the environment though there are inevitably problems in extrapolating such indices from the laboratory to the field and from short-term experiments to longer-term monitoring programmes. Biological target monitoring in the field can involve measurements of biochemical changes within the tissues of an organism; physiological or performance changes of individual organisms; changes in population size and structure; changes in distribution; and changes in the functions, performance or distributions of communities and ecosystems (Holdgate, 1979). The effects of an environmental stress may build up through these organizational levels, altering or overcoming homeostatic mechanisms that would otherwise, as indicated in section 2.2, sustain the system in the face of the stress.

In the laboratory, environmental conditions (including levels of the pollutant under investigation) can be controlled, and there is emphasis on using fast-growing, rapidly-reproducing organisms. In the field, other criteria become important. It is obvious that an indicator species must possess characteristics that truly reflect the environmental stress to which it is being subjected. Benthic invertebrates are often used because they are relatively sedentary and therefore representative of local conditions; they also have sufficiently long life-spans to provide an integrative record of water quality (Metcalfe, 1989). Concentrations of a pollutant must reflect environmental concentration, and one must be aware of how the plant or animal takes up the pollutant (in its various physical and chemical forms), how it stores it in different tissues, and how it accumulates, biomagnifies or detoxifies it. The effects of such physical stresses as temperature or moisture deficiency must be known. Often, longevity and persistence are valuable characteristics, so that external environmental factors can be ascertained from growth rings (for example in trees or molluscs), or so that the organism represents a measurable accumulation (intake minus subsequent loss) which reflects environmental concentrations and availability.

Where species have been shown, by their tolerance limits and, especially, environmental preferences, to represent particular environmental conditions, they may be used as indicator species: their presence, absence or abundance indicates the extent of environmental stress, in particular pollution (Newman, 1979). One can classify water quality, for example, using the saprobic approach, which reflects the dependence of an organism on decomposing organic substances as a food source and is based on the presence of indicator species. Alternatively, one can use measures of community diversity.

Most freshwater biotic indices have evolved from the Trent Biotic Index (TBI), which is based on the sensitivity of key groups to pollution and on the number of component groups in a sample. The TBI, however, ignores abundance, an omission rectified by the Chandler's Score System, which in turn was used as the basis of the Biological Monitoring Working Party's (BMWP) Score System, which simplified the level of taxonomic identification required. Karr (1991) advocated use of an index of biological integrity (IBI), defined as the ability to support and maintain 'a balanced, integrated, adaptive community of organisms having a species composition, diversity, and functional organization comparable to that of the natural habitat of the region'

(Karr and Dudley, 1981), noting that an IBI based on fish community characteristics has been applied widely in the USA. Such indices are attractive to those responsible for pollution control, who are often not biologists, because they produce a single figure that can be used as a criterion for judging whether acceptable limits have been breached. However,

> Careful interpretation is necessary by trained ecologists. In one case, a team of non-ecologists suspected a pollution incident when they received a series of poor BMWP Scores. We were asked to investigate. When the raw data were examined, it was apparent that a natural saltwater inundation event had occurred and marine species had entered the freshwater site. These are not allowed for in the scoring system which returned low values. (Trett, 1992).

The problems associated with the use of simple indices are not confined to the study of pollutants but apply generally in environmental monitoring. Use of a relatively simple index of environmental quality or condition is generally inappropriate in complex landscapes or ecosystems, a point emphasised by Kimmins (1990) who argues that only when we have a 'temporal fingerprint' of the ecosystem under study, so that normal patterns of disturbance and variation can be identified, may ecological indicators of environmental change be used with confidence. Noss (1990) suggests the use of a hierarchical approach to monitoring biodiversity, placing the primary attributes of biodiversity (community composition, structure and function) into a nested hierarchy that incorporates elements of each attribute at four levels of organization: regional landscape, community–ecosystem, population–species, and genetic. Indicators of each attribute at each level of organization can be identified for environmental monitoring purposes. By beginning with an inventory of landscape pattern, habitat structure, vegetation and species distribution, then overlaying data on stress levels using GIS techniques, one can direct research and monitoring programmes to high-risk ecosystems.

1.2.6 Bio-indicators and biomonitors

It is probably useful to distinguish between two types of biological detection. Animals can be used as crude indicators of an unexpected environmental problem: the equivalent of an alarm that is triggered by a critical level of a substance or an event outside normal limits of variation. For example, the damaging effects of DDT on wildlife as a consequence of its bioaccumulation and amplification through the food chain first came clearly to the attention of biologists because it was noticed that birds of prey were suffering unprecedented problems with reproduction, and in particular were producing thin-shelled eggs (as a consequence of physiological effects of DDE, the stable metabolite of DDT) (Ratcliffe, 1967). It was the observation that bird of prey numbers were declining and breeding success was impaired by breakage of thin-shelled eggs that led to the elucidation of the DDT pollution problem. In this case, birds of prey acted as bio-indicators.

Use of indicator species to detect environmental stress has also been undertaken with plants, particularly in relation to air pollution (Manning and Feder, 1980), and a variety of terrestrial animals (see, for example, Martin and Coughtrey, 1982). Similar attempts to use bird indicator species or diversity indices of environmental quality (Ohi *et al.*, 1974, 1981; Ohlendorf *et al.*, 1986) have been less frequently made than have direct measurements of contamination, such as in body fat, eggs and feathers, or such indirect measurements of environmental stress as changes in eggshell thickness, reproductive success and distribution. These measures of contamination can, of course be used as the basis for identifying indicator species. Gilbertson (1990), for example, examined changes in population numbers and reproductive success of bald eagles *Haliaeetus leucocephalus* and osprey *Pandion haliaetus* in relation to organochlorine compounds, arguing that these species are useful indicator organisms since they accumulate high levels of compounds such as DDT and polychlorinated biphenyls, and are sensitive to their chemical properties, so that if these species are present and maintaining their populations then the quality of the aquatic environment is probably satisfactory. Environmental changes that are more complex than the occurrence of pollutants may require a suite of indicator species. This was shown by Beintema (1983) in respect of the use of birds as indicators of the intensiveness of management of Dutch meadows, since each species occurred only within a certain range of management intensity. A good understanding of the ecology of the indicator species is necessary to establish and interpret such a system.

By establishing long-term surveillance programmes such as the Common Birds Census in Britain, line transect censusing of birds in Finland (Järvinen and Väisänen, 1979), standardized trapping of migrant passerines by bird observatories and equivalent research stations, as for example in Germany (Bairlein, 1981) and California (DeSante and Geupel, 1987), the extensive information obtained makes it possible to identify abnormal population changes. One example of this was the sudden decrease of the British population of the whitethroat *Sylvia communis* detected by the Common Bird Census and later attributed to the Sahel drought in Africa (Winstanley *et al.*, 1974). Another example is given by DeSante and Geupel (1987), who

18   *Can birds be used to monitor the environment?*

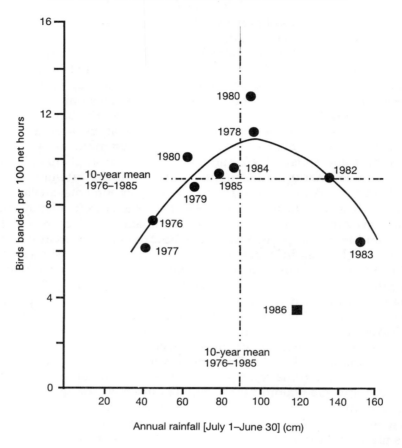

Figure 1.3 Numbers of young birds caught per 100 net-hours' trapping effort during the period May 10 to August 17 as a function of total rainfall in the previous year (July 1 to June 30) for the 11 years 1976 to 1986. Also shown is the smoothed curve for the 10 'normal' years 1976–85 (After DeSante and Geupel, 1987.)

analysed numbers of young and adult birds caught at Point Reyes Bird Observatory's Palomarin Field Station over an 11-year period. They found a relationship between the number of young birds banded per 100 net-hours of trapping and the amount of winter rainfall during the previous season, with bird productivity being high in years of average or slightly above average rainfall, and low in years with very low or very high rainfall. In 1986 the numbers of young birds deviated greatly from the normal range established over the previous ten years (Figure 1.3), whether or not rainfall was taken into account, most species falling three to eight standard errors below the predicted value based on 1976–85 data. This study provides an example of both the strength and weakness of such a bioindicator. Without this long-term surveillance study the abnormal nature of the 1986 bird breeding season might not have been noticed. However, the cause of the abnormal situation could not be determined from the data. It happens that there was an exceptionally warm and stormy period in February 1986 but there is no reason to think that this would be likely to have affected bird breeding later in 1986. Also, in early May 1986 the radioactive 'cloud' from the nuclear accident at Chernobyl passed over northern California and this coincided with rainfall but, again, there is no evidence to suggest that this caused the reproductive failure of the birds (indeed the concentrations of radionuclides involved are thought to have been far too low for any such effect to arise). The bird surveillance data indicate that something had an exceptional influence on bird breeding production in 1986 but the cause remains a matter of conjecture.

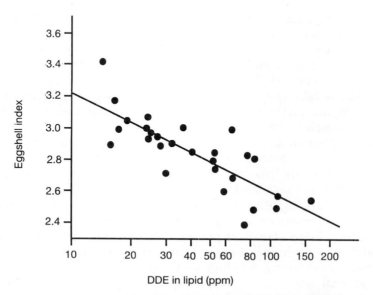

Figure 1.4 Gannet eggshell thickness index in relation to the DDE concentration in the lipid in the egg. (After Parslow and Jefferies, 1977.)

However, with appropriate design, biomonitoring can be taken much further than simply the use of sensitive species as early indicators of harmful environmental change. Biomonitor organisms can be used to quantify problems or changes and can then be monitored as a proxy for the damaging influence itself. For example, the chemical analysis of DDE levels in egg contents is somewhat expensive and requires sophisticated gas-liquid chromatographic equipment that is not readily available in many countries. The analysis is not possible in most 'Third-World' countries, where DDT is now most extensively used. However, it has been demonstrated that for any particular bird species the amount of eggshell thinning is closely and linearly correlated with the DDE level in the egg (Figure 1.4) and is not influenced by levels of other pollutants such as dieldrin or PCBs. Once such a relationship is established one could use the very easily measured shell thicknesses of regular samples of eggs to monitor the level of DDT pollution in a region.

There are many situations where biomonitoring might provide a less expensive or better means of assessing long-term trends in pollution or other forms of environmental change. Some of these are considered in section 1.2.8, and many are examined in depth in later chapters.

1.2.7 The sensitivity and reliability of monitoring programmes

The preceding sections have indicated that the sensitivity and reliability of a monitoring programme based on birds (or other organisms) depend on the relationship between the bird data and the environmental factor being monitored: in particular, how close it is and how well it is understood. These matters vary from case to case. Sensitivity and reliability can also be increased, as indicated above, by optimizing the design of the sampling programme (see, e.g., Millard and Lettenmaier, 1986).

The established statistical approach in biological investigations is that of hypothesis testing. In monitoring terms, this would typically mean that one took 'No change' or 'No departure from the standard' as the null hypothesis and only registered concern if any departure was statistically significant. Since failure to detect a departure may have serious consequences, it is important that the monitoring scheme should have high statistical power – that is, it should have a high probability of detecting changes that actually occur. Unfortunately, affordable monitoring schemes may have rather low power.

As is well known, one way to increase power is to accept less stringent levels of statistical significance than the conventional 5%. But this reduces the probability of Type II error (accepting a false null hypothesis) at the expense of increasing the probability of Type I error (rejecting a correct null hypothesis – i.e. raising the alarm unnecessarily, in the context of monitoring). There has been much discussion of the best course to adopt in these circumstances (Hinds, 1984; Nicholson and Fryer, 1992; Petermann and M'Gonigle, 1992; Schrader-Frechette and McCoy, 1992), but no consensus has yet emerged. Partly in reaction to the low power of many monitoring programmes and many studies of the effects of potentially damaging processes, some have argued the widespread adoption of the 'precautionary principle': an environmental change should only be allowed after it has been proved to be harmless. This is discussed by Gray (1990), by Gray *et al.* (1991), and by Johnston and Simmonds (1990). Such an approach would probably be widely unacceptable on cost grounds. Furthermore, firm proof of harmlessness might only be obtainable by actually instituting the environmental change and monitoring its effects. In any case, unexpected effects can never be ruled out until the change is actually instituted. Thus monitoring will always be necessary, so the problem of how to balance power and statistical significance remains. We can offer no solution to the problem here but urge those responsible for monitoring schemes to maximize their power, to estimate their power both before and after they are established, and to quote power as well as significance levels when publishing results. Failure to reject a null hypothesis of 'no effect' should never be taken as positive confirmation of that hypothesis, unless the test is sufficiently powerful.

1.2.8 Global, regional and national biomonitoring programmes

Many international agencies rely on global schemes of environmental surveillance to provide information to guide their member governments towards action against pollution and other environmental stresses. Global schemes (as with regional and national ones) build up a general picture from a synthesis of sample observations. The Scientific Committee on the Problems of the Environment (SCOPE), for example, was established by the International Council of Scientific Unions (ICSU), a non-governmental group of scientific organizations. SCOPE's mandate includes the requirement to 'assemble, review and assess the information available on man-made environmental changes and the effects of these changes on man; to assess and evaluate the methodologies of measurements of environmental parameters; [and] to provide an intelligence service on current research'. The means to achieve these aims were discussed by Munn (1973), who also recommended a suite of substances and 'environmental stress indicators' to be covered by UNEP's Global Environmental Monitoring System.

In 1988, as part of the International Geosphere Biosphere programme (IGBP), long-term environmental monitoring schemes were established at a network of sites on a worldwide basis. Not only are these intended to help identify patterns in the functioning of the biosphere, but the ground-level data collected in this way should interface with other programmes in global monitoring. Bruns et al. (1991) described such integrated ecosystem and pollutant monitoring at baseline sites in Alaska, Wyoming and Chile, where adoption of a systems approach includes evaluation of source–receptor relationships, air–water–soil–biota monitoring of contaminant pathways, and use of selected ecosystem parameters to detect human influences.

There are many national biomonitoring programmes. The US National Oceanic and Atmospheric Administration, for example, has been monitoring spatial and temporal trends of chemical contamination and biological responses through its National Status and Trends Program, with annual analyses for trace elements and organic contaminants made on sediments and on livers of benthic fish (beginning in 1984) and bivalves (from 1986) sampled from 175 sites around the coast (O'Connor and Ehler, 1991). This programme complements the Mussel Watch, a scheme established under contract with the US Environmental Protection Agency in 1976, with over a hundred stations, where oysters and mussels are used as sentinel organisms to monitor levels of heavy metals, petroleum hydrocarbons, synthetic organics and radionuclides (Goldberg, 1978). A similar scheme has been established in parts of Asia (Sivalingham, 1985). Long-term changes in US National Parks have been examined through watershed research and monitoring, started in 1980, to examine how the changing chemical environment, in particular atmospheric inputs of acidifying materials, is affecting their nature (Herrmann and Stottlemyer, 1991).

Climatic change has been recognized as an environmental problem which needs to be monitored at a global level, and the UNEP–WMO–ICSU World Climate Impact Programme was established within the framework of the IGBP. While again comprising information pulled together from national monitoring programmes, a number of supernational projects have also been initiated, for example the Landscape Ecological Impact of Climatic Change, whose brief is to examine the potential effects of future climatic change on terrestrial ecosystems and landscapes in Europe (Boer and de Groot, 1990). Similarly in Europe, the Regional Acidification Information and Simulation (RAINS) system has been mapping trans-boundary air pollution using GIS (Alcamo, 1987). Forest health, with debate centering on the relative effects of climatic change, air pollution, nutrient deficiencies, pests and pathogens (Bucher and Bucher-Wallin, 1989), is being monitored in a variety of ways, for example in North America (USDA Forest Service, 1989; Lefohn and Lucier, 1991), UK (Innes and Boswell, 1991), Germany (Krupa and Arndt, 1990) and Switzerland

(Sanasilva, 1989). The biological effects of atmospheric pollution, of course, are not confined to forest ecosystems, and monitoring programmes have also been established to examine their impact on, for instance, wetlands and agricultural ecosystems (Hutchinson and Meema, 1987).

These are just a few of the many biomonitoring programmes that have recently been established to examine the nature and consequences of pollution and general environmental change. Other monitoring programmes have been initiated in order to examine changes in the distribution and status of particular biotic groups, including birds (Sauer and Droege, 1990). The British Trust for Ornithology (BTO), for example, began the Common Birds Census (CBC) in 1962, and the Waterways Bird Survey (WBS) in 1974, in which annual changes in population size (Figure 1.5) are calculated as percentages from summed territory counts for all plots that are covered in the same way in consecutive years (Greenwood and Baillie, 1991). Population changes are published annually for 74 species (CBC) and 19 species (WBS). An annual Heronries Census and periodic surveys of individual species or groups of species are also undertaken. A Seabird Colony Register for Britain and Ireland was established in 1984 by the UK Seabird Group and the Nature Conservancy Council, incorporating

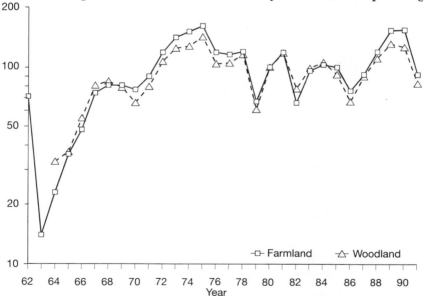

Figure 1.5 Trends in breeding population density of the wren *Troglodytes troglodytes* in farmland and in woodland Common Birds Census plots in the United Kingdom from 1962 to 1988. Numbers are given relative to those of an arbitrarily selected base year whose value is set at 100. Note the drop after the especially cold winter of 1962–63 and a rapid subsequent recovery of numbers. (Data from British Trust for Ornithology.)

24 *Can birds be used to monitor the environment?*

historical as well as new data. In addition to these studies of population size and change is the BTO Nest Record Scheme, initiated in 1939, which has been used since the 1988 breeding season to monitor annual changes in breeding performance of 68 species. The BTO Constant Effort Sites Scheme uses summer mist-netting at woodland, scrub and wetland sites to provide indices of adult and juvenile abundance and survival. The BTO Ringing Scheme, in operation since 1909, provides data on movements and migration, with sufficient data to make it possible also to monitor annual survival rates of at least 28 species.

Monitoring schemes should ideally provide information on the stages of the life cycle at which changes are taking place, and indicate probable causes of such change. To this end, the BTO's Integrated Population Monitoring Programme has been developed, which involves the collection of data on numbers, productivity and survival rates and their interpretation using population modelling techniques (Baillie, 1990, and Chapter 7 in this volume). Pienkowski (1991) also discusses the integration of measures of breeding performance and survival with those of

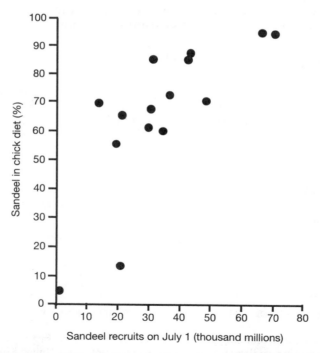

Figure 1.6 The percentage of sandeels *Ammodytes* in regurgitates produced by great skua *Catharacta skua* chicks on Foula, Shetland, between 1 and 15 July each year from 1975 to 1989 in relation to the number of sandeels recruited in Shetland waters in the previous year. (After Hamer *et al*, 1991.)

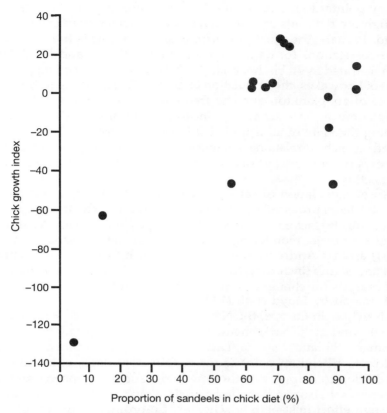

Figure 1.7 Great skua chick growth index in relation to the proportion of sandeel in the diet, for each year 1975 to 1989. The chick growth index is a comparison between actual weights of chicks and predicted weights in relation to their wing lengths based on growth studied in 1976. (After Hamer et al., 1991.)

population size, and the use of such information in providing warning of environmental change, provided one can distinguish between real environmental trends and natural environmental variation.

Many individual, detailed, long-term population studies of particular species have been undertaken in the British Isles, such as a 41-year-long study of fulmars *Fulmarus glacialis* on Eynhallow, Orkney Islands (Dunnet, 1991b), a 38-year-long study of the demography and breeding behaviour of kittiwakes *Rissa tridactyla* in NE England (Coulson and Thomas, 1985; Aebischer et al., 1990), and a 20-year study of skua and other seabird numbers, breeding and feeding ecology in Shetland (Hamer et al., 1991). Such long-term studies can provide vital evidence for environmental changes with major implications for conservation (Figures 1.6 and 1.7), and can also provide correlations that give a strong pointer to the causal mechanisms behind such changes, even though the data sets may not have been collected with such aims in mind. Perhaps the most powerful example of this is the time-series inter-correlations for data sets from the North Sea and NE Atlantic demonstrated by Aebischer *et al.* (1990) between the breeding performance of kittiwakes, the population of herring *Clupea harengus*, the abundance of zooplankton and the frequency of westerly winds. These correlations at once seem to point to an environmental factor determining the state of all trophic levels of this ecosystem. Hypotheses based on such correlational evidence need, of course, to be tested either by experiment or by deeper studies of the postulated causal connections.

An encapsulation of census material for breeding species in the British Isles is provided by Marchant *et al.* (1990), who include discussion of (likely) factors responsible for changes in numbers. A synthesis of Britain's major monitoring programmes is given by Stroud and Glue (1991) and by Andrews and Carter (1993). Information on distribution and status (including the global situation), population numbers and reasons for changes in number are given for seabirds in Britain and Ireland by Lloyd *et al.* (1991).

Elsewhere in Europe, the Netherlands, Sweden and Germany, like the UK, undertake bird censuses using territory mapping. In Sweden, Denmark, Finland, Estonia, Latvia, Czech Republic, Slovakia, France and part of Belgium point count methods are used, either alone or in addition to mapping. Other methods, such as transects and area searches, are also used, but less widely. France and Germany operate constant effort mist-netting schemes. Co-ordination of census work is achieved through the European Bird Census Council (Hustings, 1988, 1992) and of ringing data through the European Union for Bird Ringing. Analysis of ringing data can often indicate the effects of environmental change on birds; for example fluctuating ringing figures since records began at Ottenby, Sweden, in 1946 have been interpreted in terms of climatic change and man-made environmental changes (Hjort and Pettersson, 1990).

In North America, the *Journal of Field Ornithology* annually publishes results from the national Breeding Bird Census and the Audubon Christmas Bird Count (formerly published in *American Birds*). The Colonial Bird Register provides a database on colonially-nesting birds throughout North and Central America and the Caribbean, with data on colony location, species composition, habitat characteristics and numbers of breeding birds (Pettingill, 1985). Long-term population studies in the US have been less species-orientated than in Britain and the rest of Europe. Work has nevertheless been undertaken on potential uses of monitoring data, correlates of regional and national declines in bird populations, the relative value of longitudinal and cross-sectional studies in analysing slow ecological processes, and the significance of episodic events in shaping population patterns (O'Connor, 1991). Also, a number of long-running and geographically broad survey programmes have been undertaken on North American waterfowl, providing information used in making management decisions (Nichols, 1991).

A country-wide census scheme has recently started in Australia, using area-search methods.

1.3 ADVANTAGES AND DISADVANTAGES OF BIRDS AS MONITORS

A number of advantages and disadvantages of birds as biomonitors can be noted. They are easy to identify and their classification and systematics are well established, so there is little risk of monitoring being confounded by uncertainties regarding the identities of, or relationships between, the species being studied. Birds are particularly well known organisms, with much research carried out into their ecology and behaviour, and this background knowledge of biology enhances their usefulness as biomonitors, especially by reducing the risk of misinterpretations. Birds tend to be high in the food chain; thus they may be particularly suitable as monitors of any signal that accumulates through the chain, as with persistent organochlorines, but they may also be sensitive to many diverse factors affecting the food chain. Their long life-span means that birds integrate the effects of environmental stresses over time, providing the possibility of measuring, for example, pollution over a year or more, but may make it more difficult to establish any short-term perturbations. Similarly, the mobility of birds can allow monitoring over a broad spatial scale, the breadth depending on the species chosen, but migratory habits can render birds much less suitable as biomonitors because individuals may differ in their migrations to an uncertain extent and make it difficult to determine the spatial scale they represent. Furthermore, mobility can interact with temporal variation, with populations of different origins passing through the same place at different times of year and so potentially confusing a monitoring programme based on sampling at one site. Bird numbers tend to be regulated by density-dependent processes, and so their population sizes may be somewhat buffered against impacts of environmental changes. A similar problem of buffering may be evident at behavioural and physiological levels and may render birds less satisfactory as biomonitors than lower animals. For example, birds are able to regulate tissue concentrations of many metals, and body reserves of fat, to a much greater extent than invertebrates can, and so birds may less readily reflect environmental stresses. One might decide from this list that a sedentary invertebrate would be a much better biomonitor than a bird; in many situations this is true, but by no means in all, for birds have many other advantages.

One of the most compelling reasons for using birds as biomonitors is quite pragmatic. It is that they are relatively easy to study and that large amounts of data have already been gathered for bird populations. Colonial species, with easily detected, visible, and traditional breeding sites at which large amounts of data may be gathered, are particularly easy to study (though one should beware of basing studies on just one colony). There is a considerable public interest in birds and so amateur efforts can be directed into useful monitoring programmes. Such amateur work can often cover a wide geographical area, thus providing broadly-based monitoring. Those used to the high quality of work by amateur ornithologists need to be aware, however, that colleagues in the physical sciences (or even in other biological sciences) may need to be convinced of its reliability.

Because the wide interest in birds is of long standing, the most extensive and longest time series of data on populations of wildlife are from studies of birds; these provide surveillance data that can rather easily be adapted to fulfil a monitoring role. In order to establish such biomonitoring using other groups of animals it would usually be necessary to start from the present without any prospect of data covering the period, often of critical interest, before perceived changes had an influence on the biota. Indeed, it may be necessary with many other groups of animals for several years of study to be carried out to establish the necessary background biology before a surveillance or monitoring programme can even be set up. In addition, museums contain many collections of bird study skins and eggshells obtained over the last 150 years or so. These present unique opportunities for monitoring of a few particular pollutants. Although it would undoubtedly be possible, with hindsight, to design better pollutant biomonitoring sampling programmes, the fact is that bird collections exist and historical collections of sedentary invertebrates that could be used in the same way generally do not.

The widespread public interest in birds results in the further advantage that evidence of deleterious environmental changes is particularly powerful politically when it involves them. This has nothing to do with science but, if it is to achieve its ultimate goal of better management of the environment and resources, monitoring must not only be firmly based in the relevant socio-economic context but must have the administrative and political links that ensure that its findings will be translated into action.

1.4 SUBJECTS CONSIDERED IN DETAIL IN THIS BOOK

Chapter 2 of this book reviews environmental changes that are taking place. The remaining chapters deal with particular examples of the use of birds as monitors. Perhaps the best known and most firmly established use of birds as biomonitors is their use in pollutant monitoring. This is considered in a broad context in Chapter 3, and in the specific cases of monitoring radionuclides (Chapter 4) and of water quality (Chapter 5). Birds as monitors of fish stocks and of changes in the structure of the marine ecosystem are reviewed in Chapter 6. The importance of taking an integrated approach to the monitoring of bird populations (predominantly terrestrial birds) in order to be able to make some interpretations of the causal mechanisms behind changes is discussed in Chapter 7. Such population monitoring may be used to monitor the effects of a known environmental change or as a means of detecting unexpected changes. It is also important because birds themselves are a significant part of the human environment, though the extent to which their value can be quantified and whether they have intrinsic value in their own right raise philosophical questions that we cannot cover here (Diamond and Fillion, 1987; Naess, 1986; Norton, 1987).

1.5 OTHER USES OF BIRDS AS MONITORS

It is obviously impossible to deal in depth with all of the possible uses of birds as monitors, and the chapters outlined above cover only some of the areas of current research effort. Some topics not dealt with in depth are considered briefly below.

1.5.1 Habitat monitoring

The fact that bird populations and communities change as habitats are altered is not a matter of dispute. However, it seems that the changes in bird communities are often no easier to study than the directly visible changes to the habitat. To some extent bird breeding-densities may be buffered against small changes in habitat through density-dependent processes creating a floating population of non-breeders that could mask small detrimental effects on productivity. Furthermore, changes in bird communities could occur for reasons not related to habitat, and so could suggest habitat change when none had occurred. We can only agree with Morrison (1986) who succinctly concluded in a review of birds as monitors of habitat change:

> Although birds certainly respond to change, birds seldom respond in distinctly different ways to specific changes. The

problem with using birds as indicators, then, is separating the myriad of factors that can cause changes in bird populations. Birds can usually only be used to monitor the effects of a known perturbation if this monitoring is conducted in a controlled experimental design. This is a subtle point, but quite different from using birds to directly identify a specific change in the environment ... That birds, in most cases, appear to be only good indicators of general changes in the quality and quantity of habitats should not inhibit attempts to improve their usefulness. Improvements in the standardization and scope of surveys such as the Breeding Bird Survey and the Christmas Bird Count coupled with more thorough periodic examination of trends may be useful in detecting subtle changes in the quantity or quality of habitat that may be significant over large geographical areas.

This challenge is partly taken up by the Integrated Monitoring Programme of the BTO described in Chapter 7. It was also addressed in earlier work by Järvinen and Väisänen (1979), who examined the use of line transect data on bird abundance as a monitor of the effects of habitat change on birds and as a means of assessing the extent of habitat change in the Nordic countries. It was known that the area of old forests had decreased by about 15% in Finland between 1950 and 1975 and it was predicted that this would be reflected in reduced populations of forest birds. Such a trend was apparent in the line transect census data (Table 1.1). Furthermore, it was thought likely that the removal of trees would have taken out a particularly high proportion of the oldest trees and that this might have a particular

Table 1.1 Population trends of selected species favouring old forests in Finland based on line transect censuses showing the average density of the species as percentages of the density in Finland in 1973–77 (from Järvinen and Väisänen, 1979)

Species	Population index in		
	1936–49	1952–63	1973–77
Redstart *Phoenicurus phoenicurus*	185	220	100
Capercaillie *Tetrao urogallus*	220	155	100
Mistle thrush *Turdus viscivorus*	245	265	100
Black woodpecker *Dryocopus martius*	310	450	100
Siberian jay *Perisoreus infaustus*	310	225	100
Siberian tit *Parus cinctus*	900	210	100
Pine grosbeak *Pinicola enucleator*	185	80	100
Three-toed woodpecker *Picoides tridactylus*	485	65	100
Waxwing *Bombycilla garrulus*	55	135	100

impact on certain birds such as woodpeckers. This also seems to be indicated by the rates of decline of the different bird species (Table 1.1). The authors suggest that line transect monitoring of bird populations in the Nordic countries could be used to monitor the ecosystem effects of forest harvesting and in land-use planning, where bird data could be used in the evaluation of the conservation importance of different habitats. This can also be developed to allow predictions of the changes in bird numbers that might result from various scenarios of change in land use being considered in future management (Saunders et al., 1985). These approaches are frequently adopted in central Europe (see, e.g. Flade, in press; Schifferli, in press).

1.5.2 Weather forecasting

Bird behaviour is affected by weather (Elkins, 1983) and has traditionally been used to forecast it. For example, the Montagnais, North American native people who inhabited Newfoundland in the early seventeenth century, believed that one of the calls of the common loon (great northern diver) *Gavia immer* forecast a windy day (Harper, 1964). In Shetland, the red-throated diver *Gavia stellata* is known as the 'rain goose' because its call is supposed to predict rain. The latter is, perhaps, rather a safe prediction, akin to the Shetland joke 'If you can see Fair Isle its going to rain soon; if you can't see Fair Isle it's raining already'. The flight of rooks *Corvus frugilegus* is also supposed to be an indicator of the weather: 'when it hangs about home or flies up and down or especially low, rain or wind may be expected, when it "tumbles" or drops in its flight it is taken as a sure sign of rain; while when going home to roost if they fly high, the next day will be fair' (Swann, 1913). Swifts *Apus apus* and swallows *Hirundo rustica* flying high to catch insects are a sign of good and settled weather whereas low flying is supposed to indicate a change to poor weather (Swann, 1913). Virgil alluded to the signs of coming rain in his *Georgics* (30 BC): 'the swallow skims the river's wat'ry face'.

It is easy to see that many of these signals may be given by birds responding to the present conditions. Aerial insectivores can be expected to feed at the height above ground where insect prey is most abundant and it is well established that insects tend to fly less and at lower heights in damp and cool conditions. It is less clear if birds can, in fact, predict coming weather and many people dismiss these old folklores as being unscientific and unreliable.

Perhaps surprisingly, virtually no critical scientific study has been made of the abilities of bird behaviour to make the weather predictions claimed for them in folklores and ancient natural history. However, there is no doubt that at least some of them make sense on the basis of what we know about the ecology and behaviour of the species. For example, the natives of St Kilda used to use the behaviour of fulmars as a predictor of the arrival of stormy weather, which could endanger their lives if they put to sea for a long period in their small boats. They considered that the numbers of fulmars sitting on the cliffs at potential nest sites could be used to indicate the distance of depressions in the Atlantic that would bring strong westerly winds, especially during the winter. When large numbers of fulmars were sitting on their cliff sites and few sites were unattended the St Kildans considered that there was little risk of storms in the near future. This does make sense: the fulmar attends its nest site during most months of the year; it is present to an extent that varies from day to day through December to the start of laying in May; and, although the detailed patterns of colony attendance have not been related by scientists to day-to-day weather patterns, it is known that the numbers at colonies do show a tendency to be high during periods of settled winter weather (Fisher, 1952). Recent studies of the foraging of albatrosses and fulmars can go a long way to explaining this. Wandering albatrosses *Diomedea exulans* rely almost entirely on the wind to allow them to glide at high speed over the southern oceans in search of food and they can become becalmed by high pressure systems. Although they tend to avoid these, if they are caught by one they rest on the sea surface until winds return (Jouventin and Weimerskirch, 1990). In much the same way, fulmars can fly more easily in windy weather. When the wind is strong they can glide and soar using the energy from the wind to progress in any direction, whereas in calm conditions they are forced to flap their wings hard to make any progress at all (Figure 1.8). Their wings are designed for efficient high-speed gliding and have a small surface area for the weight of the bird. Consequently flapping flight must be costly in terms of energy expenditure and it would make excellent sense for fulmars to spend periods of settled windless weather attending their nest sites since they will be able to forage with a much greater net rate of energy gain when conditions are windy. Furthermore, we know that fulmars can travel long distances from their breeding sites in order to feed, especially during the winter when they are not constrained in the same way by the need to provision the chick or return to incubate. This provides a sensible biological basis for the observation that a high proportion of fulmars attending the colony can be taken to indicate that there is slack wind for a considerable distance.

Folklore also suggests that long-range weather forecasts can be based on bird behaviour. Thus rooks are said to build their nests low down in advance of poor summer weather, while overwintering thrushes are often said to arrive earlier and in greater numbers in

Other uses of birds as monitors 33

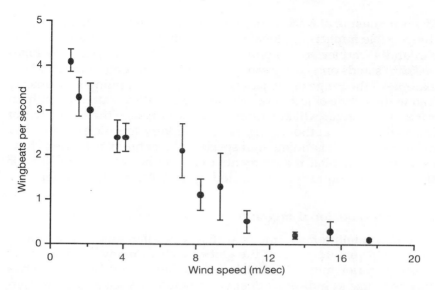

Figure 1.8 The rate of wing flapping by fulmars *Fulmarus glacialis* in level flight over the open sea under different wind speed conditions. (Data collected at Foula, Shetland by R.W. Furness in summers 1990 and 1991.)

Britain in years when the winter weather will be severe in north-west Europe. Such correlations seem unlikely, given that current models indicate that weather systems are chaotic to a degree that prevents reliable forecasting beyond about ten days, though some meteorologists believe that general forecasting is possible on a 90-day scale.

1.5.3 Monitoring climate change

Climatic change will certainly affect bird populations, though its precise effects are difficult to predict (Austin *et al.*, 1993; Hudson, 1990; Marquiss and Newton, 1990). It is, however, a good example of a change for which it is much more appropriate to measure physical variables than to study bird numbers or biology as a proxy measure. But the monitoring of certain birds might provide a measure of the impact of climate change on the ecosystem. This has been suggested by Botkin *et al.* (1991) who pointed out that Kirtland's warbler *Dendroica kirtlandii*, an endangered species which nests only in the lower peninsula of Michigan, may provide an important example of the possible impacts of global warming on endangered species. The Kirtland's warbler breeds only in a habitat of young jack pine *Pinus banksiana* stands on a coarse sandy soil. This occurs only at the southern extreme of the range of the jack pine, so global warming may quickly lead to the failure of jack pine regeneration on these extreme southern parts of its range, which would appear to presage the extinction of the warbler. The authors suggest that this very specific example may in fact be typical of endangered species in general, which often show very specific habitat requirements that may be sensitive to climatic change of the rapid type predicted by global warming models.

1.5.4 Transequatorial migrants as monitors of the tropics

Although satellite mapping now allows detailed analysis of habitat changes in remote parts of the globe, such as many tropical areas, transequatorial migrants that breed in Europe or North America may be of use as indicators of environmental change in their wintering areas or migration routes. Population census data such as the BTO Common Birds Census, or standardized trapping by bird research stations or observatories may be able to detect such changes to the extent that they affect numbers of transequatorial migrants but not other bird species. Such changes were detected as a consequence of the rapid fall in numbers of whitethroats in Britain due to drought conditions in the Sahel (Winstanley *et al.*, 1974). Subsequently, den Held (1981) showed that the numbers of at least some European populations of purple herons *Ardea purpurea*, night herons *Nycticorax nycticorax*, and squacco herons *Ardeola ralloides* (all of which are trans-Saharan migrants) are correlated with Sahelian rainfall. Overwinter survival of purple herons (Cavé, 1983), white storks *Ciconia ciconia* (Kanyamibwa *et al.*, 1990), sedge-warblers and whitethroats (Baillie and Peach, 1990; Peach *et al.*, 1991) show similar correlations. Survival of swallows *Hirundo rustica* is correlated with March rainfall in their wintering range in southern Africa (Møller, 1989).

All this suggests that these birds could provide a useful biomonitoring role of the ecological effects of Sahel droughts. Other evidence is, however, less clear-cut. Thus, although Marchant *et al.* (1990) could point to several British passerines whose populations seemed to be adversely affected by poor rainfall in the Sahel, more formal analyses of suites of Scandinavian and west European species failed to show that this was the preponderant pattern (Svensson, 1985; Marchant, 1992).

In the neotropics, although considerable concerns were raised earlier (e.g. Terborgh, 1980), it was the widespread decline of migrants on their breeding grounds in North America (e.g. Robbins *et al.*, 1989) that drew widespread attention to the problem of deforestation. It is likely that the species involved are suffering not only from neotropical deforestation but also from changes on the breeding grounds (for a recent review see Morton, 1992) but this example shows how evidence obtained about migrant birds in the developed nations can focus attention on environmental problems in faraway countries.

REFERENCES

Aebischer, N.J., Coulson, J.C. and Colebrook, J.M. (1990) Parallel long-term trends across four marine trophic levels and weather. *Nature*, **347**, 753–5.
Alcamo, J. (1987) Acidification in Europe: a simulation model for evaluating control strategies. *Ambio*, **16**, 232–45.
Andrews, J.H. and Carter, S.P. (eds) (1993) *Britain's Birds in 1990–91: The Conservation and Monitoring Review*, British Trust for Ornithology/Joint Nature Conservation Committee, Thetford.
Atkinson, D. (1992) Interactions between climate and terrestrial ecosystems. *Trends in Ecology and Evolution*, **7**, 363–5.
Austin, A., Clark, N.A., Greenwood, J.J.D. and Rehfisch, M.M. (1993) An analysis of the occurrence of rare birds in Britain in relation to weather. *BTO Research Report 99*, British Trust for Ornithology, Thetford.
Avery, M.I. and Haines-Young, R. (1990) Population estimates for the dunlin *Calidris alpina* derived from remotely sensed satellite imagery of the Flow Country of northern Scotland. *Nature*, **344**, 860–2.
Baillie, S.R. (1990) Integrated population monitoring of breeding birds in Britain and Ireland. *Ibis*, **132**, 151–66.
Baillie, S.R. (1991) Monitoring terrestrial breeding bird populations, in *Monitoring for Conservation and Ecology* (ed. F.B. Goldsmith), Chapman & Hall, London, pp. 112–32.
Baillie, S.R. and Peach, W.J. (1990) Population limitation in Palaearctic-African migrant passerines. *Ibis*, **134** (suppl. 1), 120–32.
Baines, L.M. (1988) The application of remote sensing to the management of urban wildlife habitats. Ph.D. thesis, University of Aston.
Bairlein, F. (1981) Ökosystemanalyse der Rastplätze von Zugvögeln. *Okol. Vögel*, **3**, 7–137.
Batty, L. (1989) Birds as monitors of marine environments. *Biologist*, **36**, 151–4.
Beintema, A. (1983) Meadow birds as indicators. *Environmental Monitoring and Assessment*, **3**, 391–8.
Boer, M.M. and de Groot, R.S. (eds) (1990) *Landscape-Ecological Impact of Climatic Change*, IOS Press, Amsterdam.
Botkin, D.B., Woodby, D.A. and Nisbet, R.A. (1991) Kirtland's warbler habitats: a possible early indicator of climatic warming. *Biol. Conserv.*, **56**, 63–78.
Bruns, D.A., Wiersma, G.B. and Rykiel, E.J. Jr. (1991) Ecosystem monitoring at global baseline sites. *Environ. Monit. Assess.*, **17**, 3–31.
Bucher, J.B. and Bucher-Wallin, I. (1989) *Air Pollution and Forest Decline*, Eidgenossische Anstalt für das forstliche Versuchswesen, Birmensdorf.
Budd, J.T.C. (1991) Remote sensing techniques for monitoring land-cover, in *Monitoring for Conservation and Ecology* (ed. F.B. Goldsmith), Chapman & Hall, London, pp. 33–59.

Cairns, D.K. (1987) Seabirds as indicators of marine food supplies. *Biol. Oceanogr.*, **5**, 261-71.

Calow, P. and Sibly, R.M. (1990) A physiological basis of population processes: ecotoxicological implications. *Funct. Ecol.*, **4**, 283-8.

Cavé, A.J. (1983) Purple Heron survival and drought in tropical west-Africa. *Ardea*, **71**, 217-24.

Clarke, R. (ed.) (1986) *The Handbook of Ecological Monitoring*, Clarendon Press, Oxford.

Colebrook, J.M. (1986) Environmental influences in long-term variability in marine plankton. *Hydrobiologia*, **142**, 309-25.

Colebrook, J.M., Warner, A.M., Proctor, C.A. *et al.* (1991) *60 years of the Continuous Plankton Recorder Survey: a celebration*, The Sir Alister Hardy Foundation for Ocean Science, Plymouth.

Coulson, J.C. and Thomas, C.S. (1985) Changes in the biology of the kittiwake *Rissa tridactyla*: a 31-year study of a breeding colony. *J. Anim. Ecol.*, **54**, 9-26.

Croxall, J.P., McCann, T.S., Prince, P.A. and Rothberry, P. (1988) Reproductive performance of seabirds and seals at South Georgia and Signy Island, South Orkney Islands, 1976-1987: implications for Southern Ocean monitoring studies, in *Antarctic Ocean and Resources Variability* (ed. D. Sahrhage), Springer-Verlag, Berlin, pp. 261-85.

den Held, J.J. (1981) Population changes in the Purple Heron in relation to drought in the wintering area. *Ardea*, **69**, 185-91.

DeSante, D.F. and Geupel, G.R. (1987) Landbird productivity in central coastal California: the relationship to annual rainfall, and a reproductive failure in 1986. *Condor*, **89**, 636-53.

Diamond, A.W. and Filion, F. (eds) (1987) *The Value of Birds*, ICBP Tech. Publ. 6. International Council for Bird Preservation, Cambridge.

Dickson, R.R., Kelly, P.M., Colebrook, J.M. *et al.* (1988) North winds and production in the eastern North Atlantic. *Journal of Plankton Research*, **10**, 151-69.

Duarté, C.M., Cebrián, J. and Marbà, N. (1992) Uncertainty of detecting sea change. *Nature*, **356**, 190.

Duncan, D.C. and Gaston, A.J. (1990) Movements of ancient murrelet broods away from a colony, in *Auks at sea*, (ed. S.G. Sealy), Cooper Ornithological Society, San Diego, pp. 109-13.

Dunnet, G. (ed.) (1991a) Long-term studies of birds, *Ibis*, **133** (suppl. 1), 1-137.

Dunnet, G.M. (1991b) Population studies of the fulmar on Eynhallow, Orkney Islands. *Ibis*, **133** (suppl. 1), 24-7.

Elkins, N. (1983) *Weather and Bird Behaviour*, T. & A.D. Poyser, Calton.

Estes, J.E. and Cosentino, M.J. (1989) Remote sensing of vegetation, in *Global Ecology: Towards a Science of the Biosphere* (eds M.B. Rambler, L. Margulis and R. Fester), Academic Press, London, pp. 74-111.

Felix, N.A. and Binney, D.L. (1989) Accuracy assessment of a Landsat-assisted vegetation map of the coastal plain of the Arctic National Wildlife Refuge. *Photogr. Eng. Rem. Sensing*, **55**, 475-8.

Fisher, J. (1952) *The Fulmar*, Collins, London.

Flade, M. (in press) The identification of indicator species for landscape planning in Germany, in *Bird Numbers 1992. Distribution, Monitoring and Ecological Aspects* (eds W. Hagemeijer and T. Verstrael), Proceedings of 12th International Conference of IBCC and EOAC, SOVON, Beek-Ubbergen.

Frank, T.D. (1988) Mapping dominant vegetation communities in the Colorado Rocky Mountain front range with LANDSAT thematic mapper and digital terrain data. *Photogrammetric Engineering and Remote Sensing*, **54**, 1727–34.

Furness, R.W. and Ainley, D.G. (1984) Threats to seabird populations presented by commercial fisheries, in *Status and Conservation of the World's Seabirds* (eds J.P. Croxall, P.G.H. Evans and R.W. Schreiber), International Council for Bird Preservation, Cambridge, pp. 701–8.

Garnett, S.T. (1987) Aerial surveys of waders (Aves; Charadriiformes) along the coast of north-eastern Australia. *Aust. Wildl. Res.*, **14**, 521–8.

Gilbertson, M. (1990) Freshwater avian and mammalian predators as indicators of aquatic environmental quality. *Environ. Monit. Assess.*, **15**, 219–23.

Godfray, H.C.J. and Hassell, M.P. (1992) Long time series reveal density dependence. *Nature*, **359**, 673.

Goldberg, E.D. (1978) The Mussel Watch. *Environ. Conserv.*, **5**, 101–25.

Goldsmith, F.B. (ed.) (1991) *Monitoring for Conservation and Ecology*, Chapman & Hall, London.

Gray, J.S. (1980) Why do ecological monitoring? *Marine Pollution Bulletin*, **11**, 62–5.

Gray, J.S. (1990) Statistics and the precautionary principle. *Marine Pollution Bulletin*, **21**, 174–6.

Gray, J.S., Calamari, D., Duce, R. et al. (1991) Scientifically based strategies for marine environmental protection and management. *Marine Pollution Bulletin*, **22**, 432–40.

Greenwood, J.J.D. and Baillie, S.R. (1991) Effects of density-dependence and weather on population changes of English passerines using a non-experimental paradigm. *Ibis*, **133** (suppl. 1), 121–33.

Hamer, K.C., Furness, R.W. and Caldow, R.W.G. (1991) The effects of changes in food availability on the breeding ecology of great skuas *Catharacta skua* in Shetland. *J. Zool., Lond.*, **223**, 175–88.

Harper, F. (1964) The friendly Montagnais and their neighbors in the Ungava Peninsula. *Univ. Kansas. Mus. Nat. Hist. Misc. Publ.*, **37**, 1–120.

Haslett, J.R. (1990) Geographic Information Systems: a new approach to habitat definition and the study of distributions. *Trends in Ecology and Evolution*, **5**, 214–18.

Hassell, M.P. (1989) Workshop on environmental priorities. *Biologist*, **36**, 275–80.

Hellawell, J.A. (1991) Development of a rationale for monitoring, in *Monitoring for Conservation and Ecology* (ed. F.B. Goldsmith), Chapman & Hall, London, pp. 1–14.

Herrmann, R. and Stottlemyer, R. (1991) Long-term monitoring for environmental change in U.S. National Parks: a watershed approach. *Environ. Monit. Assess.*, **17**, 51–65.

Hinds, W.T. (1984) Towards monitoring of long-term trends in terrestrial ecosystems. *Environmental Conservation*, **11**, 11–18.

Hjort, C. and Pettersson, J. (1990) Flyttfaglarnas antal och den foranderliga miljon (Changing numbers of migrant birds and the changing environment). *Calidris*, **19**, 13–23.

Holdgate, M.W. (1979) *A perspective of Environmental Pollution*, Cambridge University Press, Cambridge.

Hone, J. and Short, J. (1988) A note on the sightability of emus during an aerial survey. *Aust. Wildl. Rest.*, **15**, 647–9.

Hudson, R. (1990) Implications of a 'greenhouse climate' for British birds. *BTO Research Report* 47, British Trust for Ornithology, Tring.
Hustings, F. (1988) *European Monitoring Studies on Breeding Birds.* Samewerkende Organisaties Vogelonderzoek Nederland, Beek.
Hustings, F. (1992) European monitoring studies on breeding birds: an update. *Bird Census News*, **5**, 1–56.
Hutchinson, T.C. and Meema, K.M. (1987) *Effects of Atmospheric Pollutants on Forests, Wetland and Agricultural Ecosystems.* Springer-Verlag, New York.
Innes, J.L. and Boswell, R.C. (1991) *Monitoring of Forest Conditions in Great Britain 1990*, HMSO, London.
Inwards, R. (1869) *Weather Lore*, W. Tweedie, London.
Järvinen, O. and Väisänen, R.A. (1979) Changes in bird populations as criteria of environmental changes. *Holarctic Ecol.*, **2**, 75–80.
Jeffers, J.N.R. (1989) Environmental monitoring. *Biologist*, **36**, 171.
Johnston, P. and Simmonds, M. (1990) Precautionary principle. *Marine Pollution Bulletin*, **21**, 402.
Jouventin, P. and Weimerskirch, H. (1990) Satellite tracking of wandering albatrosses. *Nature*, **343**, 746–8.
Kanyamibwa, S., Schierer, A., Pradel, R. and LeBreton, J.D. (1990) Changes in adult annual survival rates in a western European population of the white stork *Ciconia ciconia*. *Ibis*, **132**, 27–35.
Karr, J.R. (1991) Biological integrity: a long-neglected aspect of water resource management. *Ecol. Appl.*, **1**, 66–84.
Karr, J.R. and Dudley, D.R. (1981) Ecological perspective on water quality goals. *Environ. Manage.*, **5**, 55–68.
Kimmins, J.R. (1990) Monitoring the condition of the Canadian forest environment: the relevance of the concept of 'ecological indicators'. *Environ. Monit. Assess.*, **15**, 231–40.
Koskimies, P. (1992) Monitoring bird populations in Finland. *Vogelwelt*, **113**, 161–72.
Koskimies, P. and Väisänen, R.A. (1991) *Monitoring Bird Populations. A Manual of Methods Applied in Finland*, Zoological Museum, Finnish Museum of Natural History, Helsinki.
Krupa, S.V. and Arndt, U. (1990) The Hohenheim Long-Term Experiment. *Environ. Pollut.*, **68**, 193–481.
Lefohn, A.S. and Lucier, A.A. (1991) Spatial and temporal variability of ozone exposure in forested areas of the United States and Canada: 1978–1988. *J. Air Waste Manage. Assoc.*, **41**, 694–701.
Likens, G.E. (ed.) (1989) *Long-term Studies in Ecology. Approaches and Alternatives*, Springer-Verlag, New York.
Lloyd, C., Tasker, M.L. and Partridge, K. (1991) *The Status of Seabirds in Britain and Ireland*, T. & A.D. Poyser, London.
MAFF Fisheries Laboratory (n.d.) *Monitoring the Health of our Oceans*, MAFF Fisheries Laboratory, Lowestoft.
Malingreau, J.P., Tucker, C.J. and Laporte, N. (1989) AVHRR for monitoring global tropical deforestation. *Int. J. Remote Sens.*, **10**, 855–67.
Manning, W.J. and Feder, W.A. (1980) *Biomonitoring Air Pollutants with Plants.* Applied Science Publishers, London.
Marchant, J.H. (1992) Recent trends in breeding populations of some common trans-Saharan migrant birds in northern Europe. *Ibis*, **134** (suppl. 1), 113–19.
Marchant, J.H., Hudson, R., Carter, S.P. and Whittington, P. (1990) *Population*

Trends in British Breeding Birds. British Trust for Ornithology/Nature Conservancy Council, Tring.

Marquiss, M. and Newton, I. (1990) Birds, in *The Greenhouse Effect and Terrestrial Ecosystems of the UK* (eds M.G.R. Cannell and M.D. Hooper) ITE Research Publication, HMSO, London, pp. 38–42.

Martin, M.H. and Coughtrey, P.J. (1982) *Biological Monitoring of Heavy Metal Pollution*. Applied Science Publishers, London.

McGowan, J.A. (1990) Climate and change in oceanic systems: the value of time-series data. *Trends in Ecology and Evolution*, **5**, 293–9.

Metcalfe, J.L. (1989) Biological water quality assessment of running waters based on macroinvertebrate communities: history and present status in Europe. *Environ. Pollut.*, **60**, 101–39.

Millard, S.P. and Lettenmaier, D.P. (1986) Optimal design of biological sampling programmes using the analysis of variance. *Estuarine, Coastal and Shelf Science*, **22**, 637–56.

Miller, R.I., Stuart, S.N. and Howell, K.M. (1989) A methodology for analysing rare species distribution patterns utilizing GIS technology: the rare birds of Tanzania. *Landsc. Ecol.*, **2**, 173–89.

Møller, A.P. (1989) Population dynamics of a declining Swallow *Hirundo rustica* population. *J. Anim. Ecol.*, **8**, 1051–63.

Montevecchi, W.A. and Tuck, L.M. (1987) *Newfoundland Birds: Exploitation, Study, Conservation*. Publ. Nuttall Orn. Club 21, Cambridge, Mass.

Morrison, M.L. (1986) Bird populations as indicators of environmental change. *Current Ornithology*, **3**, 429–51.

Morton, E.S. (1992) What do we know about the future of migrant landbirds?, in *Ecology and Conservation of Migrant Landbirds* (eds J.M. Hagan and D.W. Johnston), Smithsonian Institution Press, Washington.

Munn, R.E. (1973) *Global Environmental Monitoring Systems (GEMS): Action plan for Phase 1*. SCOPE Report NO 3, SCOPE, Paris.

Naess, A. (1986) Intrinsic value: Will the defenders of nature please rise?, in *Conservation Biology. The Science of Scarcity and Diversity* (ed. M.E. Soulé) Sinauer, Sunderland, Mass, pp. 504–15.

Newman, J.R. (1979) The effects of air pollution on wildlife and their use as biological indicators, in *Animals as Monitors of Environmental Pollutants*. Nat. Acad. Sci., Washington, D.C., pp. 223–32.

Nichols, J.D. (1991) Extensive monitoring programmes viewed as long-term population studies: the case of North American waterfowl. *Ibis*, **133** (suppl. 1), 89–98.

Nicholson, M.D. and Fryer, R.J. (1992), The statistical power of monitoring programmes. *Marine Pollution Bulletin*, **24**, 146–9.

Norton, B.G. (1987) *Why Preserve Natural Variety?* Princeton University Press, Princeton, New Jersey.

Noss, R.F. (1990) Indicators for monitoring biodiversity: a hierarchical approach. *Conserv. Biol.*, **4**, 355–64.

O'Connor, R.J. (1991) Long-term bird population studies in the United States. *Ibis*, **133** (suppl. 1), 36–48.

O'Connor, T.P. and Ehler, C.N. (1991) Results from the NDAA National Status and Trends Program on distribution and effects of chemical contamination in the coastal and estuarine United States. *Environ. Monit. Assess.*, **17**, 33–49.

Ohi, G., Seki, H., Akiyama, K. and Yagyo, H. (1974) The pigeon, a sensor of lead pollution. *Bull. Environ. Contam. Toxicol.*, **12**, 92–8.

Ohi, G., Seki, H., Minowa, K. et al. (1981) Lead pollution in Tokyo – the pigeon reflects its amelioration. *Environ. Res.*, **26**, 125–9.

Ohlendorf, H.M. Hoffman, D.J., Saiki, M.K. and Aldrich, T.W. (1986) Embryonic mortality and abnormalities of aquatic birds: apparent impacts of selenium from irrigation drainwater. *Sci. Total Environ*, **52**, 49–63.

Parslow, J.L.F. and Jefferies, D.J. (1977), Gannets and toxic chemicals. *Brit. Birds*, **70**, 366–72.

Peach, W.J., Baillie, S.R. and Underhill, L. (1991) Survival of British sedge warblers *Acrocephalus schoenobaenus* in relation to west African rainfall. *Ibis*, **133**, 300–5.

Peakall, D. (1992) *Animal Biomarkers as Pollution Indicators*, Chapman & Hall, London.

Petermann, R.M. and M'Gonigle, M. (1992) Statistical power analysis and the precautionary principle. *Marine Pollution Bulletin*, **24**, 231–4.

Pettingill Jr. O.S. (1985), *Ornithology in Laboratory and Field*, (5th edn), Academic Press, London.

Pienkowski, M.W. (1990) Foreword, in *Population Trends in British Breeding Birds* (eds J.H. Marchant, R. Hudson, S.P. Carter and P. Whittington) British Trust for Ornithology, Tring, pp. v–viii.

Pienkowski, M.W. (1991) Using long-term ornithological studies in setting targets for conservation in Britain. *Ibis*, **133** (suppl. 1), 62–75.

Pimm, S.L. (1991) *The Balance of Nature? Ecological Issues in the Conservation of Species and Communities*, University of Chicago Press, Chicago and London.

Ratcliffe, D.A. (1967) Decrease in eggshell weight in certain birds of prey. *Nature*, **215**, 208–10.

Robbins, C.S., Sauer, J.R., Greenberg, R.S. and Droege, S. (1989) Population declines in North American birds that migrate to the neotropics. *Proc. Natl. Acad. Sci.*, **86**, 7658–62.

Robertson, J.G.M., Eknert, B. and Ihse, M. (1990) Habitat analysis from infra-red-aerial photographs and the conservation of birds in Swedish agricultural landscapes. *Ambio*, **19**, 195–203.

Sader, A.A., Stone, T.A. and Joyce, A.T. (1990) Remote sensing of tropical forests: an overview of research and applications using non-photographic sensors. *Photogr. Eng. Rem. Sensing*, **56**, 1343–51.

Samiullah, Y. (1990) Biological monitoring: animals. *MARC Report 27*, GEMS Monitoring and Assessment Research Centre, Kings College, London.

Sanasilva (1989) *Das Programm Sanasilva 1988–1991*, Eidgenossische Forschungsantait fur Wald, Schnee und Landschaft, Birmensdorf.

Sauer, J.R. and Droege, S. (1990) Survey designs and statistical methods for the estimation of avian population trends. *Biological Report – US Fish and Wildlife Service* **90**, US Fish and Wildlife Service, Washington, D.C.

Saunders, D.A., Rowley, I. and Smith, G.T. (1985) The effects of clearing for agriculture on the distribution of cockatoos in the southwest of Western Australia, in *Birds of Eucalpyt Forests and Woodland: Ecology, Conservation, Management* (eds A. Keast, H.F. Recher, H. Ford and D. Saunders) Surrey Beatty, Chipping Norton (New South Wales), pp. 309–21.

Schifferli, L. (in press) Scientific research for application in landscape planning and habitat management, in *Bird Numbers 1992. Distribution, Monitoring and Ecological Aspects* (eds W. Hagemeijer and T. Verstrael). Proceedings of 12th International Conference of IBCC and EOAC, SOVON, Beek-Ubbergen.

Schrader-Frechette, K.S. and McCoy, E.D. (1992) Statistics, costs and rationality in ecological inference. *Trends in Evolution and Ecology*, **7**, 96–9.

Shlyakter, A.I. and Kammen, D.M. (1992) Sea-level rise or fall? *Nature*, **357**, 25.
Sivalingham, P.M. (1985) An overview of the Mussel-Watch Programme in Asia: in conjunction with the WESTPAC Programme. *Asian Environ.*, **6**, 39–50.
Spellerberg, I.F. (1991) *Monitoring Ecological Change*, Cambridge University Press, Cambridge.
Stroud, D. and Glue, D. (1991) *Britain's Birds in 1987–90: The Conservation and Monitoring Review.* British Trust for Ornithology/Nature Conservancy Council, Thetford.
Svensson, S.E. (1985) Effects of changes in tropical environments on the North European avifauna. *Ornis Fennica*, **62**, 56–63.
Swann, H.K. (1913) *A Dictionary of English and Folk-names of British Birds*, Witherby & Co., London.
Takekawa, J.E., Carter, H.R. and Harvey, T.E. (1990) Decline of the common murre in central California, 1980–1986, in *Auks at Sea* (ed. S.G. Sealy), Cooper Ornithological Society, San Diego, pp. 149–63.
Temple, S.A. and Wiens, J.A. (1989) Bird populations and environmental changes: can birds be bio-indicators? *American Birds*, **43**, 260–70.
Terborgh, J.W. (1980), The conservation status of neotropical migrants: present and future, in *Migrant Birds in the Neotropics: Ecology, Behavior, Distribution and Conservation* (eds A. Keast and E.S. Morton) Smithsonian Institution Press, Washington.
Thompson, D.W. (1910), *The Works of Aristotle. Vol. IV Historia Animalium*, Clarendon Press, Oxford, pp. 614–20.
Trett, M. (1992) Aquatic monitoring – a fishy tale? *In Practice*, **4**, 12–13.
USDA Forest Service (1989) *Patterns of Forest Condition and Air Pollution: Report of the Progress of the National Vegetation Survey*, USDA Forest Service, Athens, Georgia.
Usher, M.B. (1991) Scientific requirements of a monitoring programme, in *Monitoring for Conservation and Ecology* (ed. F.B. Goldsmith) Chapman & Hall, London, pp. 15–32.
Veitch, N. (1990) Thematic Mapper data as input to moorland breeding bird habitat prediction, in *Evaluation of Land Resources in Scotland. Proc. RSGS Symposium, Stirling* (eds. J.S. Bibby and M.F. Thomas) Macauley Land Use Research Institute, Aberdeen, pp. 89–92.
Verner, J.S. (1986) Future trends in management of non-game wildlife: a researchers viewpoint, in *Management of Nongame Wildlife in the Midwest: a Developing Art* (eds J.B. Hale, L.B. Best and R.L. Clawson), Proceedings of a Symposium held at the 47th Midwest Fish and Wildlife Conference, Grand Rapids, Michigan, December 1985, pp. 149–71.
Warren, C.E. and Davis, G.L. (1967), Laboratory studies on the feeding of fishes, in *The Biological Basis of Freshwater Fish Production* (ed. S.D. Gerking), Blackwell Scientific, Oxford, pp. 175–214.
Winstanley, D., Spencer, R. and Williamson, K. (1974) Where have all the Whitethroats gone? *Bird Study*, **21**, 1–14.
Wooler, R.D., Bradley, J.S. and Croxall, J.P. (1992) Long-term population studies of seabirds. *Trends in Ecology and Evolution*, **7**, 111–14.
Zonneveld, I.S. (1988) Reflection, absorption and transmission of light and infrared radiation through plant tissues, in *Vegetation Mapping* (eds A.W. Kuchler and I.S. Zonneveld), Kluwer, Dordrecht, pp. 223–31.

2
Environmental changes

P.J. Jarvis

2.1 INTRODUCTION

In one sense the term 'environment' represents all elements of the biosphere, but it is often of more use in the narrower sense of those parts or attributes of a habitat that are functionally significant during all or part of an organism's life. All living organisms have a range of tolerance for each environmental factor, though the limits and the optimum may vary between individuals and populations as well as between species. They may vary according to life stage or reproductive condition and because of interaction with other environmental factors. It is to environmental factors that populations and species respond, either through evolutionary change or through phenotypic plasticity. The response persists (and the population or species succeeds) where this has led to continued, indeed often enhanced, efficiency in utilizing the environment.

But organisms do not, of course, simply fine-tune their functional efficiency in a static set of environmental conditions. Environmental changes take place. Some are short-term, cyclical and predictable, such as seasonal variation in temperature or water availability. Others occur over years, centuries or millennia and produce longer-term responses and adaptations by plants and animals. Environmental changes may be natural, man-made, or a combination of the two. Changes may have direct impacts on a species or may have indirect consequences, for example by adversely affecting availability of food.

This chapter examines the nature of such longer-term environmental changes; considers how such changes might affect the behaviour,

Birds as Monitors of Environmental Change. Edited by R.W. Furness and J.J.D. Greenwood. Published in 1993 by Chapman & Hall, London. ISBN 0 412 40230 0.

The nature and scale of environmental changes

demography, ecology and distribution of birds; and identifies those elements of environmental change where trends might be quantitatively described by using birds as biological monitors. It focuses on the twentieth century, but looks ahead to possible or likely trends in the twenty-first. Emphasis is placed on climatic change, changes in habitat and land use, other changes in economic activities, species introductions and pollution.

2.2 THE NATURE AND SCALE OF ENVIRONMENTAL CHANGES

Environmental change may be examined in terms of time-scale, geographical scale, intensity of perturbation, and inherent potential of the population or community to respond or recover. All four elements are intricately linked.

The time-scale may relate to the length of time a particular change has been taking place. In the case of climatic change, whether natural or exacerbated by human activity, it is difficult to gauge when a particular effect or consequence began. With many forms of pollution, however, the time-scale is easier to ascertain; for example, the increase in lead in the environment following the introduction of tetraethyl lead as an anti-knock additive to petrol in 1921, and its recent decline with greater use of unleaded fuels. Time-scale may also refer to the recovery time after a discrete disturbance or perturbation, for instance a pollution incident. Walker and Walker (1991) encapsulated the pattern of disturbance by human activity on an Alaskan Arctic terrestrial ecosystem by plotting recovery time against spatial scale (Figure 2.1).

Perturbations may be low-level, chronic or acute. The first kind will probably have little effect on bird activity or distribution. Chronic perturbations, for example some forms of pollution or land use changes, may depress bird activity and numbers, but species might nevertheless persist. Acute perturbations will tend to exclude bird species. Intermittent acute perturbation may lead to local or regional extinction or displacement, but recovery may occur through subsequent immigration, increased reproductive effort or enhanced survival of remaining birds. There will be short, medium, and sometimes long-term effects. Synergistic, additive and compensatory interactions may be found.

How birds react to environmental changes, as individuals, populations or communities, depends on their resilience (the extent to which they can tolerate or accommodate stress or disturbance) or elasticity (the extent to which a population or community, having been perturbed, can return to its original state). These characteristics reflect the susceptibility of a population, species or community, which may

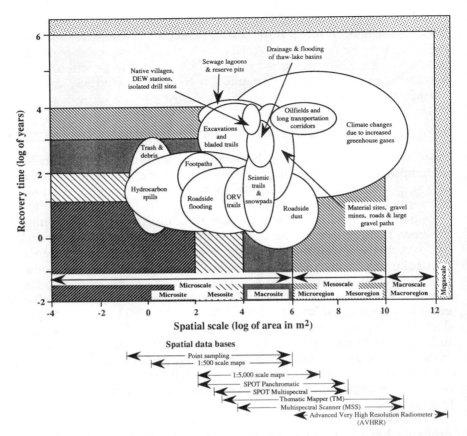

Figure 2.1 Spatial scales and estimated recovery time for human disturbances to the environment of the North Slope of Alaska. (Adapted from Walker and Walker, 1991).

be defined as the extent to which the birds are prone to perturbation together with the inherent ability to withstand or recover from that perturbation. All these ideas relate to stability, which usually refers to the tendency of a system (such as a bird population) to remain near an equilibrium point or to return to it after a disturbance. The more rapidly the system returns, with minimum fluctuation in properties, the greater its inherent stability (Holling, 1973; Orians, 1975).

The impact of environmental changes, whether as persistent perturbation, increasing stress or periodic acute disturbance, thus depends not only on the nature of the environmental factor(s) concerned but also on certain properties of the bird population, species or community. The impact on an individual will depend on its age (adult or juvenile), reproductive status and health (including such things as fat reserves and parasite load). Injury or impact may have symptoms reflecting biochemical and physiological changes, morphological changes (such as weight), behavioural changes, and possibly genetic changes. Growth and reproductive abilities may be enhanced or reduced.

Where climatic changes involve spatial shifts, as predicted for global warming (the greenhouse effect), one might imagine that the mobility of many bird species would allow them to make a simple distributional shift to a similar climatic environment. But if climatic changes are sufficiently rapid, the plant species upon which the birds depend for food, nesting or shelter may be unable to 'migrate' quickly enough to survive. Engelmann spruce *Picea engelmannii*, for example, can only migrate at about 20 km per century and for birch *Betula* species in Scotland 'the selection differentials needed for populations to change in response to climatic warming of 2°C over a period of 60 years show that, even in the most genetically variable populations, the rate of change in temperature (assuming an increase in mean annual temperature of 0.7°C every 20 years) is too rapid for native populations to evolve in the absence of gene flow from birch populations with earlier dates of budburst' (Billington and Pelham, 1991). Even where plants are able to migrate there will be a lag in their response to climatic changes (Davis, 1989), and even where plant species have been able to respond over a number of years to warmer temperatures, a single harsh period may be sufficient to lead to a reversion, as happened in pinewood in northern Sweden where defoliation and mortality due to frost and drought occurred in 1987 (Kullman, 1991).

If plant species cannot evolve or migrate quickly enough to withstand the impact of environmental changes, the same might be true of certain relatively immobile animal groups required by specialist avian predators. Such a scenario has been suggested, for example, for Australasia (Parsons, 1989). In such cases, birds would have to adapt to new habitat and food types in order to survive. Indirect as well as direct consequences of environmental changes must therefore be considered.

Climatic changes have greatest impact on birds already at the edge of their range, where changes in distribution, abundance, community structure and breeding (for instance in reproductive phenology, clutch size, clutch number per annum and reproductive success) may both be caused by and reflect environmental changes. It is necessary, therefore, to examine how environmental changes can affect bird life;

and, in turn, to invert the relationship to gauge the extent to which changes in bird life reflect environmental changes and can be used to monitor those changes.

Environmental changes take place naturally in response to geophysical, geomorphological, atmospheric, ecological and evolutionary processes, but there is no doubt that the human impact on the natural environment has become increasingly significant (Goudie, 1990). Global and regional changes in the biosphere that have occurred, been modified or been accentuated through human activities have been well-documented in recent years (for example, Botkin *et al.*, 1989; Turner *et al.*, 1990).

The nature, extent and some of the consequences of environmental changes are here examined in terms of climatic changes; changes in economic activities (in particular farming, forestry and fisheries); habitat alteration and fragmentation; introduced species; and pollution. There is inevitably overlap between many areas within these sections.

2.3 CLIMATIC CHANGES

2.3.1 Natural variation

Climatic variation is a natural phenomenon, occurring not only to a time-scale measured in millennia, as seen with the major glaciations, but also apparent over centuries and even decades. Two clear examples in the twentieth century are climatic warming, especially since the 1930s, in northern temperate regions, and the Sahelian droughts since the late 1960s. Thus, in Fennoscandia, greater warmth in the 1930s resulted in a northerly advance of many plant and animal species. Worldwide temperatures had, in fact, been increasing since the nineteenth century, and in Finland during the 1930s there was an increase of 1.5°C in the mean annual temperature. More critically, from around 1930 winters in Finland were warmer (by over 1°C) especially towards the Arctic, and remained so until 1966–1970 when there were markedly reduced winter temperatures, followed by higher temperatures again during the 1970s and 1980s. The coldest phase of winter shifted from December/January during the late 1960s to November/December during the 1980s. Higher mean temperatures during the growing season not only promoted faster plant growth but also extended the growing season for plants and the breeding season for animals. Increased productivity by vegetation was translated, through feeding relationships, into increased productivity in animals. Complementing such shifts in temperature were very variable trends in precipitation. Helsinki, for example, had increased precipitation up to the beginning of the century, followed by a general decrease (Heino, 1978).

Von Haartman (1978) noted that the spread and increase in numbers of many bird species coincided with such climatic amelioration but that the degree of spread often greatly exceeded the northward displacement of isotherms. He concluded that changes in land use probably represented a more important reason for the changes in bird distribution than did climatic change. A similar conclusion was reached by Järvinen and Väisänen (1979) in their examination of the inter-relationship between climatic changes, habitat changes and interspecific competition as explanations for the decline in Finland of Siberian tit *Parus cinctus*, for the expansion then decline of crested tit *P. cristatus*, for the expansion of chaffinch *Fringilla coelebs*, and for the general stability of brambling *F. montifringilla* populations. Climatic amelioration could account for some population changes but abandonment of forest grazing, increase of spruce *Picea abies* through plantation forestry, increased edge effect, extensive clear-cutting and other habitat changes were sufficient to explain all the long-term trends. Standard competition models were inadequate by themselves and environmental changes (affecting the carrying capacity) and population changes resulting from climatic changes had to be considered.

In the late 1960s and early 1970s, the drought in the Sahel zone of Africa received a lot of media attention. The reduction in rainfall was not as pronounced as earlier in the twentieth century but the extent of desertification, and the consequent socio-economic impact, were more severe. This was due to greater human population pressure and changes in land use during the middle years of the century, following a series of years in the 1940s and 1950s with unusually high and less variable rainfall. Although there have been serious periods of drought since, 1972 has been identified in retrospect as a critical year, a time when there were also serious droughts in India, the Soviet Union, Australia and elsewhere.

As with the Fennoscandian example, the Sahel drought has to be set in a wider geographical context. In the mid-1920s, winter and spring rains north of the Sahara and in the Middle East had decreased, to a minimum of 16% below average. Summer rainfall in the 1920s, though, was 16% above average. Since the 1920s, the pattern has been reversed: with some exceptions (such as in the 1950s), winter and spring precipitation were above average, while summer rainfall was below average. Thus North Africa has become wetter, the Sahel has become drier, and the desert lands of the Sahara have been migrating southwards at a rate that has averaged 9 km/y since 1960. The post-1960 decline in precipitation was indeed seen (even before the greenhouse effect was mooted) as part of a larger, natural climatic change which was predicted to continue until 2030 (Winstanley, 1973). The general southward shift in the Sahelian vegetation has had an impact on bird distributions. Although resident species and more general savanna species may be able to shift their activities southwards, the significance of the Sahel to Palaearctic migrants must not be forgotten. Whether these birds winter in the Sahelian savanna or use the zone as a staging post, they also need to adjust to shifts in climate and vegetation, including the greater extent of desert that must be flown over using (presumably) the same amount of stored energy. The status of a number of Palaearctic migrants has been adversely affected during the 1970s and 1980s. Peach *et al.* (1991) note that fluctuations in population size and annual adult survival rates of British sedge warblers *Acrocephalus schoenobaenus* since the late 1960s are strongly correlated with wet season rainfall in West Africa, with habitat availability in the West African winter quarters the main limiting factor.

With the increasing human population and advances in the nature and spread of technology, human impact has become an increasingly significant factor in variations in local, regional and global climate. Because human activity is only a contributory element it is rarely feasible to identify precisely the extent to which variations and changes in climate result from human activities. Goudie (1990) stressed the importance of inadvertent influence on atmospheric quality and on the albedo (relating to reflectance) of land masses.

2.3.2 The greenhouse effect

The pre-industrial level of CO_2 was around 260 ppm by volume (Wigley, 1983); the 1988 level averaged about 350 ppm. Models of climatic change use and provide different scenarios, but a level of around 600 ppm by the year 2065 is predicted by the model developed by the Goddard Institute for Space Studies (GISS) (Houghton *et al.*, 1990). Two key uncertainties that enter modelling predictions are the future production of fossil fuels (the burning of which represents the main cause of the release of CO_2 into the atmosphere) and the extent to which CO_2 will be absorbed by the oceans (Henderson-Sellars, 1990). Carbon dioxide levels in the atmosphere affect the global heat balance: CO_2 is nearly transparent to incoming solar radiation but absorbs outgoing infra-red radiation that would otherwise escape into space and result in a loss of heat from the lower atmosphere. Increased CO_2 levels cause an increase in this 'greenhouse effect' and thus an increase in surface temperatures. The situation is exacerbated by increases in some other gases, especially methane and chlorofluorocarbons (CFCs), which together probably contribute just as much as CO_2 to the greenhouse effect.

Current best estimates suggest that if greenhouse gas concentrations in the atmosphere continue to increase at present rates (Figure 2.2),

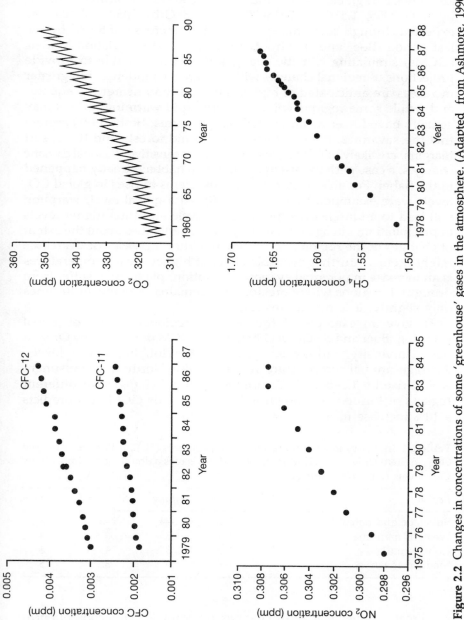

Figure 2.2 Changes in concentrations of some 'greenhouse' gases in the atmosphere. (Adapted from Ashmore, 1990.)

an increase in global mean annual surface air temperatures of 0.5°C by 1995–2005, 1.5°C by 2015–2050, and 3°C by 2050–2100 can be expected, though some models suggest an increase of 5.5°C by the last period (Boer and de Groot, 1990). General Circulation Models (GCMs) simulating climatic changes have been unable to provide projections of regional changes with any great confidence, but greater increases are anticipated at higher latitudes than nearer the equator and, while some regions will have very rapid warming, others may actually have lower mean annual temperatures. Stott (1991) argued that the savannas, lying between the humid forests and the desert margins, are likely to be the ecosystems most sensitive to global climatic changes, a case perhaps strengthened by what has already happened in the Sahel. Possible responses by ecosystems to rises in global CO_2 levels were examined by Bazzaz (1990). At a global level, warming will lead to an intensified hydrological cycle and thus higher levels of precipitation; but again there are great uncertainties about the extent of changes at the regional level, where either increases or decreases might occur. A further probable effect of higher mean temperatures is an increase in potential evapotranspiration: plants may suffer from drought in regions where precipitation remains the same or declines only slightly, if temperatures rise.

To give an example of the possible regional impact of global warming, Boer and de Groot (1990) used a GCM developed by Oregon State University, with a scenario based on a doubling of CO_2 levels. Average annual temperature for the entire Northern Hemisphere would rise by 1.5–5.5°C. Computed seasonal changes for different regions of Europe are shown in Table 2.1. This GCM also predicts a 10% increase in mean annual precipitation.

Table 2.1 Increases in seasonal average temperatures (°C) for Europe using the Oregon State University GCM scenario that envisages a doubling of atmospheric CO_2 levels (after Boer and de Groot, 1990)

Region	Winter	Spring	Summer	Autumn
Subarctic and boreal	5–8	3–4	2–3	3
Western maritime	3–4	3	3	3
Continental	4–5	3	4	4
Mediterranean	3–4	3	3	3

Climatic change will lead to changes in hydrology, with consequences particularly severe for semi-arid and sub-humid regimes, due to direct effects on stream flow and water quality and indirect effects on groundwater and soil moisture, and hence on vegetation (Falkenmark,

1990). Szaboles (1990) also predicted changes in the properties of soils, with direct effects on soil temperature, moisture content and biota, and indirect effects through changes in groundwater depth and surface vegetation. As a result of increasing mean annual temperature in boreal Europe, soil productivity is expected to increase. Increased salinization of many soils in Europe, however, is also predicted, with implications for (semi-)natural vegetation and agriculture. From examination of changes in soil acidity, redox potential of soils and sediments, and salinization, Stigliani (1990) argued that acidification will probably increase, and toxic chemicals such as heavy metals are likely to be disbound and released from environmental sinks. Likely changes in the nature and properties of tropical soils are noted by Sombroek (1990).

GCM scenarios also generally predict rises in sea level, with consequent changes in tidal regimes (especially increasing tidal amplitude, with mean high water levels rising more than low tide levels), in current patterns (with changes in sediment transport and thus coastal erosion and accretion), and in wave action (Boorman, 1990). Coastal vegetation and associated animal life will in turn be affected (Reed, 1990). In the ocean itself, climatic change and the ocean carbon cycle are closely linked, and global warming may lead to reduced CO_2 uptake, thereby exacerbating the problems of the increase in greenhouse gases (Williamson and Holligan, 1990). There are problems of establishing baselines against which to measure change but McGowan (1990) was hopeful that time-series data from the Atlantic and Pacific can be used in this way. He predicted that the marine biotic response will also be measurable, with large zooplankton serving as an indicator for the state of the rest of the system.

Climatic warming will also affect lakes and streams. Schindler *et al.* (1990) reported that already during the previous twenty years in the Experimental Lakes Area of north-west Ontario, lake temperatures had risen by 2°C and the length of the ice-free season had increased by three weeks. Rates of water renewal had decreased and concentrations of most chemicals increased. Thermoclines had deepened and cold stenothermal species such as lake trout *Salvelinus namaycush* and opossum shrimp *Mysis relicta* had decreased. Such trends will continue with global warming.

Regarding likely species responses to climatic change, newly-developed theories have been criticised by Hengeveld (1990) as being perhaps too biologically mechanistic. The palaeoecological record, indeed, clearly shows that Late Quaternary global warming caused individual species distributions to change along environmental gradients in different directions, at different rates, and over different periods, creating new community structures and dynamics

(Graham and Grimm, 1990). If the rate of warming caused by the greenhouse effect is greater than hitherto (as is expected to be the case), individual species' responses are likely to be even less predictable. Shugart (1990) and Woodward (1990) are more optimistic about incorporating known physiological responses of plants to raised CO_2 levels into ecosystem models: the effect of elevated CO_2 on transpiration and gas exchange, for instance, should increase the sensitivity of plant community structure (especially of forests) to climatic change (Peters, 1990). Even so, small differences in initial conditions or driving variables of a model can lead to large differences in the responses of a community, and Cohen and Pastor (1991) feel that there may be a fundamental limit to our ability to predict ecosystem responses to climatic change.

Many species may thus shift in abundance or distribution as a response to global warming. Others, however, might become extinct, or change genetically; but while sufficient heterozygosity may exist in some species, other taxa have little genetic variation for characters that allow a particular abundance or distribution. Indeed, climatic change can affect non-selective components of microevolution such as genetic variance and covariance, and the magnitudes of gene flow, genetic drift and susceptibility to mutation (Holt, 1990). The evolution of resistance to environmental stress, such as may be caused by rapid climatic shifts, might involve genetic changes associated with a reduction in metabolic energy expenditure. Especially where a species is at its physiological limits, stressful conditions may require such an increase in energy expenditure that survival is unlikely. Metabolic limitations would prevent spread into, or even within, climatically more extreme habitats, and species extinctions are thus probable (Parsons, 1990).

Birds are no exception to such scenarios, particularly where they have very specific food or habitat requirements or are at the edge of their range (in cases where climatic factors provide the distributional constraints). There will be many circumstances where they could act as monitors of climatic change, directly or otherwise. For example, Botkin *et al.* (1991) suggest that Kirtland's warbler *Dendroica kirtlandii* could be an early indicator species of the greenhouse effect: jack pine *Pinus banksiana* forests in central Michigan, managed as the primary nesting habitat for this endangered species, will respond to warming by growing at a significantly slower rate than recently, and may therefore become unsuitable for the warbler within 30 to 60 years. Taylor and Wilson (1990) noted that the changes in numbers of Adelie penguins *Pygoscelis adeliae* in the Ross Sea over the past ten years were correlated with physical evidence of climatic variation, and suggest that the population dynamics of this species might represent a sensitive indicator of environmental change in Antarctica. Speculating on the impact of climatic warming on marine birds in the north-west Atlantic, Brown (1991) predicted northerly shifts in species ranges off Nova Scotia and Newfoundland, caused as much by alterations in the ranges of prey species as by the direct effects of warming on the birds themselves.

Changes in land use are shown in the next section to have been a key feature of global environmental change. Shifts in climatic regimes will affect economic activities such as agriculture (Parry, 1990), which in turn will affect associated plant and animal life. Agricultural pests may expand in geographical range; have greater per generation rates of population growth and greater numbers of generations each year; and expand their range of host plants. The most inherently vulnerable regions, economically and ecologically, may well be regions of low and middle latitudes which at the moment have limited biological productivity because of poor rainfall, a problem which global warming may intensify (Parry et al., 1990).

2.3.3 Aerosols and the ozone layer

Two further aspects of climatic change need mentioning. One consequence of industrialization and increased burning of plant material, which may be of equal significance to CO_2 emission, is the increase in amount of dust or smoke particles entering the lower atmosphere. Such an increase in atmospheric aerosols might influence temperatures through scattering and absorption of solar radiation, but there is debate on the likely consequences of such an effect.

The other aspect of concern lies in the depletion of the ozone layer, the layer of relatively high concentration of ozone at a height of 16–18 km towards the poles and of around 25 km in the tropics. The ozone layer absorbs incoming solar ultraviolet radiation, affecting convective processes and atmospheric circulation, and thus strongly influencing global climate. Human activities are causing ozone depletion in the stratosphere, a consequence mainly of increased emissions of CFCs (from refrigerant systems, aerosols and foam food containers), but also from nitrogen oxides (largely from nitrogenous chemical fertilizers) and a variety of combustion products (especially from jet aircraft).

In 1984, satellite monitoring and chemical analyses identified a 50% decrease in the total column of ozone over Antarctica in the austral spring. This 'hole' in the ozone layer was associated with the Antarctic because the very cold polar winter plays a role in releasing chlorine atoms, which act as catalysts in destroying ozone; because the long summer day lengths promote photochemical processes; and because there is a well-defined circulation vortex which tends to contain the chemical reactions (Titus and Seidel, 1986). Ozone depletion, however, has since been observed over the rest of the world, and although attempts have been made to reduce output of CFCs the problem has certainly not been solved. Thus, in July 1991 it was discovered that ozone loss over Europe and North America had been occurring twice as rapidly as had previously been thought: the amount of ozone overhead in February and March 1990, for instance, had fallen by 8% between latitudes 30°N and 50°N (UK Stratospheric Ozone Research Group, 1991). Ozone depletion is expected to reach maximum severity (about twice that already experienced) around the year 2000; conditions will not return to those of the early 1970s until the mid-twenty-first century.

The ozone layer, by controlling the amount of ultraviolet radiation reaching the Earth's surface, affects the rate of photosynthesis. Oceanic phytoplankton are particularly vulnerable to increases in UV-B radiation, and any reduction in planktonic primary production would have serious consequences along the entire food chain, including fish and fish-eating birds. Increased susceptibility to carcinogenic reactions following increased UV amounts will also affect unprotected mammalian skin.

2.4 CHANGES IN ECONOMIC ACTIVITIES

2.4.1 Farming

One consequence of the Green Revolution of the 1960s has been the need for heavy applications of chemical fertilizers and pesticides, and many areas now face problems of chemical-resistant pests, altered soil structure, decreased soil fertility and contamination of water supplies. Land degradation, including greater susceptibility to erosion, is a chronic problem, particularly, but not only, in arid and semi-arid regions (WRI/IIED, 1988). In India, for example, an average of 30 t/ha of soil is redistributed each year, much of it being washed into rivers. But soil erosion has also been significant in the developed world: Iowa has lost half of its topsoil over the past two centuries and Kansas has lost 30% of its soil nitrogen and over a third of its soil organic matter since 1960 (Anon, 1988a). In intensively-farmed parts of central and eastern England, fields commonly lose 20 t/ha/y of soil and exceptional annual losses of 50 t/ha (equivalent to 3 mm of topsoil) have been noted (Arden-Clarke and Hodges, 1987). The ploughing up of many permanent pastures and the use of inorganic fertilizers tend to reduce the amount of organic matter in the soil, which in turn reduces infiltration capacity and increases surface runoff. These problems are exacerbated by expansion into marginal, often steeply-sloping land, and by greater use of heavy machinery which leads to soil compaction, again increasing surface runoff.

Another implication of increased mechanization has been the amalgamation of fields to provide larger working units and the consequential loss not just of hedgerows but of the whole hedge–ditch–verge complex that provided valuable habitat heterogeneity in many rural landscapes. Hooper (1977) estimated that an average of 8000 km of hedgerows were removed annually from England and Wales between 1945 and 1970, a figure which hides a peak of perhaps 11 000 km per annum towards the end of this period and which masks regional variability, with Norfolk, for instance, losing around half of its hedges (Baird and Tarrant, 1973). Since the early 1970s, the rate of hedgerow removal in Britain has declined, partly because the economic, amenity and ecological values of the hedge have become better appreciated, but there is still both considerable loss and marked decline in quality (Joyce et al., 1988).

The extent to which hedgerow removal has affected bird life has been discussed by O'Connor and Shrubb (1986). Hedgerows are prime breeding habitats for dunnock *Prunella modularis*, yellowhammer *Emberiza citrinella* and whitethroat *Sylvia communis*, and are important for many other passerines, especially where trees are an integral feature. Hedges provide food and shelter as well as nesting sites. Numbers of species and individual birds increase as the variety of shrubs increases (Osborne, 1984; O'Connor and Shrubb, 1986), in turn a function largely of hedgerow age (Pollard et al., 1974). Number of bird species increases with the amount of hedgerow up to a peak at 7–11 km hedge per square km (O'Connor and Shrubb, 1986). Some hedgerow removal therefore leads to a decline in bird density but not necessarily species diversity. Moreover, the proximity of woodland further influences the nature of the hedgerow bird community and there is reciprocal effect: the number of forest-interior birds present in small woodland patches is significantly increased by the density of 'connecting elements', which include hedgerows, which have either a habitat or corridor function, or both (van Dorp and Opdam, 1987).

Changes in hedgerow management also affect bird species. Mechanical cutting of hedgerows leads to a thinner, less densely structured body of hedge than does hedge laying by hand. Many local authorities in the UK trim external hedges (those bordering roads) during spring, causing much disturbance and exposing nests. And, with increases in the sowing of cereals in autumn, hedges are cut earlier than previously, with the consequence that berry crops, on which winter thrushes such as redwing *Turdus iliacus* and fieldfare *T. pilaris* greatly depend, are reduced (Lack, 1992).

Of all tilled land in England and Wales, 70% is now ploughed in autumn compared with 22% in the early 1960s. Autumn tilling contributed to an increase in stubble burning (now banned), with lower numbers of invertebrates and seeds available to birds in consequence (Edwards, 1984). Inglis *et al.* (1990) argue that the decline in woodpigeon *Columba palumbus* numbers in East Anglia is linked to a shift from spring-sown barley to autumn-sown barley and wheat during the 1970s, together with a reduction in the area of clover ley. Because cultivation exposes food, spring-sown fields generally provide better feeding in spring than do autumn-sown fields (O'Connor and Shrubb, 1986). Furthermore, some ground-nesting birds such as lapwing *Vanellus vanellus* prefer spring cultivation for nesting (Shrubb and Lack, 1991).

Increased mechanization has directly affected birds by increasing nest and brood losses in ground-nesting species, by reducing grain wastage, and by compressing the period in the farming year over which food is available to granivorous species such as rook *Corvus frugilegus*.

Pasture has been converted to arable land over much of Europe and North America. Much previously unimproved grassland has been drained, irrigated, or both; it has also been fertilized and subjected to herbicides. The distribution and status in the UK of breeding waders such as redshank *Tringa totanus* and snipe *Gallinago gallinago* have been affected by the disappearance of moist pasture and grazing marshes (Smith, 1983) and a reduction in ley pasturage has adversely affected skylarks *Alauda arvensis* (O'Connor and Shrubb, 1986).

Shrubb (1990) showed how agricultural changes have affected the nesting of lapwings in England and Wales between 1962 and 1985, noting a general increase in the use of upland grassland as breeding habitat. Lapwings do best in a mosaic of arable and pasture, as created by rotation schemes that have declined in use since the 1960s. Larger clutches were found in emergent crops than in recently-tilled land, and in unimproved compared with improved pasture. Brood size at hatching showed a progressive decline throughout the study period, with farm work and increased stocking rates causing desertion and loss. Loss of spring-tilled land to autumn cereals is once again emphasized as limiting good nesting and feeding opportunities (Shrubb and Lack, 1991).

O'Connor and Shrubb (1986) discussed how changes in farming practice have influenced the behaviour and status of 13 bird species: five were probably influenced by crop management, four by changes in the timing of tilling, four by changes in the area of grass or tillage or of cereals, and two each by stock management, grassland management or herbicides. They thus stressed the influence on bird populations of management changes as well as habitat changes, a theme developed by Lack (1992) in a book aimed at farmers who wish their practices to be both agriculturally efficient and sympathetic to bird life. Recently, some farmers have adopted a more conservation-orientated approach to farming, associated in part with economic incentives (for example the use of set-aside land in the UK), or with the development of game shoots. Even if the advantages to wildlife are incidental, they may reverse some of the trends seen in the farmland bird community in recent years.

Similar situations occur outside the British Isles. Many of the trends noted for the British hedgerow landscape, for example, are seen throughout Europe (Terrasson and Tendron, 1975). In Spain and Portugal, where most of the European populations of chough *Pyrrhocorax pyrrhocorax* occur, there is evidence of general decline due to changes in traditional farming practice, with pastoral and cereal husbandry no longer providing appropriate food or habitat. Drastic changes in farming practice are considered to be responsible for reduced numbers of corn bunting *Miliaria calandra* in the Netherlands, where the Dutch population fell from 1100–1250 pairs in 1975 to 125–175 pairs in 1989 (Hustings *et al.*, 1990). Similarly, the decline of kestrel *Falco tinnunculus* in Switzerland is attributed to crops growing more rapidly and densely, and to lower densities of prey animals in farmland (Schmidt, 1990).

One can also trace the consequences for bird life of conversion from a natural to an arable or pastoral landscape in areas with a more recent history of farming. In Western Australia, for example, clearance of native vegetation has left only scattered remnants of original woodland, scrub and grassland, though the number of water points has undoubtedly increased. Species diversity of birds in the central wheatbelt has declined over the past 80 years, though this reflects a balance between the disappearance or decline of some species and the increase in range and abundance of others (Saunders and Curry, 1990).

2.4.2 Forestry

Temperate forests account for around 57% of the world's total forest area and are expanding globally through replanting programmes. Tropical forests, however, are being rapidly destroyed, with an estimated 7.3 m ha cleared each year for agriculture and a further 4.4 m ha being degraded through selective logging and ecologically inappropriate reforestation (WRI/IIED, 1988; section 2.5). Forests not only provide habitat, but also play a critical role in the regional (and probably global) climate, in hydrology, in biogeochemical cycles and as a protection against soil erosion (Myers, 1983). Clearance of forest thus leads to habitat loss, land-use change and habitat fragmentation, but often where forest remains, and certainly when it is replanted, there are also important changes in species composition, species diversity and forest structure, with consequent effects on animal life. In much of North America and Europe there must also be added the consequences of extensive forest decline (*Waldsterben*) caused by a combination of air pollution, soil nutrient status, pests and disease.

At the beginning of the twentieth century there were some 1.2 M ha of woodland in Great Britain. By 1990 this figure had increased to 2.1 M ha (nearly 10% of the land surface) of which 1.7 M ha was high forest. This increase was largely a result of coniferous plantation by the Forestry Commission, which had been established in 1919. Between 1947 and 1980, coniferous forest increased more than threefold to over 1.2 M ha, while broadleaved woodland actually decreased by 29% to just over 0.8 M ha (Peterken and Allison, 1989). Where there had originally been extensive tracts of mixed oak woodland and, in the north, Scots pine *Pinus sylvestris*, high forest in the UK is now 28% planted with Sitka spruce *Picea sitchensis*; 13% Scots pine; 9% oak (*Quercus robur* and *Q. petraea*); 7% lodgepole pine *Pinus contorta*; 6% Norway spruce *Picea abies*; and 6% Japanese or hybrid larch *Larix* × *eurolepis* (Forestry Trust, 1991). Where the tree species are introductions, woodland structure is often also changed, partly because of management, partly because structure largely reflects dominant tree species, their ages, and the nature and cyclicity of natural or man-induced disturbance. Thus mature beechwood and, especially, coniferous forests both have a sparse shrub and field layer because of their shade and slowly-decaying litter. Mixed oakwood generally has a structurally more complex and floristically richer understorey.

The impact of woodland type and structure on bird life is inevitably complex (Avery and Leslie, 1990: Simms, 1971; Yapp, 1962). Some species are essentially confined to one particular kind of woodland: for example, the dependence of capercaillie *Tetrao urogallus*, crossbill *Loxia curvirostra*, siskin *Carduelis spinus* and crested tit on coniferous forest. Other woodland bird species may range more widely in habitat but nevertheless have a strong preference for coniferous areas: for example goldcrest *Regulus regulus* and coal tit *Parus ater*.

Natural treefalls lead to relatively small gaps in the forest canopy, altering the light regime and microclimate, and allowing plants to regrow or colonize in such a way as to increase species diversity and structural complexity, a modification that persists for many years. With clear-felling and replanting, however, the resulting woodland tends to be much simpler in age, floristic content and physiognomy. In Britain, through the 1960s into the early 1980s, coniferous 'panels' were commonly planted in otherwise deciduous woodlands, though this practice has declined following the Government's 'Broadleaves Policy' of 1985. The greater habitat heterogeneity of deciduous woodland tends to support greater numbers and a greater diversity of birds, and certainly the decline and dilution of such woods has adversely affected species such as treecreeper *Certhia familiaris* and nuthatch *Sitta europaea*. It must be admitted, however, that other bird species apparently benefit from a deciduous–coniferous compartment mix, for example sparrowhawk *Accipiter nisus*, tree pipit *Anthus trivialis*, and redpoll *Carduelis flammea* (Fuller, 1982).

Coppicing is where trees are cut back to their stumps to produce new shoots. Young coppice supports high numbers of warblers, finches, buntings and other summer migrant species; thrushes and tits increase in numbers with coppice age; but bird species diversity declines after about seven years' regrowth, with canopy closure (Fuller and Moreton, 1987). Mixed-species coppice supports more species than single-species coppice, and stand density, length of rotation and the spatial structure of the compartments are all variables that also affect bird life. Over the past century, coppicing has greatly declined as a means of woodland management, though there are signs of a resurgence in south-eastern England.

There was a loss of 25% of native Scots pine forest in Scotland over the period 1950–1986. Hill *et al*. (1990) showed how bird density in natural pinewood increases in response not only to canopy closure and increasing tree height, but also to the number of snags and the amount of dead wood on the ground. All these are features that tend to be absent from plantations, especially young ones. Thus the 40% of the Scottish crested tit population that inhabits conifer plantations (the rest occurs in natural pinewoods) is mainly in plantations that are over twenty years old (Cook, 1982). As plantations have matured, during the last few decades, the crested tit has increased its range. Many passerines benefit from afforestation (though often they are species that are already common in other habitats) and it has been argued that the greater numbers of songbirds associated with afforestation are exploited by raptors such as merlin *Falco columbarius* (Bibby, 1986, 1987; Parr, 1991) and sparrowhawk (Newton, 1986).

Where comparable habitats are examined, it may be that introduced tree species do not differ significantly in their associated avifauna. Rose (1979), for example, showed that for at least one study site the bird communities in plantations of Scots pine were only slightly more species-rich than in those of Corsican pine *Pinus nigra* var. *maritima* and Reed (1982) reported that even native pines may have a lower species richness and density of birds than do spruces or larches. Reed, however, is sceptical of some of the supposed adaptations by birds to coniferous plantations. That merlins breed in plantation glades, for example, is offset by the need for this species to commute between nesting sites and the nearest areas of moorland for feeding. Sparrowhawks studied in the south of Scotland used lowland territories more often and generally bred most successfully in larch (introduced), followed in order by pine (some introduced), spruce (introduced) and native broadleaves; in upland areas and in forestry plantations, however, the relationship between territory use and tree species was not statistically significant. Also, some aspects of breeding performance (laying date, nestling growth and nest success) deteriorate with increase in the extent of woodland around the nesting territory (Newton *et al.*, 1979). Larch and pine, being more open in structure, might provide better visibility for detection of prey and intruders. Larch might also provide better nesting material. On the other hand, larch and pine give poorer protection against rain, and are associated with lower prey densities than around other tree species.

Petty and Avery (1990) identify a number of trends regarding forest bird communities in Great Britain. Species richness is greater in larger woods; the typical bird communities differ with successional stage and species of tree; and the number of species and of individual birds tends to be different at the junction of two habitats from those typical within the habitats themselves (an ecotone or edge effect). These trends are clearly affected by environmental changes wrought by human activity.

General patterns of density, species richness and relative abundance of breeding birds have been examined in woodlands in other parts of the world. Taking a wide sample of North American forests, James and Wamer (1982) demonstrated that young forests in secondary succession and mature deciduous forests can be equally rich in bird species; coniferous forests and dense successional stands comprising one or two tree species have fewest species. Forests rich in tree species and those with high canopies held most birds. Area of woodland is also an important variable regarding the bird community: this is examined in section 2.5.

The dependence by birds on a variety of features found within woodland was demonstrated by Loyn (1985) for eucalypt-dominated vegetation in Australia, where shrub structure, successional stage, presence of nectar-rich flowering shrubs, presence of arborescent wattles, abundance of tree hollows, infestation by canopy insects and the nature of the bark and exudates of the eucalypts themselves all affect the structure of the avian community. Milledge and Recher (1985), however, found that neither species richness nor abundance of birds was correlated with eucalypt woodland structure in southeastern Australia and indeed some sites with a simple structure had particularly rich avifaunas; what was important was the floristic diversity.

Logging activity and afforestation in Australia, as elsewhere, would be expected to affect the nature of the bird community. That this is not always the case was shown by Shields et al. (1985) for northern New South Wales where, despite greater plant species diversity and a greater extent of mixed woodland and open habitat, there were no significant differences in bird species diversity or abundance between logged and unlogged sites. More commonly, however, adverse impacts are demonstrable. Kavanagh et al. (1985), for example, found few bird species and individuals on logged compared with unlogged coupes and, while bird life recovered as vegetation regenerated, only 78% of the original number of bird species were found four years after felling; the absence of hole-nesting birds, of canopy feeders and of litter foragers was particularly notable. In woodchip logging areas, Smith (1985) found that 10–15 year forest regeneration on ridges supported a smaller bird community than mature eucalypt forest, while 10–15 year regeneration in gullies supported a bird community of similar size but different composition from that in mature forest.

Rymer (1981) estimated that by 2010 there would be 1.4 m ha of forestry plantation in Australia, principally of Monterey pine *Pinus radiata*. Such afforestation will continue to have a pronounced impact on the avifauna. Driscoll (1977) found that mature pine plantation in upland New South Wales had as many or indeed more birds than native forest, but that the latter had greater species diversity, and the exotic and indigenous vegetation types had rather different bird faunas.

The area of exotic forestry in New Zealand is around 750 000 ha, with 83% of plantations growing *P. radiata*, and a further 7% Douglas fir *Pseudotsuga menziesii*. Bull (1981) noted two major limitations for pine plantations as a habitat for native fauna: that most are monocultures, and that forest is usually clear-felled on a relatively short rotation, *P. radiata* providing sawlogs in 25–30 years, and pulp timber even sooner. Nevertheless, several native birds have become widespread in pine forests; indeed Gibbs (1961) discovered that whitehead *Mohoua albicilla*, pied tit *Petroica macrocephala* and New Zealand robin *P. australis* were as numerous in one pine forest as in almost any native forest, and a good deal more numerous than in most. Brown kiwi *Apteryx australis* has been found in Waitangi State Forest, Northland, at one of the highest densities (around one kiwi per hectare) ever found in any forest type. In Nelson, brown creeper *Finschia novaeseelandiae* and New Zealand robin were abundant in *P. radiata* and Douglas fir plantations, but were rare or absent in other habitat types (Clout, 1980).

It is not only coniferous plantations that have had an adverse impact. Teak *Tectona grandis*, for example, has been introduced into Papua New Guinea. Bell (1979) showed that the bird community in teak plantation 11 and 19 years old was a depauperate version of that in the original rainforest: less than half of the original species were found in teak and both numbers and biomass of birds were much lower. In forestry areas of young *Pinus caribaea*, *Albizia falcateria* and other introduced timber species in East Kalimantan, both diversity and density of bird and mammal species declined drastically compared with native forest (Wilson and Johns, 1982).

Clearly there are some benefits to bird life to be gained from the global expansion of exotic plantations, and these can be optimized by forest management that includes both sympathetic silviculture (Avery and Leslie, 1990) and explicit conservation methods (Bull, 1981). Generally, however, such forests do not support as much or as varied a wildlife as does native forest.

2.4.3 Fish and fisheries

The influence of oceanic circulation and chemistry on fish growth is profound, and the impact of periodic anomalies such as the El Nino–Southern Oscillation event of 1982–1983 is seen throughout the whole marine trophic system of the Indo–Pacific region (Glynn, 1988, 1990). Such influences have consequences for both seabirds (Duffy, 1990) and land birds (Gibbs and Grant, 1987; Grant and Grant, 1987) through their impact on food supply, breeding phenology and reproductive success. Whether El Nino events constitute an environmental change in the sense adopted in this volume is questionable. What is undeniable is the importance of deep currents for the global carbon cycle, and their influence on the greenhouse effect.

The oceanic fish stock is thus affected by oceanographic changes, but fishery effort is perhaps even more critical. The world marine fish catch has increased from 30 Mt in 1958 to nearly 90 Mt in 1986 (WRI/IIED, 1988). The global annual catch is approaching maximum sustainable yield (estimated by FAO to be 100 Mt) and indeed certain fisheries have exceeded their maximum sustainable yields.

Seabirds with different foraging methods are affected differently by changes in fish prey numbers and availability, such effects being strongly linked not only with survival but also with breeding success. Baird (1990) described how the diets of black-legged kittiwake *Rissa tridactyla*, glaucous-winged gull *Larus glaucescens* and tufted puffin *Fratercula cirrhata* were affected by changes in the numbers, species, frequency of occurrence and body length of prey in Alaskan waters over 1977–1978. The surface-feeding gulls showed a reduction in breeding success by up to 90% over this period, while the puffin, which hunts by diving, showed no such trend. During 1978 sampling of fish prey populations showed stratification and a reduction in stocks of the species predominantly taken during 1977 (capelin *Mallotus villosus*), possible explanations lying with lower sea temperatures in 1978 and seawater dilution in 1977. Similarly, Murphy et al. (1991) showed how the reproductive performance of kittiwakes fluctuated in western Alaska during 1975–1989, a period when year to year variability in spring air temperatures was much more pronounced than during the previous 68 years. Spring air temperature is correlated with the time at which sea ice breaks up, which affects sea temperature and salinity, and consequent availability of prey such as sandeel *Ammodytes hexapterus*. Aebischer et al. (1990) have noted the similarities in trends of abundance of phytoplankton, zooplankton and herring *Clupea harengus*, variations in kittiwake breeding performance, and variations in weather patterns over 1955–1987 in the north-west North Sea.

The dependence of bird species on particular prey and their ability to switch to alternative prey have been shown in a number of studies (e.g. Barrett and Furness, 1990; Furness and Barrett, 1991; Hamer et al., 1991). The complexities of changes in population size and structure, reproductive success, food supply and behaviour suggest the difficulties in using seabirds as indicators or monitors of changes in the fish stock until the relationship between stock abundance and the ecological responses of sensitive bird species is elucidated (see Chapter 6). Even then one needs to consider the intermeshing of other elements of environmental change such as pollution, predator–prey interactions and human disturbance (Dunnet et al., 1990). If seabird populations can respond quickly to such fluctuations in prey availability it will be salutary to consider what changes might result from such longer-term environmental changes as global warming and overfishing.

2.5 HABITAT ALTERATION AND FRAGMENTATION

Habitat loss through conversion to other habitat types is the greatest threat to species survival and the maintenance of diversity, and not only in the developed world. Recent studies have shown that in South-East Asia 68% of the original habitats for wildlife have been lost, and in sub-Saharan Africa the figure is 65% (MacKinnon and MacKinnon, 1986a, 1986b). Much attention has been paid to deforestation, particularly of tropical moist and dry forests (section 2.4.2) which carry anything from 50% to 90% of all known species on a dwindling 6% of the Earth's land surface. Yet 28% of the tropical forests (448 M ha) have already been destroyed. For the world as a whole, tropical deforestation is most extensive in South-East Asia and Central America, with 98% of the dry tropical forest of the Pacific coast of Central America and Mexico having been converted to other land uses (WRI/IIED, 1988). A more recent survey has confirmed that the rate of deforestation in tropical regions is continuing to increase (Sayer and Whitmore, 1991).

Some other terrestrial habitats have almost disappeared. For example, the tallgrass prairies of the United States have been reduced by 98% (Myers, 1979) and up to 50% of the world's swamps and marshes have been destroyed (WRI/IIED, 1987). What remains of such habitats is often degraded and impoverished. Further discussion in this section focuses on the UK.

Habitat alteration has already been discussed in the context of farming and forestry, with some consequences of recent changes in land use and management having been noted. In a more general context, the impact that habitat alteration has on bird life depends on what has been lost and what has been substituted, on the area involved, and on the spatial relationships that exist in the new landscape, including the extent of connectivity and fragmentation.

Lowland heath, for example, represents an important habitat for birds in north-west Europe, although it is ironic that this vegetation type itself largely results from human activity, a plagioclimax that has persisted through burning and grazing following prehistoric and subsequent forest clearance. In turn, various pressures have been placed on such heathlands, particularly since the mid-eighteenth century when improvements in agricultural techniques permitted reclamation for farming purposes. In the present century heathland has been transformed not only into arable land and pasture but also into forestry plantations, golf courses, residential overspill, industrial estates, land for military purposes and gravel extraction pits. In east Dorset, for instance, the area of heathland declined between the 1750s and 1978 by some 80% (from around 39 960 ha to 7900 ha) and, just as critically, became heavily fragmented, so that by 1960 a near-continuous block had been broken into over a hundred pieces, and by 1980 there were only 18 areas of heathland over 100 ha (Chapman *et al.*, 1989; Moore, 1962; Webb and Haskins, 1980). Similar losses have been noted in continental Europe (Noirfalise and Vanesse, 1976).

Webb (1989) noted a negative correlation between the diversity of the invertebrate community and the area of the surrounding heathland parcel, in turn related to degree of isolation; diversity was also related to the floristic composition and structure of the sample fragment. Such a relationship also appears to hold for other animal groups, including birds. The impact of habitat loss and fragmentation on vulnerable species was shown by Bibby (1978) in his study of Dartford warbler *Sylvia undata*; he also discussed the adverse consequences of permitting encroachment of birch, bracken and pine. Dartford warblers are at the edge of their range in southern England, yet the major threat to them lies with habitat loss rather than climatic change.

The consequences to birds of the conversion of heathland to forestry plantation have been examined at Thetford Chase in the Breckland region of East Anglia (Lack, 1933, 1939; Lack and Lack, 1951). The average density of heathland birds was estimated to be nearly 1.25 birds/ha, and that of coniferous forest birds to be 4.5 birds/ha. Consequently, the 20 000 ha or so of former heathland planted to conifers in the Breckland by the 1960s meant that each year there were some 25 000 fewer heathland birds and 90 000 more birds associated with conifers present in the area. Such figures extrapolated throughout north-west Europe indicate the drastic decline that must have taken place in numbers of heathland birds.

Not only have there been changes in total numbers of birds, but also in the nature of the bird communities themselves. As heathland is ploughed up prior to conifer planting, so species such as wheatear *Oenanthe oenanthe* and stone curlew *Burhinus oedicnemus* leave. Birds found on well-grown heath, such as skylark and meadow pipit *Anthus pratensis*, persist until the trees are about four years old but then decline to zero by the ninth or tenth year of planting. Species such as stonechat *Saxicola torquata* and whinchat *S. rubetra* increase in the fourth and fifth year, then decrease. Other scrub species (e.g. willow warbler, whitethroat and wren *Troglodytes troglodytes*) are common in years 5–12, then decline. Species such as blackbird *Turdus merula*, chaffinch and goldcrest, begin to invade by year seven. By the fifteenth year or so, when the trees are cleared of smallwood, the scrub bird community disappears altogether, leaving a truly woodland bird community.

Yapp (1962) and Simms (1971) also recognized this successive change in the bird community, though they rightly emphasized tree height rather than age. Yapp, unlike Lack, found no sudden changes in bird numbers, but rather a gradual shift, possibly because the former's work concentrated on afforestation of grass moorland where interspecific competition may not have been so pronounced. Other studies of moorland afforestation were summarized in Petty and Avery (1990), who noted that in all cases bird species richness, diversity and overall density were all greater at the end than at the beginning of forest growth, although actual peak values were often associated with the intermediate scrub stage. The net gain in woodland birds, however, is at the expense of (often rare) heathland species.

Similar threats from habitat alteration face birds of moorland and wetland. Afforestation of moorland may adversely affect relatively common birds, such as meadow pipit and skylark, though the latter's UK population halved during the 1980s largely as a consequence of changes in the farming landscape. More significantly, numbers of ring ousel *Turdus torquatus* and twite *Acanthis flavirostris*, both essentially upland species, are dwindling, and even rarer species are under threat, for instance golden eagle *Aquila chrysaetos* and raven *Corvus corax*. Thus Marquiss *et al.* (1978) noted a decline in the raven population of southern Scotland and northern England dating from the 1960s, so that by 1974–75 only 55% of formerly-used sites remained occupied. Both the timing and the geographical distribution of this decline coincided with upland afforestation. Indeed, with time, the adverse effects of afforestation appeared to become disproportionately greater, possibly because the remaining, dwindling area of moorland in a territory became increasingly critical for survival and reproductive success. Changes in sheep husbandry might also have reduced the food available as carcasses to the birds, and Mearns (1983) argued that a trend towards bringing sheep down to shelter for winter and lambing was perhaps more important than afforestation in affecting raven numbers. Newton *et al.* (1982) found no decline in the raven in Wales, despite afforestation, though they noted that sheep were more available and trees younger than in the Scottish borders.

Before conversion to forest or farming, many upland areas require draining and such schemes pose threats to many birds. Likewise, drainage schemes in lowland areas, generally for agricultural purposes, can adversely affect the bird community. As with heathland and moorland, a wide range of species is involved, but it is the already rare species that are especially vulnerable, for example bittern *Botaurus stellaris*, water rail *Rallus aquaticus* and black-tailed godwit *Limosa limosa*, which have all suffered from reductions in wetlands ranging from reedswamp to periodically flooded meadow.

Floristic changes associated with wetland may represent another adverse factor, although sometimes these reflect increased flooding rather than draining. For instance, the extensive spread of reedgrass *Glyceria maxima* swamp on the Ouse Washes, eastern England, during the last two decades (Burgess *et al.*, 1990) has led to a reduction of food quality for grazing wildfowl such as wigeon *Anas penelope* and swans *Cygnus* spp. which are found there in internationally important numbers in winter (though this effect may have been offset by an increased availability of *Agrostis stolonifera*). Replacement of species-rich grassland by reedgrass has reduced the number of seed-producing plants, affecting the food supply for wintering seed-eating ducks (Thomas, 1982). Reedgrass also provides poorer nesting habitat for wetland birds.

Habitat fragmentation not only reduces the area available to species but extends the amount of edge and increases the changes of population isolation. Habitat 'islands' lend themselves to analysis using island biogeography theory. The species–area relationship is such that in general terms

$$S = cA^z$$

where S = number of species, A = area, and c and z are constants, c providing an estimate of the number of species per unit area and z indicating how quickly new species are added with increasing area. The relationship does not always hold and even where it does the values of c and z differ between regions. However, a rough rule of thumb would be that an order of magnitude increase in area results in an approximate doubling of number of species present. Habitat heterogeneity is also an important variable, though often a larger area is actually associated with greater habitat diversity.

Several studies have shown that the number of bird species found in forest fragments does indeed reflect area (Ambuel and Temple, 1983; Blake and Karr, 1984; Forman *et al.*, 1976; Lynch and Whigham, 1984; Moore and Hooper, 1975; Whitcomb *et al.*, 1981). Area itself, however, tends to be an indicator of species diversity rather than a cause; Robbins (1980) and Lynch and Whigham (1984), for instance, found that floristic diversity and canopy height tended to be greater in larger areas; they also noted that isolation between woodland parcels is important. Varying sampling effort can lead to bias in analysis and interpretation (Woolhouse, 1983).

Whatever the actual quantitative relations, in qualitative terms smaller forests generally possess fewer bird species than larger ones. Small forests, too, will contain smaller populations of particular species, and this in turn will tend to lead to genetic drift, inbreeding depression and loss of heterozygosity (genetic diversity), especially where habitat islands are physically isolated from each other. The net result is often that population numbers fall below a critical threshold, and so, for genetic reasons as well as because of the greater vulnerability to physical disturbance of small populations in small areas, a bird population may become locally extinct.

Rolstad (1991) argued that because of their high mobility and large home ranges, birds usually perceive fragmented forests in a fine-grained manner, embracing several woodlots within their area of activity. On a regional scale, however, these fine-grained habitat components are aggregated into a coarser-grained environment, and bird populations may be separated into isolated demes. Distance–area effects, such as an increased island effect and decreasing habitat island size, directly prevent dispersal and reduce population size. Landscape effects, such as reduced habitat fragment/matrix ratio and interior/edge ratios (discussed later), increase the pressure from surrounding predators, competitors, parasites and disease. At a certain point, non-fragmented areas become so widely spaced out that regional distance–area effects appear, giving rise to metapopulation dynamics. The importance of context in examining habitat islands is stressed by Loman and von Schantz (1991), who noted that bird density was highest in small habitat islands and in islands close to other islands in Swedish farmland. Birds found in small fragments actually utilize resources from neighbouring sites, whether similar types or the overall farmland matrix.

The increased extent of 'edge habitat' associated with habitat fragmentation may be an advantage or disadvantage to birds. The edge is often a habitat zone in its own right, incorporating habitat features of both surrounding habitats and often possessing features unique to itself. It may thus be distinguished as an ecotone and may be richer in species content than the surrounding areas, containing some species characteristic of each of these plus some characteristic only of the ecotone itself (Ranney *et al.*, 1981). Levenson (1981) demonstrated how the area of edge of a particular unit width increases with an increase in island site: in small blocks the whole block will comprise 'edge' community but as the block increases so 'interior' conditions begin to form. Woodland edge, for example, may support a greater number of berry-bearing trees and shrubs than interior forest, and may have a higher arthropod abundance, these conditions in turn encouraging frugivorous and insectivorous birds, respectively (Hansson, 1983). Many studies have indicated the increase (and decrease) of woodland bird species at the forest edge (reviewed by Petty and Avery, 1990) and clearly the size, shape and degree of isolation of habitat islands and their edges affect the nature and number of birds present. The disadvantage of increasing edge habitat, especially where this reduces the extent of core habitat, lies in the concomitant increase of a zone liable to disturbance, especially from human activity. Fragmentation of the Dorset–Hampshire heathland, for example, has increased the vulnerability of species such as Dartford warbler to physical disturbance, especially when on or near the nest.

2.6 INTRODUCED SPECIES

Plant and animal species have been introduced into, and have become naturalized in, new geographical areas through deliberate or accidental human activity. Some species introductions are ecologically or economically beneficial. Most are neutral and persist in low populations in a limited area. Many, however, are harmful, economically (becoming weeds and pests) or ecologically (causing reductions of native species). Ecological characteristics of introductions are summarized by Jarvis (1979) and extensive accounts are provided by the various publications emanating from the SCOPE programme, 'The ecology of biological invasions', initiated in 1982 and brought together by Drake *et al.* (1989).

Introduced birds have themselves become naturalized throughout the world (Lever, 1987; Long, 1981) and many have become pests. Adverse impact on the native avifauna includes transmission of disease, interspecific competition (both interference and exploitative), and predation. Diseases affecting other bird species include Newcastle disease, avian malaria, blackhead and bird pox. Regarding competition, the European starling *Sturnus vulgaris* has contributed to the decline of several North American hole-nesters, such as the eastern bluebird *Sialis sialia*, and the Indian mynah *Acridotheres tristis* has acquired notoriety by its aggressive behaviour wherever it has been introduced, usurping both nest sites and food. There are few naturalized raptors which have had a serious impact on other bird species, though barn owls *Tyto alba* have numerically reduced or made extinct a dozen or so island endemic birds, and on Macquarie Island, wekas *Gallirallus australis* prey on ground-nesting seabirds and have contributed to the decline of nine species of Procellariidae and Pelecanidae (Lever, 1987). Frequently, as with all introduced species, the success of an introduced bird has been facilitated or even permitted through habitat modification by humans.

Other introduced animals have had serious, sometimes disastrous, effects on bird populations. King (1984), for example, describes the impact that introduced mustelids, rats (*Rattus* spp.) and cats *Felis catus* have had on the New Zealand avifauna. Having evolved in the absence of predatory mammals, many New Zealand birds are ground-nesting species, and quite a few have little or no ability to fly, making them particularly vulnerable to introduced carnivores. Such a history, together with the adverse consequences of habitat alteration (with 71% of the original forest being lost), largely explains why some twenty endemic species of land birds have become extinct in New Zealand during the past two hundred years, and why 11% of the 318 rare or endangered bird species listed in the IUCN Red Data Book are New Zealand endemics.

The avifaunas of small islands are particularly vulnerable to the introduction of predators. The impact of feral cats on island bird life, for instance, is explored by Fitzgerald (1988). Reptile predators can also be important: thus the snake *Boiga irregularis* that 'ate Guam' exterminated seven of eleven native bird species (plus seven introductions) and the remaining four indigenous species are critically endangered (and indeed may now also have gone) (Pimm, 1987; Savidge, 1987).

The impact that alien plants can have on bird life has already been indicated using the example of forestry plantations. On the whole such impacts are adverse, though there are occasions where an introduced plant species has been beneficial to birds. For example, the most important food items of the endangered Seychelles black parrot *Coracopsis nigra barklyi* are, together with an endemic palm, the introduced *Averrhoa bilimbi* and *Mangifera indica*; indeed, nine out of sixteen food plant species for this bird are introductions (Evans, 1979). Introduced animals may likewise become significant prey items for native birds: thus, in Australia, the diet of wedge-tailed eagle *Aquila audax* includes such introductions as rabbit *Oryctolagus cuniculus*, red fox *Vulpes vulpes*, feral cat and feral goat as well as the non-indigenous sheep. Other examples of the benefits as well as costs of introduced species are reviewed by Jarvis (1983a) in the context of nature conservation.

2.7 POLLUTION

Pollution may be considered as 'the introduction by man into the environment of substances or energy liable to cause hazards to human health [and] harm to living resources and ecological systems' (Holdgate, 1979). Pollutants can affect birds directly, through lethal or sublethal stress, and indirectly through habitat alteration.

The last fifty years have seen a number of critical changes in the nature of pollution. First there is the pervasiveness of pollutant material, which is now found wherever there is human activity and even in places where there is none. Secondly, a number of chemicals have been developed that are persistent and are transferred through components of the ecosystem, often with biomagnification *en route*. Thirdly, there is the problem of radiation, with increases of damaging and long-lasting artificial radionuclides through the testing of nuclear weapons and through accidents such as occurred at Chernobyl in 1986 (see Chapter 4). Fourthly, there are various non-biodegradable products, in particular plastics. Fifthly, despite safety precautions, there is an increased chance of major accidents simply from the sheer volume of material that is being transferred, as has been seen with oil spills. And sixthly, there is the threat of using the environment as part of the theatre of war, as seen with the use of herbicides in Vietnam and deliberate spillage and burning of oil in the Gulf crisis of 1991.

The following sections give a brief account of major hazards: oil pollution, pesticides and related chemicals, heavy metals, plastics and acidification.

2.7.1 Oil pollution

Deliberate discharge, such as by the washing of oil tanks, accounts for most marine oil pollution. Accidents, however, may have severe local results. Petroleum fractions interact with marine biota in many ways at many levels. Non-volatile fractions are degraded by microorganisms, degradation being more rapid at higher temperatures and in more nutrient-rich waters; oil therefore tends to persist for longer in polar and temperate waters than in the tropics. Major incidents cause an immediate large-scale mortality among seabirds but chronic oil pollution may have a more serious long-term impact on the marine ecosystem. Prolonged exposure to oil, for example, will lead to a decline in primary productivity and shift in the phytoplankton composition; pelagic food chains are modified; and medusae increase at the expense of fish. Seabirds are affected not only by a reduction in preferred prey or prey biomass but also through exposure to the physical, carcinogenic and mutagenic effects of oil components (Holdgate et al., 1982).

The acute impact of oil on birds arises through external contamination, ingestion and embryotoxicity (Leighton et al., 1985). The major effect of external petroleum oil is to reduce plumage integrity and water repellency. Oiled feathers become matted and waterlogged, and birds may lose their buoyancy and drown. The insulating properties of the feathers are also reduced. The effect of external oil depends on dose and circumstance: a scavenging gull in summer may tolerate a substantial oil burden, while an auk fishing in sub-zero waters may die from a small dose. In attempting to preen, birds ingest oil, which may be toxic or may clog up the digestive system. Inability to digest efficiently, to forage effectively and perhaps to find uncontaminated food lead to rapid depletion of fat reserves and energy. Reproductive impairment, depressed growth in young birds, changes in osmoregulation (related to excretion of excess dietary salts), depressed adrenal gland function, anaemia and other stress and toxic effects have also been noted.

Numbers of birds killed through major oil pollution incidents are generally high, but depend on a range of circumstances. Some 30 000 birds are estimated to have died as a result of the *Torrey Canyon* disaster off the Scilly Isles in March 1967, in which 120 000 t of oil were spilled (Bourne, 1976; Bourne et al., 1967), but 40 000 birds died as a result of only 200 t of oil being discharged into the Dutch Wadden Sea in February 1969 (Samiullah, 1990). The grounding of the *Exxon Valdez* in March 1989 led to the loss of 35 000 t of crude oil, the largest spillage ever in American waters. The consequences for wildlife from Prince William Sound down to the tip of the Alaska Peninsula some 900 km away, and the problems associated with clean-up and restoration procedures, illustrate the risk that will always remain, however much care is taken with oil extraction and transport. Such incidents also pose danger to terrestrial birds: by August 1989, 109 bald eagles had been found dead from scavenging oil-polluted carcasses and few pairs produced any young, the parents abandoning the nest, not producing eggs, or the eggs not hatching (Pain, 1989).

That spills are an inevitable consequence of the oil industry is suggested by the fact that, despite precautions, there was a record 791 reported incidents around Britain's coast in 1990 (though none major), according to the Advisory Committee on the Protection of the Sea, the number having risen for the fifth successive year to double the 1985 level. On the other hand, the US National Research Council Marine Board reported that the amount of oil entering oceans from shipping operations had been cut by 60% since 1981 (*The Guardian* newspaper, 13 August, 1991).

It is often difficult to distinguish mortality of seabirds due to oil pollution from that through natural causes (Dunnet, 1982) and most populations of most species appear to be able to recover from crashes due to oil-induced mortality. But there are real risks to particular colonies and to some rare species (Bourne, 1982).

Oil spills also occur in freshwater systems. Large numbers of ducks, geese and herons were killed following a spill on the St Lawrence River in 1979, for example, and many oiled anatids were noted after an incidents on the Monongahela River, Pennsylvania in 1988 (Shales *et al.*, 1989).

2.7.2 Pesticides and related chemicals

Chemical agents are often target-specific and are deliberately released into the environment to control pests, weeds and diseases. The nature and strong persistence of some of these chemicals in the inorganic environment and in living (and dead) organisms, however, has commonly led to side-effects which establish the compounds as pollutants. Since the use of birds as monitors of pesticides and related chemicals is reviewed in Chapter 3, I deal only briefly with chemical pollution and pesticide use. Pesticides that have caused greatest concern as pollutants are synthetic organic compounds, which have been introduced into the environment in large quantities since their development in the 1940s. Organophosphate compounds and their metabolites affect the nervous system as powerful inhibitors of the enzyme acetylcholinesterase. Organochlorine compounds affect the nervous system by entering the nerve membrane and interfering with system functions. Organochlorines are solids of very low water solubility and, being resistant to breakdown, are highly persistent within animals, in contrast to organophosphates which are liquids of often high water solubility and which are generally rapidly metabolized and excreted by animals. A major organochlorine is DDT, one component of which is DDD, and with a metabolite DDE, all of which have a pronounced environmental persistence. DDT, for example, has a half life of 2–5 years in soil and it takes 5–25 years for a 95% reduction in concentration. In pigeons, half lives are 28 days for DDT, 24 days for DDD and 250 days for DDE. For most bird species the 96-hour LC_{50} (the concentration that kills half the individuals in 96 h) for DDT is more than 500 ppm (Walker, 1975). The chlorinated cyclodiene insecticide aldrin (which can be converted to dieldrin) and heptachlor (to heptachlor epoxide) have a number of biochemical features in common with DDT, though they are generally more toxic to animals; the 96-hour LC_{50} for dieldrin in pigeon is 67 ppm. The persistence of organochlorines and the toxic hazards posed to wildlife by pesticides have in many cases led to their restriction. By 1976, DDT, aldrin and dieldrin had been banned for most uses in the USA, and they were phased out in many European countries by the end of the decade, being largely replaced by organophosphates and carbamates.

Samiullah (1990) summarizes trends in organochlorine residue concentrations in birds sampled during national or international surveys. Thus the US National Pesticides Monitoring Program, using European starling, shows that geometric mean levels of DDE ($\mu g/g$ fresh weight) fell from 0.58 in 1968 to 0.17 in 1979, and dieldrin from 0.08 to 0.01. While use of many of the more persistant organochlorines has been banned or severely restricted in the developed world, most are still in use by the developing nations. Broad reviews of the impact of pesticides on bird populations are given by Hall (1987) and Samiullah (1990).

2.7.3 Heavy metals

Birds may acquire heavy metals from preening contaminated material off their feathers, and by taking in food or water containing industrial effluent, polluted agricultural runoff or air pollution fallout. The most serious heavy metals for birds, however, come from two other sources – mercury from mercurial seed dressings and lead from lead shot. Once a metal enters the body it may be stored or excreted. Tissue metal concentrations directly reflect the degree of environmental contamination. Birds eliminate some heavy metals, especially methylmercury, by depositing them during periods of rapid feather growth.

During the 1940s the practice of treating seeds with mercurial fungicides became widespread in many countries, and granivorous birds and their predators commonly ingested toxic quantities. Borg *et al.* (1969) provided a comprehensive survey of mercury contamination in Swedish birds from the 1950s to 1963. Of seed-eaters, highest mercury concentrations were found in pheasant *Phasianus colchicus*, partridge *Perdix perdix*, pigeons and a number of corvids. Of seed-eaters found dead, 48% had wet weight liver mercury concentrations above 2 µg/g fresh weight, and 13% had over 20 µg/g. For *Accipiter* species, concentrations of 2 µg/g and 20 µg/g were exceeded by 78.8% and 17.8% of specimens respectively; for buzzards *Buteo buteo*, the corresponding percentages were 67.2% and 4.8%, and for various owl species, 46.2% and 9.7%. These raptors fed on granivorous birds or seed-eating rodents. Following the ban on alkylmercury seed dressing in Sweden in 1966, mercury levels in exposed birds fell dramatically. In Alberta mercury seed dressing was common and birds contained high mercury levels, but in Saskatchewan such a practice was rare and the birds had very much lower mercury levels (Fimreite *et al.*, 1970). That mercury remains in the environment was shown by Lindberg and Odsjo (1983), who studied the influence of diet on mercury levels in peregrine feathers, and by Lindberg (1984) for gyrfalcon *Falco rusticolus*, in Sweden. Peregrines were at risk from elevated mercury concentrations in waders that had migrated from contaminated areas elsewhere.

The toxic significance to wildfowl of lead from shotgun discharges and discarded angling weights is considerable. The birds swallow the spent shot and grind it in their gizzards, releasing lead into the bloodstream. Annual losses due to such lead poisoning in the USA may be 2–3% of the autumn population of all freshwater aquatic birds (Sanderson and Bellrose, 1986). Some success has recently been made with (part-)substitution of steel shot. Tin and tungsten are now commonly-used substitutes for lead angling weights in the UK, especially following 1987 legislation restricting sale of angling lead shot, and a ban on use of lead by eight of the ten water authorities in England and Wales. Lead weights may persist for some time in the environment: in 1987, following reduction in their use, 24% of all mute swans *Cygnus olor* rescued or given autopsies in the Thames area were victims of lead poisoning (Anon, 1988b). However, there are indications that English mute swan populations, hard hit by lead poisoning, had begun to recover by 1990 (J.J.D. Greenwood and S. Delaney, pers. comm.).

Other sources of mercury and lead exist, and other heavy metals also have toxic effects on birds. Annotated bibliographies and reviews are included in Jarvis (1983b) and Samiullah (1990). The use of birds as monitors of environmental pollution by heavy metals is reviewed in Chapter 3.

2.7.4 Plastics

A problem with plastics is their low biodegradability. Careless anglers deposit plastic as well as lead, primarily as nylon monofilament line.

In South Wales, 25.9 km of fishing line were collected from a 400 m length of Abethaw beach over 12 months, and 341 m from 225 m of bank at Llanishen Reservoir after the first two weeks of the fishing season. Llanishen anglers each left an average of 41 fragments, totaling 6.6 m, on the bank during each visit (Cryer et al., 1987). Such material readily entangles the feet and bills of water birds, and can damage feathers. The plastic yokes used to hold cans of drinks can get around waterbirds' necks and prevent them from feeding, as can plastic netting. Such material can even affect nestlings, if used for nest building (Montevecchi, 1991).

Ingestion of plastic appears to have little or no adverse impact on digestion and hence on body mass or condition (Furness, 1985a; Ryan, 1987; Ryan and Jackson, 1987). However the chemicals in plastic may be toxic to birds. Fragments of manufactured plastic items ('user plastics') and the spherules of the raw material produced by industry ('industrial plastics') were of about equal abundance in the stomachs of fulmars *Fulmarus glacialis* found dead on the Dutch coast and collected from Arctic colonies, ingestion of the former suggesting a stronger impact of toxic chemicals from plastics than is generally assumed (van Franeker, 1985). Procellariiformes may be susceptible to a build-up of plastic because they have small, constricted gizzards (Furness, 1985b). Plastics are also found in Antarctic birds, mostly originating from wintering areas outside the continent, though highest incidence has been noted in chicks of Wilson's storm-petrel *Oceanites oceanicus* that died before fledging (van Franeker and Bell, 1988). An overview of pollution of marine environments by plastics is provided by Coleman and Wehle (1984), and this subject is considered further in Chapter 3.

2.7.5 Acidification

Acidification has been increasing in many developed and in some developing countries, largely as a result of increased amount of atmospheric sulphur dioxide and nitrogen oxides (NO_x) which fall as acid precipitation (pH below 5.6), but also in some places as a consequence of land-use changes, especially following coniferous afforesttion on upland acidic soils. Annual global emissions of man-made SO_2 are 160–180 M t, over 80% coming from fossil fuel combustion; annual global emissions of NO_x (as NO_2) are around 150 M t, 40% being accounted for by fossil fuels and a further 25% by burning biomass. The wet deposition rate of sulphur ranges from 0.3 g/m^2/y in Scandinavia to above 3 g/m^2/y in Central Europe, with a value of 1.0 g/m^2/y to the south and south-east of the Great Lakes in North America. The rate for nitrogen ranges from 0.1 /m^2/y in Scandinavia to 2.0 in Central Europe with a deposition of 0.5 g/m^2/y to the south and south-east of the Great Lakes. Dry deposition probably reaches similar levels to wet deposition over Europe and eastern America (Samiullah, 1990).

In aquatic ecosystems, all trophic levels contain organisms that are sensitive to acidification; generally, number and diversity of species decline. Research on the impact of this on bird life in Europe and North America is reviewed by Diamond (1989), who notes, for example, that piscivorous birds in Ontario are scarcer and breed less successfully in areas of high acidic deposition. Populations of osprey *Pandion haliaetus* (Eriksson, 1984) and dipper *Cinclus cinclus* may have been adversely affected by eggshell thinning and reduced egg mass (Ormerod *et al.*, 1986, 1988). Dippers spend a significantly greater proportion of their day foraging, swimming and flying at acidic streams, and less time resting, with mean energy expenditure 4.5–7.0% greater on acidic than circumneutral streams, yet with less time available for self-maintenance, predator avoidance and breeding (O'Halloran *et al.*, 1990). These issues are considered in greater detail in Chapter 5.

2.8 CONCLUSIONS

The International Council for Bird Preservation's Twentieth World Conference was held at Hamilton, New Zealand, in November 1990. With conservation as their focus, delegates provided evidence of why there is, in general, a continuing decline in bird populations and species diversity (Noon and Young, 1991). The vulnerability of islands was stressed, but the general themes that emerged emphasized the significance of the loss and fragmentation of habitat through human activity, pollution, predation and competition (especially from introduced species), disease and overhunting. Most changes of status cannot be attributed to a single factor but are the expression of several factors working simultaneously or sequentially, and often synergistically. These represent most of the themes examined in this chapter.

The consequences of such environmental changes do not adversely affect all bird species, for some do indeed gain from habitat changes or from a weakening in competition for resources following a perturbation that has led to a reduction of species diversity. The net consequences, though, are of loss. To some extent 'aggressive conservation' (an approach advocated by the ICBP conference) can offset, mitigate or occasionally redress such losses. The one theme not addressed by the ICBP was climatic change. It is doubtful whether there is the political will or the conservation expertise that can reduce or offset the likely consequences to birds of global warming and reduced ozone levels. And of course it is the whole biosphere and the whole world economy that will be affected by such environmental changes.

REFERENCES

Aebischer, N.J., Coulson, J.C. and Colebrook, J.M. (1990) Parallel long-term trends across four marine trophic levels. *Nature*, **347**, 753–5.
Ambuel, B. and Temple, S.A. (1983) Area-dependent changes in the bird communities and vegetation of southern Wisconsin forests. *Ecology*, **64**, 1057–68.
Anon (1988a) Soil erosion tops crisis list. *New Scientist*, **117** (1600), 29.
Anon (1988b) Ban on lead shot gives swans new lease of life. *New Scientist*, **118** (1607), 18.
Arden-Clarke, C. and Hodges, D. (1987) *Soil Erosion in Britain*, Soil Association, Bristol.
Ashmore, M. (1990) The greenhouse gases. *Trends in Ecology and Evolution*, **5**, 296–7.
Avery, M. and Leslie, R. (1990) *Birds and Forestry*, Poyser, London.
Baird, P.H. (1990) Influence of abiotic factors and prey distribution on diet and reproductive success of three seabird species in Alaska. *Ornis Scand.*, **21**, 224–35.
Baird, W.N. and Tarrant, J.R. (1973) *Hedgerow Destruction in Norfolk 1946–1970*, Centre for East Anglian Studies, UEA, Norwich.
Barrett, R.T. and Furness, R.W. (1990) The prey and diving depths of seabirds on Hornøy, North Norway after a decrease in the Barents Sea capelin stocks. *Ornis Scand.*, **21**, 179–86.
Bazzaz, F.A. (1990) The response of natural ecosystems to the rising global CO_2 levels. *Annual Review of Ecology and Systematics*, **21**, 167–96.
Bell, H.L. (1979) The effects on rain forest birds of plantings of teak, *Tectona grandis*, in Papua New Guinea. *Australian Wildlife Research*, **6**, 305–18.
Bibby, C.J. (1978) Conservation of the Dartford warbler on English lowland heaths: a review. *Biological Conservation*, **13**, 229–307.
Bibby, C.J. (1986) Merlins in Wales: site occupancy and breeding in relation to vegetation. *Journal of Applied Ecology*, **23**, 1–12.
Bibby, C.J. (1987) Foods of breeding merlins *Falco columbarius* from Wales. *Bird Study*, **34**, 64–70.
Billington, H.L. and Pelham, J. (1991) Genetic variation in the date of budburst in Scottish birch populations: implications for climate change. *Functional Ecology*, **5**, 403–9.
Blake, J.G. and Karr, J.R. (1984) Species composition of bird communities and the conservation benefit of large versus small forests. *Biological Conservation*, **30**, 173–87.
Boer, M.M. and de Groot, R.S. (eds) (1990) *Landscape-ecological Impact of Climatic Change*, IOS Press, Amsterdam.
Boorman, L.A. (1990) Impact of sea level changes on coastal areas, in *Landscape-ecological Impact of Climatic Change* (eds M.M. Boer and R.S. de Groot), IOS Press, Amsterdam, pp. 379–91.
Borg, K., Wanntorp, P.H., Erne, K. and Hanko, E. (1969) Alkylmercury poisoning in terrestrial Swedish wildlife. *Viltrevy*, **6**, 307–79.

Botkin, D.B., Caswell, M.F., Estes, J.E. and Orio, A.A. (1989) *Changing the Global Environment: Perspectives on Human Involvement*, Academic Press, London.

Botkin, D.B., Woodby, A. and Nisbet, R.A. (1991) Kirtland's warbler habitats: a possible early indicator of climatic warming. *Biological Conservation*, **56**, 63–78.

Bourne, W.R.P. (1976) Seabirds and pollution, in *Marine Pollution* (ed. R. Johnston), Academic Press, London, pp. 403–502.

Bourne, W.R.P. (1982) Concentrations of Scottish sea birds vulnerable to oil pollution. *Marine Pollution Bulletin*, **13**, 270–3.

Bourne, W.R.P., Parrack, J.D. and Potts, G.R. (1967) Birds killed in the *Torrey Canyon* disaster. *Nature*, **215**, 1123–5.

Brown, R.G.B. (1991) Marine birds and climatic warming in the northwest Atlantic. *Occasional Paper, Canadian Wildlife Service*, **68**, 49–54.

Bull, P.C. (1981) The consequences for wildlife of expanding New Zealand's forest industry. *New Zealand Journal of Forestry*, **26**, 210–31.

Burgess, N.D., Evans, C.E. and Thomas, G.J. (1990) Vegetation changes on the Ouse Washes wetland, England, 1972–1988 and effects on their conservation importance. *Biological Conservation*, **53**, 173–89.

Chapman, S.B., Clarke, R.T. and Webb, N.R. (1989) The survey and assessment of heathland in Dorset, England, for conservation. *Biological Conservation*, **47**, 137–52.

Clout, M.N. (1980) Comparison of bird populations in exotic plantations and native forest. *New Zealand Journal of Ecology*, **3**, 159–60.

Cohen, Y. and Pastor, J. (1991) The responses of a forest model to serial correlations of global warming. *Ecology*, **72**, 1161–5.

Coleman, F.C. and Wehle, D.H.S. (1984) Plastic pollution: a worldwide oceanic problem. *Parks*, **9**, 9–12.

Cook, M.J.H. (1982) Breeding status of the crested tit. *Scottish Birds*, **12**, 97–106.

Cryer, M., Corbett, J.J. and Winterbotham, M.D. (1987) The deposition of hazardous litter by anglers at coastal and inland fisheries in south Wales. *Journal of Environmental Management*, **25**, 125–35.

Davis, M.B. (1989) Lags in vegetation response to greenhouse warming. *Climatic Change*, **15**, 75–82.

Diamond, A.W. (1989) Impacts of acid rain on aquatic birds. *Environmental Monitoring and Assessment*, **12**, 245–54.

Drake, J.A., Mooney, H.A., di Castri, F. *et al.* (1989) *Biological Invasions, SCOPE 37*, Wiley, Chichester.

Driscoll, P.V. (1977) Comparison of bird counts from pine forests and indigenous vegetation. *Australian Wildlife Research*, **4**, 281–8.

Duffy, D.C. (1990) Seabirds and the 1982–1984 El Nino-Southern Oscillation, in *Global Ecological Consequences of the 1982–83 El Nino-Southern Oscillation* (ed. P.W. Glynn), Elsevier, Amsterdam, pp. 395–418.

Dunnet, G.M. (1982) Oil pollution and seabird populations. *Philosophical Transactions of the Royal Society, London, B*, **297**, 413–27.

Dunnet, G.M., Furness, R.W., Tasker, M.L. and Becker, P.H. (1990) Seabird ecology in the North Sea. *Netherlands Journal of Sea Research*, **26**, 387–425.

Edwards, C.A. (1984) Changes in agricultural practice and their impact on soil organisms, in *Agriculture and the Environment* (ed. D. Jenkins), Institute of Terrestrial Ecology, Cambridge, pp.56–65.

Eriksson, M.O.G. (1984) Acidification of lakes: effects on water birds in Sweden. *Ambio*, **13**, 260–2.

Evans, P.G.H. (1979) Status and conservation of the Seychelles black parrot. *Biological Conservation*, **16**, 233–40.
Falkenmark, M. (1990) Hydrological shifts as part of landscape ecological Impacts of Climatic Change, in *Landscape-ecological Impact of Climatic Change* (eds M.M. Boer and R.S. de Groot), IOS Press, Amsterdam, pp. 194–217.
Fimreite, N., Fyfe, R.W. and Keith, J.A. (1970) Mercury contamination of Canadian prairie seed eaters and their avian predators. *Canadian Field Naturalist*, **84**, 269–76.
Fitzgerald, B.M. (1988) Diet of domestic cats and their impact on prey populations, in *The Domestic Cat: the Biology of its Behaviour* (eds D.C. Turner and P. Bateson), Cambridge University Press, Cambridge, pp. 123–47.
Forestry Trust (1991) *The Forestry Trust Information Leaflet*, Forestry Trust for Conservation and Education, Reading.
Forman, R.T.T., Galli, A.E. and Leck, C.F. (1976) Forest size and avian diversity in New Jersey woodlots with some land use implications. *Oecologia*, **26**, 1–8.
Fuller, R.J. (1982) *Bird Habitats in Britain*, Poyser, Calton.
Fuller, R.J. and Moreton, B.D. (1987) Breeding bird population of Kentish sweet chestnut *Castanea sativa* coppice in relation to the age and structure of the coppice. *Journal of Applied Ecology*, **24**, 13–27.
Furness, R.W. (1985a) Plastic particle pollution: accumulation by procellariiform seabirds at Scottish colonies. *Marine Pollution Bulletin*, **16**, 103–6.
Furness, R.W. (1985b) Ingestion of plastic particles by seabirds at Gough Island, South Atlantic Ocean. *Environmental Pollution (Series A)*, **38**, 261–72.
Furness, R.W. and Barrett, R.T. (1991) Seabirds and fish declines. *National Geographic Research and Exploration*, **7**, 82–95.
Galbraith, H. and Furness, R.W. (1983) Habitat and distribution of waders breeding on Scottish agricultural land. *Scottish Birds*, **4**, 98–107.
Gibbs, H.L. and Grant, P.R. (1987) Ecological consequences of an exceptionally strong El Nino event on Darwin's finches. *Ecology*, **68**, 1735–46.
Gibbs, J.A. (1961) Ecology of the birds of Kaingaroa Forest. *Proceedings of the New Zealand Ecological Society*, **8**, 29–38.
Glynn, P.W. (1988) El Nino-Southern Oscillation 1982–1983: nearshore population, community, and ecosystem responses. *Annual Review of Ecology and Systematics*, **19**, 309–45.
Glynn, P.W. (ed.) (1990) *Global Ecological Consequences of the 1982–83 El Nino-Southern Oscillation*, Elsevier, Amsterdam.
Goudie, A. (1990) *The Human Impact on the Natural Environment*, 3rd edn, Basil Blackwell, Oxford.
Graham, R.W. and Grimm, E.C. (1990) Effects of global climatic change on the patterns of terrestrial biological communities. *Trends in Ecology and Evolution*, **5**, 289–92.
Grant, P.R. and Grant, B.R. (1987) The extraordinary El Nino event of 1982-3: effects on Darwin's finches on Isla Genovesa, Galapagos. *Oikos*, **49**, 55–66.
Hall, R.J. (1987) Impact of pesticides on bird populations, in *Silent Spring Revisited* (eds G.J. Marco, M. Hollingworth and W. Durham), American Chemical Society, Washington, D.C., pp. 85–111.
Hamer, K.C., Furness, R.W. and Caldow, R.W.G. (1991) The effects of changes in food availability on the breeding ecology of great skuas *Catharacta skua* in Shetland. *Journal of Zoology, London*, **223**, 175–88.
Hansson, L. (1983) Bird numbers across edges between mature conifer forest and clearcuts in central Sweden. *Ornis Scandinavica*, **14**, 97–103.

Heino, R. (1978) Climatic change in Finland during the last hundred years. *Fennia*, **150**, 3–13.

Henderson-Sellars, A. (1990) Modelling and monitoring 'greenhouse' warming. *Trends in Ecology and Evolution*, **5**, 270–5.

Hengeveld, R. (1990) Theories on species responses to variable climates, in *Landscape-ecological Impact of Climatic Change* (eds M.M. Boer and R.S. de Groot), IOS Press, Amsterdam, pp. 274–89.

Hill, D., Taylor, S., Thaxton, R., Amphlet, A. and Horn, W. (1990) Breeding bird communities of native pine forest, Scotland. *Bird Study*, **37**, 133–41.

Holdgate, M.W. (1979) *A Perspective of Environmental Pollution*, Cambridge University Press, Cambridge.

Holdgate, M.W., Kassas, M. and White, G.F. (1982) *The World Environment 1972–1982*, UNEP/Tycooly, Dublin.

Holling, C.S. (1973) Resilience and stability of ecological systems. *Annual Review of Ecology and Systematics*, **4**, 1–23.

Holt, R.D. (1990) The microevolutionary consequences of climatic change. *Trends in Ecology and Evolution*, **5**, 311–15.

Hooper, M.D. (1977) Hedgerows and small woodlands, in *Conservation and Agriculture*, (eds J. Davidson and R. Lloyd), Wiley, Chichester, pp. 45–57.

Houghton, J.T., Jenkins, G.J. and Ephraums, J.J. (1990) *Climatic Change, The IPCC Scientific Assessment*, Cambridge University Press, Cambridge.

Hustings, F., Post, F. and Schepers, F. (1990) Verdwijnt de Grauwe Gors *Miliaria calandra* als broedvogel uit Nederland? (Are corn buntings *Miliaria calandra* disappearing as breeding birds in the Netherlands?) *Limosa*, **63**, 103–11.

Inglis, I.R., Isaacson, A.J., Thearle, R.J.P. and Westwood, N.J. (1990) The effect of changing agricultural practice upon woodpigeon *Columba palumbus* numbers. *Ibis*, **132**, 262–72.

James, F.C. and Wamer, N.O. (1982) Relationships between temperate forest bird communities and vegetation structure. *Ecology*, **63**, 159–71.

Järvinen, O. and Väisänen, R.A. (1979) Climatic changes, habitat changes, and competition: dynamics of geographical overlap in two pairs of congeneric bird species in Finland. *Oikos*, **33**, 261–71.

Jarvis, P.J. (1979) The ecology of plant and animal introductions. *Progress in Physical Geography*, **3**, 187–214.

Jarvis, P.J. (1983a) Introduced species and nature conservation. *Department of Geography, Working Paper* **23**, University of Birmingham, Birmingham (UK).

Jarvis, P.J. (1983b) *Heavy Metal Pollution, an Annotated Bibliography: 1976–1980*, Geobooks, Norwich.

Joyce, B., Williams, G. and Woods, A. (1988) Hedgerows: still a cause for concern. *RSPB Conservation Review*, **2**, 34–7.

Kavanagh, R.P., Shields, J.M., Recher, H.F. and Rohan-Jones, W.G. (1985) Bird populations of a logged and unlogged forest mosaic in the Eden Woodchip Area, in *Birds of Eucalypt Forests and Woodlands: Ecology, Conservation, Management* (eds A. Keast, H.F. Recher, H. Ford and D. Saunders), Surrey Beatty, Chipping Norton (New South Wales), pp. 273–81.

King, C. (1984) *Immigrant Killers. Introduced Predators and the Conservation of Birds in New Zealand*, Oxford University Press, Auckland.

Kullman, L. (1991) Cataclysmic response to recent cooling of a natural boreal pine (*Pinus sylvestris* L.) forest in northern Sweden. *New Phytologist*, **117**, 351–60.

Lack, D. (1933) Habitat selection in birds, with special reference to the effects of afforestation on the Breckland avifauna. *Journal of Animal Ecology*, **2**, 239–62.

Lack, D. (1939) Further changes in the Breckland avifauna caused by afforestation. *Journal of Animal Ecology*, **8**, 277–85.

Lack, D. and Lack, E. (1951) Further changes in bird life caused by afforestation. *Journal of Animal Ecology*, **20**, 173–9.

Lack, P. (1992) *Birds on Lowland Farms*, HMSO, London.

Leighton, F.A., Butler, R.G. and Peakall, D.B. (1985) Oil and Arctic marine birds: an assessment of risk, in *Petroleum Effects in the Arctic Environment* (ed. F.R. Engelhardt), Elsevier, London, pp. 183–215.

Levenson, J.B. (1981) Woodlots as biogeographic islands in southeastern Wisconsin, in *Forest Island Dynamics in Man-dominated Landscapes*, (eds R.L. Burgess and D.M. Sharpe), Springer-Verlag, New York, pp. 13–39.

Lever, C. (1987) *Naturalized Birds of the World*, Longman, Harlow.

Lindberg, P. (1984) Mercury in feathers of Swedish gyrfalcons *Falco rusticolus* in relation to diet. *Bulletin of Environmental Contamination and Toxicology*, **12**, 227–32.

Lindberg, P. and Odsjo, T. (1983) Mercury levels in feathers of peregrine falcon *Falco peregrinus* compared with total mercury content in some of its prey species in Sweden. *Environmental Pollution (Series B)*, **5**, 297–318.

Loman, J. and von Schantz, T. (1991) Birds in a farmland – more species in small than in large habitat island. *Conservation Biology*, **5**, 176–88.

Long, J.L. (1981) *Introduced Birds of the World*. David & Charles, Newton Abbot.

Loyn, R.H. (1985) Ecology, distribution and density of birds in Victorian forests, in *Birds of Eucalypt Forests and Woodland: Ecology, Conservation, Management* (eds A Keast, H.F. Recher, H. Ford and D. Saunders), Surrey Beatty, Chipping Norton (New South Wales) pp. 33–46.

Lynch, J.F. and Whigham, D.F. (1984) Effects of forest fragmentation on breeding bird communities in Maryland, USA. *Biological Conservation*, **28**, 287–324.

MacKinnon, J. and MacKinnon, K. (1986a) *Review of the Protected Area System in the Afrotropical Realm*, IUCN/UNEP, Gland.

MacKinnon, J. and MacKinnon, K. (1986b) *Review of the Protected Area System in the Indo-Malayan Realm*, IUCN/UNEP, Gland.

Marquiss, M., Newton, I. and Ratcliffe, D.A. (1978) The decline of the raven (*Corvus corax*) in relation to afforestation in southern Scotland and northern England. *Journal of Applied Ecology*, **15**, 129–44.

McGowan, J.A. (1990) Climate and change in oceanic ecosystems: the value of time-series data. *Trends in Ecology and Evolution*, **5**, 293–9.

Mearns, R. (1983) The status of the raven in southern Scotland and Northumbria. *Scottish Birds*, **12**, 211–18.

Milledge, D.R. and Recher, H.F. (1985) A comparison of forest bird communities on the New South Wales south and mid-north coasts, in *Birds of Eucalypt Forests and Woodland: Ecology, Conservation, Management* (eds A. Keast, H.F. Recher, H. Ford and D. Saunders), Surrey Beatty, Chipping Norton (New South Wales), pp. 47–52.

Montevecchi, W.A. (1991) Incidence and types of plastic in gannets' nests in the northwest Atlantic. *Canadian Journal of Zoology*, **69**, 295–7.

Moore, N.W. (1962) The heaths of Dorset and their conservation. *Journal of Ecology*, **50**, 369–91.

Moore, N.W. and Hooper, M.D. (1975) On the number of bird species in British woods. *Biological Conservation*, **8**, 239–50.

Murphy, E.C., Springer, A.M. and Roseneau, D.G. (1991) High annual variability in reproductive success of kittiwakes (*Rissa tridactyla* L.) at a colony in western Alaska. *Journal of Animal Ecology*, **60**, 515–34.

Myers, N. (1979) *The Sinking Ark*, Pergamon, Oxford.

Myers, N. (1983) Tropical moist forests: over-exploited and under-utilized? *Forest Ecology and Management*, **6**, 59–79.

Newton, I. (1986) *The Sparrowhawk*, Poyser, Berkhamsted.

Newton, I., Davis, P. and Davis, J. (1982) Ravens, buzzards and land-use in Wales. *Journal of Applied Ecology*, **19**, 681–706.

Newton, I., Marquiss, M. and Moss, D. (1979) Habitat, female age, organochlorine compounds and breeding of European sparrowhawks. *Journal of Applied Ecology*, **16**, 777–93.

Noirfalise, E. and Vanesse, R. (1976) *Study on the Heathlands of Western Europe*, Council of Europe, Brussels.

Noon, B.R. and Young, K. (1991) Evidence of continuing worldwide declines in bird populations: insights from an international conference in New Zealand. *Conservation Biology*, **5**, 141–3.

O'Connor, R.J. and Shrubb, M. (1986) *Farming and Birds*, Cambridge University Press, Cambridge.

O'Halloran, J., Gribbin, S.D, Tyler, S.J. and Ormerod, S.J. (1990) The ecology of dippers *Cinclus cinclus* (L.) in relation to stream acidity in upland Wales: time–activity budgets and energy expenditure. *Oecologia*, **85**, 271–80.

Orians, G.H. (1975) Diversity, stability and maturity in natural ecosystems, in *Unifying Concepts in Ecology*, (eds W.H. van Dobben and R.H. Lowe-McConnell), Dr W. Junk, The Hague, pp. 139–50.

Ormerod, S.J., Allinson, N., Hudson, D. and Tyler, S.J. (1986) The distribution of breeding dippers in relation to stream acidity in upland Wales. *Freshwater Biology*, **16**, 501–7.

Ormerod, S.J., Bull, K.R., Cummins, C.P., Tyler, S.J. and Vickery, J.A. (1988) Egg mass and shell thickness in dippers *Cinclus cinclus* in relation to stream acidity in Wales and Scotland. *Environmental Pollution*, **55**, 107–21.

Osborne, P. (1984) Bird numbers and habitat characteristics in farmland hedgerows. *Journal of Applied Ecology*, **21**, 63–82.

Pain, S. (1989) Alaska has its fill of oil. *New Scientist*, **123** (1677), 34–40.

Parr, S.J. (1991) Occupation of new conifer plantations by merlins in Wales. *Bird Study*, **38**, 103–11.

Parry, M. (1990) *Climatic Change and World Agriculture*, Earthscan, London.

Parry, M.L., Porter, J.H. and Carter, T.R. (1990) Agriculture, climatic change and its implications. *Trends in Ecology and Evolution*, **5**, 318–22.

Parsons, P.A. (1989) Environmental stresses and conservation of natural populations. *Annual Review of Ecology and Systematics*, **20**, 29–49.

Parsons, P.A. (1990) The metabolic cost of multiple environmental stresses: implications for climatic change and conservation. *Trends in Ecology and Evolution*, **5**, 315–17.

Peach, W., Baillie, S. and Underhill, L. (1991) Survival of British sedge warblers *Acrocephalus schoenobaenus* in relation to West African rainfall. *Ibis*, **133**, 300–5.

Peterken, G.F. and Allison, H. (1989) Woods, trees and hedges: a review of changes in the British countryside. *Focus on Nature Conservation* **22**, Nature Conservancy Council, Peterborough.

Peters, R.L. (1990) Effects of global warming on forests. *Forest Ecology and Management*, **35**, 13–33.

Petty, S.J. and Avery, M.I. (1990) Forest bird communities. *Forestry Commission Occasional Paper 26*, HMSO, London.
Pimm, S.L. (1987) The snake that ate Guam. *Trends in Ecology and Evolution*, **2**, 293–5.
Pollard, E., Hooper, M.D. and Moore, N.W. (1974) *Hedges*, Collins, London.
Ranney, J.W., Bruner, M.C. and Levenson, J.B. (1981) The importance of edge in the structure and dynamics of forest islands, in *Forest Island Dynamics in Man-dominated Landscapes* (eds R.L. Burgess and D.M. Sharpe). Springer-Verlag, New York, pp. 67–95.
Reed, D.J. (1990) The impact of sea-level rise on coastal salt marshes. *Progress in Physical Geography*, **14**, 465–81.
Reed, T.M. (1982) Birds and afforestation. *Ecos*, **3**, 8–10.
Robbins, C.S. (1980) Effects of forest fragmentation on breeding bird populations in the Piedmonts of the mid-Atlantic region. *Atlantic Naturalist*, **33**, 31–6.
Rolstad, J. (1991) Consequences of forest fragmentation for the dynamics of bird populations: conceptual issues and the evidence. *Biological Journal of the Linnean Society*, **42**, 149–63.
Rose, C.I. (1979) Observations on the ecology and conservation value of native and introduced tree species. *Quarterly Journal of Forestry*, **73**, 219–28.
Ryan, P.G. (1987) The effects of ingested plastic on seabirds: correlations between plastic load and body condition. *Environmental Pollution*, **46**, 119–25.
Ryan, P.G. and Jackson, S. (1987) The lifespan of ingested plastic particles in seabirds and their effect on digestive efficiency. *Marine Pollution Bulletin*, **18**, 217–19.
Rymer, L. (1981) Pine plantations in Australia as a habitat for native animals. *Environmental Conservation*, **8**, 95–6.
Samiullah, Y. (1990) Biological Monitoring: Animals. *MARC Report 27*, GEMS Monitoring and Research Centre, Kings College, London.
Sanderson, G.C. and Bellrose, F.C. (1986) A review of the problems of lead poisoning in waterfowl. *Illinois Natural History Society Survey, Special Publication 4*, Illinois Natural History Society.
Saunders, D.A. and Curry, P.J. (1990) The impact of agricultural and pastoral industries on birds in the southern half of Western Australia: past, present and future. *Proceedings of the Ecological Society of Australia*, **16**, 303–21.
Savidge, J.A. (1987) Extinction of an island forest avifauna by an introduced snake. *Ecology*, **68**, 660–8.
Sayer, J.A. and Whitmore, T.C. (1991) Tropical moist forests: destruction and species extinction. *Biological Conservation*, **55**, 199–213.
Schindler, D.W., Beaty, K.G., Fee, E.J. *et al.* (1990) Effects of climatic warming on lakes of the central boreal forest. *Science*, **250**, 967–70.
Schmidt, H. (1990) Die Bestandsentwicklung des Turmfalken *Falco tinnunculus* in der Schweiz (The status of the kestrel *Falco tinnunculus* in Switzerland). *Ornithologische Beobachter*, **87**, 327–49.
Shales, S., Thake, B.A., Frankland, B. *et al.* (1989) Biological and ecological effects of oil, in *The Fate and Effects of Oil in Freshwater* (eds J. Green and M.W. Trett), Elsevier, London, pp. 81–171.
Shields, J.M., Kavanagh, R.P. and Rohan-Jones, W.G. (1985) Forest avifauna of the Upper Hastings River, in *Birds of Eucalypt Forests and Woodland: Ecology, Conservation, Management* (eds A. Keast, H.F. Recher, H. Ford

and D. Saunders), Surrey Beatty, Chipping Norton (New South Wales), pp. 55–64.

Shrubb, M. (1990) Effects of agricultural change on nesting lapwings *Vanellus vanellus* in England and Wales. *Bird Study*, **37**, 115–27.

Shrubb, M. and Lack, P.C. (1991) The numbers and distribution of lapwings *V. vanellus* nesting in England and Wales in 1987. *Bird Study*, **38**, 20–37.

Shugart, H.H. (1990) Using ecosystem models to assess potential consequences of global climatic change. *Trends in Ecology and Evolution*, **5**, 303–7.

Simms, E. (1971) *Woodland Birds*, Collins, London.

Smith, K.W. (1983) The status and distribution of waders breeding on wet lowland grasslands in England and Wales. *Bird Study*, **30**, 177–92.

Smith, P. (1985) Woodchip logging and birds near Bega, New South Wales, in *Birds of Eucalypt Forests and Woodlands: Ecology, Conservation, Management* (eds A. Keast, H.F. Recher, H. Ford and D. Saunders), Surrey Beatty, Chipping Norton (New South Wales), pp. 259–71.

Sombroek, W.G. (1990) Soils on a warmer Earth: the tropical regions, in *Soils on a Warmer Earth* (eds H.W. Scharpenseel *et al.*), Elsevier, Amsterdam, pp. 18–28.

Stigliani, W.M. (1990) Climate change and its potential effects on the retention and release of hazardous chemicals in soils and sediments, in *Landscape-ecological Impact of Climatic Change* (eds M.M. Boer and R.S. de Groot), IOS Press, Amsterdam, pp. 361–78.

Stott, P. (1991) Recent trends in the ecology and management of the world's savanna formations. *Progress in Physical Geography*, **15**, 18–28.

Szaboles, I. (1990) Effects of predicted climatic changes on European soils, with particular regard to salinization, in *Landscape-ecological Impact of Climatic Change* (eds M.M. Boer and R.S. de Groot), IOS Press, Amsterdam, pp. 177–93.

Taylor, R.H. and Wilson, P.R. (1990) Recent increase and southern expansion of Adelie penguin populations in the Ross Sea, Antarctica, related to climatic warming. *New Zealand Journal of Ecology*, **14**, 25–9.

Terrasson, F. and Tendron, G. (1975) *Evolution and Conservation of Hedgerow Landscapes in Europe*, Council of Europe, Brussels.

Thomas, G.J. (1982) Autumn and winter feeding ecology of waterfowl at the Ouse Washes, England. *Journal of Zoology, London*, **197**, 131–72.

Titus, J.G. and Seidel, S. (1986) Overview of the effects of changing the atmosphere, in *Effects of Changes in Stratospheric Ozone and Global Climate* (ed. J.G. Titus), UNEP/USEPA, Washington, D.C., pp. 3–19.

Turner II, B.L., Clark, W.C., Kates, R.W. *et al.* (eds) (1990) *The Earth as Transformed by Human Action: Global and Regional Changes in the Biosphere Over the Past 300 Years*, Cambridge University Press, Cambridge.

UK Stratospheric Ozone Research Group (1991) *Stratospheric Ozone*, HMSO, London.

van Dorp, D. and Opdam, P.F.M. (1987) Effects of patch size, isolation and regional abundance on forest bird communities. *Landscape Ecology*, **1**, 59–73.

van Franeker, J.A. (1985) Plastic ingestion by the North Atlantic fulmar. *Marine Pollution Bulletin*, **16**, 367–9.

van Franeker, J.A. and Bell, P.J. (1988) Plastic ingestion by petrels breeding in Antarctica. *Marine Pollution Bulletin*, **19**, 672–4.

von Haartman, L. (1978) Changes in the bird fauna in Finland during the last hundred years. *Fennia*, **150**, 25–32.

Walker, C.(1975) *Environmental Pollution by Chemicals*, 2nd edn, Hutchinson, London.
Walker, D.A. and Walker, M.D. (1991) History and patterns of disturbance in Alaskan Arctic terrestrial ecosystems: a hierarchical approach to analysing landscape change. *Journal of Applied Ecology*, **28**, 244–76.
Webb, N.R. (1989) Studies on the invertebrate fauna of fragmented heathland in Dorset, UK, and the implications for conservation. *Biological Conservation*, **47**, 153–65.
Webb, N.R. and Haskins, L.E. (1980) An ecological survey of heathland in the Poole Basin, Dorset, England, in 1978. *Biological Conservation*, **17**, 281–96.
Whitcomb, R.F., Robbins, C.S., Lynch, J.F. *et al.* (1981) Effect of forest fragmentation on avifauna of the eastern deciduous forest, in *Forest Island Dynamics in Man-dominated Landscapes* (eds R. Burgess and D.M. Sharp), Springer-Verlag, New York, pp. 125–205.
Wigley, T.M.L. (1983) The pre-industrial carbon dioxide level. *Climatic Change*, **5**, 315–20.
Williamson, P. and Holligan, P.M. (1990) Ocean productivity and climatic change. *Trends in Ecology and Evolution*, **5**, 299–303.
Wilson, W.L. and Johns, A.D. (1982) Diversity and abundance of selected animal species in undisturbed forest, selectively logged forest and plantations in East Kalimantan, Indonesia. *Biological Conservation*, **24**, 205–18.
Winstanley, D. (1973) Rainfall patterns and general atmospheric circulation. *Nature*, **245**, 190–4.
Woodward, F.I. (1990) Global change: translating plant ecophysiological responses to ecosystems. *Trends in Ecology and Evolution*, **5**, 308–11.
Woolhouse, M.E.J. (1983) The theory and practice of the species-area effect applied to breeding birds of British woods. *Biological Conservation*, **27**, 315–32.
WRI/IIED (1987) *World Resources, 1987–88*, World Resources Institute/International Institute for Environment and Development, Basic Books, New York.
WRI/IIED (1988), *World Resources, 1988–89*, World Resources Institute/International Institute for Environment and Development, Basic Books, New York.
Yapp, W.B. (1962) *Birds and Woods*. Oxford University Press, Oxford.

3
Birds as monitors of pollutants

R.W. Furness

3.1 INTRODUCTION

Several authors of books on the monitoring of pollution have advocated the use of animals as monitors in terrestrial and aquatic environments (e.g. Phillips, 1980; Schubert, 1985). Such studies tend to emphasize the use of sedentary invertebrate animals as biomonitors. By comparison, birds suffer from several apparent drawbacks. They are mobile, so pollutants will be picked up from a wide, often ill-defined, area; they are long-lived, so pollutant burdens may be integrated in some complex way over time; and they have more complex physiology, and so may regulate pollutant levels better then invertebrates. Furthermore, birds tend to be more difficult to sample, and killing birds may be unacceptable for conservation or ethical reasons. However, some of these characteristics may at times be positively advantegeous. Integrating pollutant levels over greater areas or time-scales or over food webs, may be useful, provided that species are chosen carefully. Less sampling may be necessary if birds can reflect pollutant levels in the whole ecosystem or over a broad area. In addition, since they are high in food chains, birds may reflect pollutant hazards to humans better than do most invertebrates. It is also significant that birds are extremely popular animals with the general public, so pollutant hazards to them are likely to receive greater attention than threats to invertebrates.

Gilbertson *et al.* (1987) point out that an important consideration in selection of monitoring species is the coefficient of variation of the pollutant level within the population. They provide some evidence

Birds as Monitors of Environmental Change. Edited by R.W. Furness and J.J.D. Greenwood.
Published in 1993 by Chapman & Hall, London. ISBN 0 412 40230 0.

Introduction

indicating that organochlorine levels in seabirds show lower coefficients of variation than in fish or marine mammals, allowing as tight a confidence interval to be obtained from analysis of seabirds as from a larger sample of fish or mammals. This makes sampling seabirds more cost effective in the case of organochlorines. Similar considerations may apply to other pollutants, such as metals, and in particular to the choice of tissue, blood, egg, feather or chick as sampling unit. Advantages of different tissues for pollutant monitoring are discussed in the sections on pesticides, PCBs, heavy metals, and radionuclides, as appropriate.

Most ornithologists would agree that the clearest example of the use of birds as monitors of the environment is their use over the last three decades as qualitative indicators and quantitative monitors of pesticides in food webs. Birds have been particularly successful in this regard because organochlorines are not easily biodegraded and are lipophilic, with very low solubility in water. These characteristics result in bioaccumulation and bioamplification of residues through the food chain. Animals that ingest organochlorines assimilate and store them in their lipid-rich tissues and, because the rate of breakdown of organochlorines in animal tissues is slow, the concentration in the body will increase to a high dynamic equilibrium or even increase throughout the life of the animal (bioaccumulation). Thus organochlorine levels in animals are generally much higher than in their food. Since these animals are themselves food for the next trophic level, organochlorines become concentrated through each step in the food chain (bioamplification). As a result, top predators accumulate the highest concentrations of the pollutant. As it happens, birds of prey have been found to be especially sensitive to the toxic effects of organochlorines and their populations were affected before any other signs of the increasing levels of organochlorines in ecosystems were manifest. The deaths of birds of prey, the thinning of eggshells, and failure to hatch eggs were qualitative signs, or indicators, of the organochlorine problem. Subsequent research has shown that some responses, such as eggshell thinning, can be related quantitatively to organochlorine burdens and so can be used as quantitative monitors of pesticide levels in bird of prey populations.

The success of research into organochlorine pollution in birds has led to a general view that top predators are the best organisms for monitoring of pollutants. This is certainly not universally so. Many pollutants are water-soluble and not lipophilic, so their concentrations do not increase up the food chain. For these substances there are no good reasons to look to top predators as monitors and there are many reasons to use animals at lower trophic levels.

This chapter reviews the use of birds as monitors or indicators for a wide range of different pollutants. The value of birds as indicators of pesticide problems in the environment, as monitors of residue levels and toxic effects, especially of the organochlorines, is so well known and so well reviewed in many books (Diamond and Filion, 1987; Moriarty, 1988; Newton, 1979, 1986; Ratcliffe, 1980) that the chapter will deal only briefly with this subject and devote more space to aspects of pollution that have received less attention. The aim is to consider critically the contribution that birds can make to pollutant monitoring and the limitations and assumptions of such an approach. The particular subjects of birds as monitors of radionuclide pollution and birds as monitors of freshwater quality (especially acidification) are dealt with in chapters by Brisbin (Chapter 4) and by Ormerod and Tyler (Chapter 5) and are mentioned only briefly here.

Surveillance or monitoring of pollutants may be for various reasons: to measure the level of environmental contamination or the rate of change in contamination, to assess the rate of release into the environment of a pollutant, to assess the biological effects of pollutants on species or communities, or to assess the hazard to man. In practice, many monitoring programmes attempt to examine several of these simultaneously, which may obscure the main aim of the monitoring programme (Mejstrik and Pospisil, 1989; Moriarty, 1988). Unfortunately, a programme that is optimally designed for one purpose is often unsuitable for others. For example, the MAFF monitoring of pollutant levels in fish around the British Isles has as a main aim the assessment of hazard to humans through consumption of fish (Franklin, 1990, 1991). As a result, pollutant levels are measured in the edible portion of fish, rather than in the organs in which the pollutants may be preferentially stored, and attention is focused on obtaining small numbers of fish from all areas rather than looking particularly at areas where pollution levels may be expected to be high, such as close to industrial estuaries or sewage dumping grounds. Environmental biologists interested in the uptake of pollutants in food chains in contaminated areas would need to use a different monitoring protocol.

The concept of 'sentinel organisms' is common in pollution studies. Sentinel species are usually chosen because they are especially sensitive to a particular pollutant and so provide the earliest warning of pollution in an ecosystem. Such sentinel organisms can be used as indicator species to detect increasing pollution. As discussed in Chapter 1, it can be useful to draw a distinction between such qualitative indicators of a problem and the use of organisms to provide a quantitative monitor of pollutant levels. The latter is inherently far more difficult since organisms will respond to the combined influences of various different pollutants and to other environmental stresses. For example, a bird of prey suffering unusually low breeding success may have high levels of DDE in its tissues, but levels of other organochlorines tend to be correlated with DDE levels, as may levels of alkyl-mercury. Furthermore, a substantial part of breeding failure may result from factors such as food supply or weather, which may well interact with pollutant stresses. Perhaps the greatest challenge in the use of birds as monitors of pollutants lies in disentangling such interactions.

I shall deal with each category of pollutant in turn, although the same principles often apply to more than one type of pollutant.

3.2 PESTICIDES

3.2.1 Organochlorine insecticides

The insecticide DDT was introduced in the 1940s and quickly became widely used throughout the areas of intensive agriculture in the UK and elsewhere. In the late 1950s cyclodiene organochlorines (dieldrin, aldrin and heptachlor) became commonly used compounds, dieldrin particularly in sheep dips and all three as seed dressings against soil pests. Cyclodiene organochlorines have been used much more extensively in the UK than in North America, whereas the use of DDT was widespread in both continents in the 1950s and early 1960s. Although the organochlorine insecticides have a high contact toxicity for insects, their contact toxicity for humans and other vertebrates is low, especially in the case of DDT. This, their stability, and their low cost made them particularly popular as insecticides during the post-war boom in agricultural intensification. Harmful environmental side-effects were not anticipated, particularly from the more toxic cyclodienes because these were targeted directly at pests by application as seed dressings rather than disseminated as sprays.

In 1961 the British Trust for Ornithology carried out a national peregrine *Falco peregrinus* survey in Britain, in response to complaints to the Nature Conservancy from pigeon racers that peregrines were killing too many of their birds. This survey showed that peregrine breeding numbers had fallen dramatically from a fairly stable level of about 800 pairs before 1940: site occupancy in 1961 was only 68% of the earlier level, in 1962 56%, and in 1963 only 44% with the decreases most dramatic in the south and east. The story is told in detail by Ratcliffe (1980). No evidence could be found to link this decline with weather, food, habitat changes, disease or human persecution, but the finding of unexpectedly high residues of DDE (the stable metabolite of DDT) and the dieldrin and aldrin metabolite

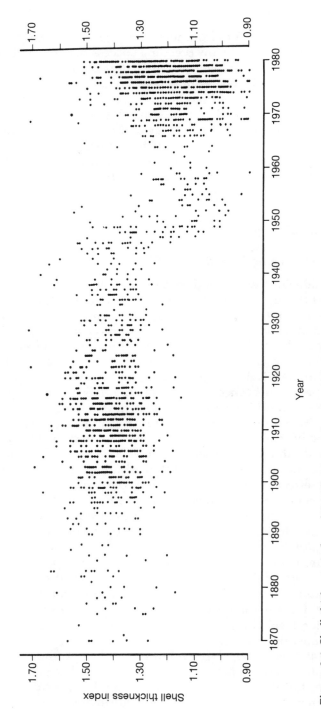

Figure 3.1 Shell thickness index of British sparrowhawks. (After Newton, 1986.)

HEOD in peregrine tissues and eggs (Moore and Ratcliffe, 1962) suggested a toxic effect, particularly since residue levels were higher in areas of greatest peregrine decline and similar to levels shown experimentally to be toxic to some species of laboratory birds. Further evidence for such a link was soon found. By comparing fresh samples of eggs with museum specimens, Ratcliffe (1967, 1970) demonstrated that shell thinning had begun at the same time as the first widespread use of DDT in 1947 (Figure 3.1). It later became clear that the degree of shell thinning in birds could be related to DDE residues in the egg contents and laying female (but see Bunck et al., 1985) although other organochlorines did not cause shell thinning. The dose–response relationship (Figure 3.2) varied among bird species, with birds of prey being particularly sensitive to this effect of DDE (in addition to tending to accumulate the highest levels because of food chain bioamplification). Shell-thinning caused eggshell breakage and hence reduced breeding success, and this was detected both by studies of breeding performance within bird of prey populations and by retrospective analysis of the British Trust for Ornithology's nest record cards, which showed a reduction in brood sizes after the introduction of DDT.

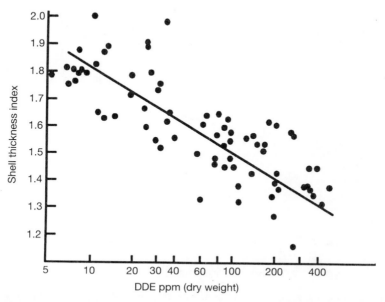

Figure 3.2 Shell thinning in relation to egg DDE levels in peregrines. (After Peakall et al., 1975.)

Evidence was gained in the Netherlands, UK and USA of increased mortality of adult birds of prey because of high accumulated burdens of organochlorines (Newton, 1979). Declines in particular populations gave further evidence for the effects of organochlorines. For example, peregrine numbers and productivity in Alaska, where DDT usage was never extensive, declined only after 1967, correlated with increased use of DDT in South America, where the Alaskan peregrines spend the winter (Cade *et al.*, 1971; Springer *et al.*, 1984).

In the light of strong, albeit circumstantial, evidence, use of DDT, dieldrin and other organochlorine insecticides was progressively reduced in Western Europe and North America, both by voluntary controls and by legislation. Subsequent monitoring of residues in tissues and eggs, and of the numbers and breeding success of birds of prey, has shown clear recoveries after the reduction of organochlorine levels in the food chain (Figure 3.3, and Newton and Wyllie, 1992).

Some details of the effects of organochlorines on birds of prey have been a matter of dispute and the mechanisms remain to be fully elucidated. For example, more recent work indicates that DDT can cause feminization of bird embryos (Fry and Toone, 1981). It now seems that in Britain a decline in the numbers of peregrines and other birds of prey was due predominantly to dieldrin poisoning and not

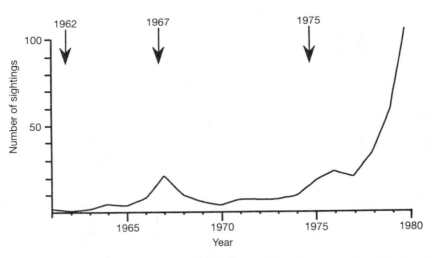

Figure 3.3 Recovery of sparrowhawk numbers after the restrictions in use of organochlorine pesticides in Hertfordshire, England. Dates of major restrictions in the use of aldrin and dieldrin are marked with arrows. (After Newton, 1986.)

to the effects of DDE. Numbers crashed just after the widespread use of dieldrin and aldrin, which are many times more toxic than DDE. In North America, where the cyclodienes were not so extensively used, bird of prey populations declined more slowly, probably as a result of the effect of DDE on breeding success rather than adult mortality (Newton, 1986).

Peregrines and other birds of prey acted as 'sentinel' species for the harmful environmental effects of organochlorines. Had baseline population survey data for birds of prey not existed, and volunteer ornithological survey efforts not been directed into a peregrine census in 1961, the effects of organochlorines on wildlife would not have been clearly evident until the environment had become even more severely polluted. The background information on birds and the strong interest in them combined, by good fortune, to point to the existence of a problem. The possibility of turning to museum collections to investigate long-term changes in eggshells was invaluable in helping to provide clear evidence to convince sceptics that organochlorine usage was the cause of the problem.

Although the use of birds of prey to indicate a DDT pollution problem is an outstanding example of the value of birds as biological indicators of environmental change, there are a number of major limitations to this approach. Because birds of prey are scarce (as are most top predators) it is not possible to take large samples of birds or of eggs to monitor spatial variation or long-term temporal changes in organochlorine levels. Instead, tissues of adults are obtained from birds fortuitously found dead, which include birds that have starved. Since starvation causes the mobilization of fat reserves, in which organochlorines are held, the depletion of fat reserves in a starving bird leads to an increase in the concentration of organochlorines in soft tissues, especially the liver (Bogan and Newton, 1977). Highly emaciated birds may use up almost all of their body fat and a substantial amount of muscle protein before they die, giving concentrations of organochlorines an order of magnitude higher than before starvation sets in. If it is assumed that the proportion of emaciated birds remains about the same over time, then this effect simply increases the variance of organochlorine levels within samples, making trends more difficult to confirm statistically. If only birds whose cause of death is known are used for analysis, the variance is diminished but the already small sample size is severely reduced.

Even by purposely killing birds for organochlorine analysis, problems of defining organochlorine concentrations are not completely overcome. Because of seasonal changes in body composition, involving lipid stores in particular, organochlorine levels fluctuate seasonally as well as in relation to age, sex and diet (Anderson and Hickey, 1976;

Anderson *et al.*, 1984; Clark *et al.*, 1987; Newton, 1986). Another problem is that some analysts present levels per unit fresh weight, others in relation to dry weight, and others per unit weight of lipid. Such values are often not readily interchangeable because conversion ratios change seasonally, vary between populations according to their feeding conditions, and particularly vary among species.

One solution to the problems of sampling adult tissues is to use eggs instead. However, organochlorine levels may vary through the clutch sequence, generally being higher in last-laid eggs than in the first laid, apparently because birds produce their first eggs mostly from recent dietary uptake whereas body reserves contribute more to later eggs (Mineau, 1982). Furthermore, regular egg collection from most bird of prey populations is unacceptable, so analyses are often restricted to addled eggs remaining in the nest after the chicks have hatched. Addled eggs may not have pollutant levels typical of the whole population as they are more likely to be produced by young birds, birds of poor quality, or those adversely affected by toxic substances. However, although this problem should be borne in mind, some comparisons

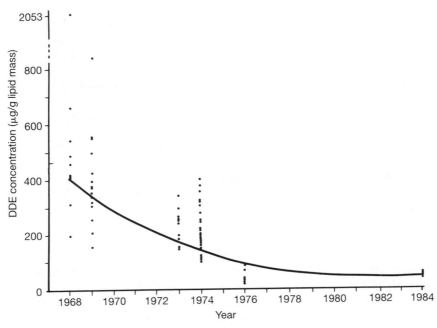

Figure 3.4 Exponential decrease in DDE levels in eggs of gannets from Bonaventure Island, Canada, following restrictions in the use of DDT in North America in the 1960s. (After Chapdelaine *et al.*, 1987.)

suggest that pollutant levels in addled eggs may not differ much from those in fertile developing eggs. The use of addled eggs is made slightly more difficult by the fact that eggs lose water during incubation, and hence eggs collected from nests long after hatching was due tend to be severely dehydrated, and sometimes highly infected with bacteria.

Although birds of prey have remained an important focus for surveillance of organochlorine levels in environments, many investigators have preferred to avoid the difficulties of small sample sizes by studying more abundant bird species, usually slightly lower in the food chain. Becker et al. (1985a, 1988) and Becker (1991) chose eggs of common terns *Sterna hirundo*, herring gulls *Larus argentatus* and other common coastal birds to monitor organochlorine contamination of the North Sea coast. Substantial variation was seen in the degree of contamination of populations close to polluted rivers and the overall contamination had declined in recent years due to reduced use of organochlorines. Chapdelaine et al. (1987) showed that DDE and dieldrin levels in gannet *Sula bassana* eggs from Bonaventure Island decreased from 1968 to 1984 (Figure 3.4), and that this change corresponded to a period of increased hatching success from abnormally low levels (for gannets) in 1966–70 attributed to organochlorine toxicity (Table 3.1).

Table 3.1 Hatching success of gannets on Bonaventure Island, Quebec, in relation to organochlorine levels in the eggs (µg/g lipid) (from Chapdelaine et al., 1987)

Year	Number of samples	Organochlorine data					Breeding data	
		% lipid	DDE mean	(S.D.)	Dieldrin mean	(S.D.)	Number of nests	Hatching success (%)
1966	0						437	40
1967	0						507	36
1968	10	5.3	435.6	(44.1)	17.1	(2.8)	–	–
1969	19	5.3	409.9	(36.0)	14.8	(1.3)	–	–
1970	0						261	37
1973	10	3.9	234.9	(18.4)	8.3	(1.3)	–	–
1974	30	4.8	222.7	(14.3)	6.9	(0.5)	503	58
1976	6	4.6	48.3	(8.2)	6.8	(0.5)	474	85
1979	0						489	89
1984	6	4.9	29.2	(2.6)	3.8	(0.3)	531	78

The US Fish and Wildlife Service chose to monitor organochlorine and other pollutant levels in bald eagles *Haliaeetus leucocephalus*, mallards *Anas platyrhynchos*, black ducks *Anas rubripes*, and starlings *Sturnus vulgaris* (Cain and Bunck, 1983; Dustman et al., 1971). Bald eagles were chosen because they are top predators most likely to suffer toxic effects of organochlorines (Frenzel and Anthony, 1989); mallards and black ducks because their combined ranges cover almost all of the USA and samples were readily available through sport shooting; starlings because they are abundant and considered a pest so that their killing in large numbers for tissue analysis was unlikely to cause adverse public reaction.

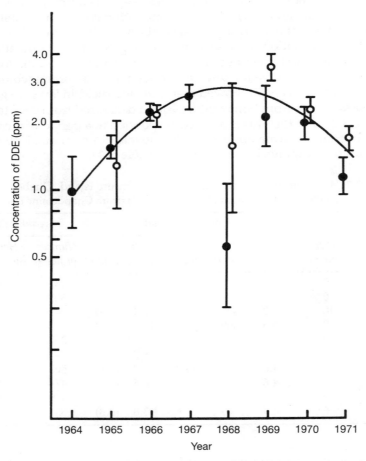

Figure 3.5 Concentrations of DDE in shag eggs from the Farne Islands (closed circles) and Isle of May (open circles). Values plotted are geometric means, and a quadratic model provided the best fit regression as a function of time (excluding anomalous points from 1968 when levels were affected by a preceding red tide). (After Coulson et al., 1972.)

Organochlorines may be transported considerable distances from sites of use. Transport can be atmospheric, oceanic, or by the migrations of contaminated animals. Studies of organochlorine residues in seabirds that are resident at high latitudes have provided evidence of the transport of these pollutants even to the polar regions (Nettleship and Peakall, 1987).

Much of the organochlorine surveillance has been directed at establishing long-term trends in pollutant levels. Coulson *et al.* (1972) gave one of the first and most striking examples, showing a decrease in DDE and dieldrin levels in shag *Phalacrocorax aristotelis* eggs in northeast England and south-east Scotland following restrictions in the use of these pesticides in the late 1960s. However, their analyses also highlighted one problem. Levels of both DDE and dieldrin were very much lower in shag eggs in 1968 than in other years, coinciding with a 'red tide' (dinoflagellate bloom) which had profound effects on the birds breeding in 1968. For unknown reasons, this seems to have affected levels of DDE (Figure 3.5) and dieldrin in the eggs that year.

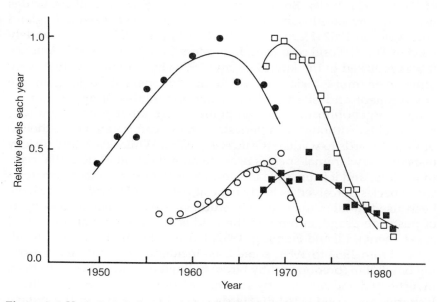

Figure 3.6 Variations in known industrial production of DDT and PCB and the relative levels of DDT and PCB in guillemot eggs. Production of DDT (closed circles); DDT in eggs (open circles); PCB production (closed squares); PCB in eggs (open squares). Maximum annual levels are 81 000 t for DDT production; 30 000 t for PCB production; 640 µg/g extractable lipid for DDT in eggs; 430 µg/g extractable lipid for PCB in eggs. (After Olsson and Reutergardh, 1986.)

Guillemot eggs have been used to show the rapid loss of DDT from food chains following the restriction of its use (Figure 3.6).

Although the main aim of most surveillance of organochlorine is to establish trends in environmental contamination, the question of toxic effects is also important even if more difficult to address. Correlations between organochlorine levels and breeding success or population trends are suggestive, but difficult to attribute to the effects of a single organochlorine because of correlation between different organochlorines, and because of possible additive or synergistic effects. Newton (1988) discusses an approach to this that relates pollutant levels in populations to population trends, in order to assess the likely threshold level that will lead to population decline. He points out that this critical level may bear little relation to the LC_{50}s of laboratory toxicity tests but provides a much better indication of the threshold at which pesticide levels in the environment become unacceptable.

3.2.2 Organophosphorus, carbamate and synthetic pyrethroids

The use of pesticides has altered dramatically since the introduction of DDT in the 1940s. Both the variety and the amount of pesticides used have increased; some 16 708 t of pesticides were used in England and Wales in 1982 (Hardy et al., 1987). Although DDT is now widely used in Third-World countries, in western Europe and North America it was replaced in the late 1960s and 1970s by compounds that were often even more toxic but which did not persist so long in the environment and often had more specific modes of application or action: organophosphorus compounds, carbamates and synthetic pyrethroids. Although less persistent, these compounds nonetheless present hazards to wildlife (Grue et al., 1983). While laboratory toxicity tests on a few animal species provide an indication of potential risks to wildlife, and it is possible in some cases to perform laboratory tests with species perceived as likely to be at risk (over 50 bird species have been used in laboratory toxicity studies), surveillance of the effects of pesticide usage on birds in the wild is an essential part of hazard assessment (Hill and Fleming, 1982; Hill and Hoffman, 1984; Powell, 1984). From 1964 to 1986, some 2500 incidents of suspected poisoning of wildlife (predominantly birds) in England and Wales have been investigated and reported by scientists of the Ministry of Agriculture, Fisheries and Food (MAFF) (Hardy et al., 1987). Many of these have involved suspected poisoning by pesticides. Such surveillance has been valuable in permitting some hazardous chemicals to be identified and withdrawn from use. For example, carbophenothion seed treatment was found to result in many deaths of grey geese *Anser* spp., and it has now been restricted in consequence (Hardy et al., 1987).

Surveillance of pesticide levels in dead birds found fortuitously by the public, especially birds of prey and aquatic top predators such as the grey heron *Ardea cinerea* and kingfisher *Alcedo atthis*, has been a strategy of the NERC-funded laboratories of the Institute of Terrestrial Ecology at Monks Wood. However, cost permits the quantification of residue levels for only a small fraction of the vast numbers of chemicals now in use. Furthermore, organophosphorus pesticides are rapidly metabolized and may not be detectable in birds even though toxic effects may be evident at the biochemical level (Peakall, 1992). The magnitude of the problem is shown by the estimate that 30 000 different man-made chemicals (not all of them pesticides) enter the Great Lakes in North America and some of these are suspected of being toxic to birds (Fox and Weseloh, 1987). One approach to this problem is to combine residue analysis and regular monitoring of eggs or tissues with studies of eggshell thinning (Fox, 1976; Jorgensen and Kraul, 1974), monitoring of breeding performance (Peakall *et al.*, 1980; Weseloh *et al.*, 1983), embryotoxicity (Ellenton *et al.*, 1983; Hoffman *et al.*, 1987), analysis of the behaviour of adult birds, genotoxicity (sister chromatid exchange rates (Ellenton and McPherson, 1983)), congenital abnormality rates (Svecova, 1989) and biochemical parameters (Busby *et al.*, 1983; Ellenton *et al.*, 1985; Walker *et al.*, 1987; Zinkl *et al.*, 1979, 1984), especially blood enzyme levels (Boersma *et al.*, 1986). Such an approach is advocated and discussed by Fox and Weseloh (1987), Peakall and Boyd (1987) and Peakall (1992). It is likely to reveal the toxic effects of pollutants in birds but may not enable the causal agent to be identified because several chemicals may be involved and may interact in unknown ways; levels of lipophilic pollutants tend to be intercorrelated. A slightly different approach is advocated by Mineau *et al.* (1987). They point out that organophosphates, carbamates and synthetic pyrethroids are too numerous to be readily quantified individually as residues in birds and that such measurements are of limited value because the compounds are short-lived, their residue levels being a poor reflection of the impact of the chemicals. They advocate a more experimental approach to discovering the effect of pesticides by means of detailed ecological research (in their particular case on ducks in relation to pesticide usage on the Canadian prairies). They cite as their model for such an approach the extensive field-manipulation studies of the effects of pesticides, herbicides and habitat management on the population biology of the grey partridge *Perdix perdix*, studied by the Game Conservancy in England (Hudson and Rands, 1988; Potts, 1986).

3.2.3 Rodenticides

Although warfarin has a low toxicity to birds that eat poisoned rodents (Townsend et al., 1981), more recently developed rodenticides can be highly toxic to birds (Newton et al., 1990). Those likely to be at risk are some owls, some birds of prey and some gulls, and laboratory toxicity testing has taken this into account when selecting test species (Hardy et al., 1987). Given the known hazard, especially to a scarce and declining species such as the barn owl *Tyto alba*, studies of numbers and breeding have been started in order to monitor any geographical patterns or long-term trends that may be related to patterns of rodenticide usage in the UK. Similar concerns exist elsewhere, for example in the Netherlands, where owl mortality resulting from secondary poisoning by rodenticides is thought to be significant. The existence of owl surveillance data such as nest record cards, and recordings of the breeding distribution and density of populations over long periods before the introduction of new rodenticides, although not planned for this purpose, provides an invaluable baseline with which to compare ecological data gathered over the period of rodenticide use.

3.3 PCBs

Polychlorinated biphenyls (PCBs) have been used since the early 1930s in paints, plastics, electrical equipment (as dielectrics in capacitors for example), as transmission fluids in mechanical equipment, and in a wide variety of other roles. They exist as many different chemical species, differing in number and position of chlorine atoms, and are used in thousands of different mixtures. Their properties and toxicity vary but all are chemically stable and resistant to attack. They are not pesticides and were not thought to be an environmental hazard. The discovery that they are hazardous largely resulted from routine surveillance of organochlorine pesticide residues in wildlife (Bourne and Bogan, 1972; Koeman et al., 1969). In the mid-1960s those using gas liquid chromatography (GLC) to assess levels of DDT, DDD, DDE and HEOD (dieldrin) in animal tissues increasingly found numerous unattributed peaks that were obviously due to organochlorines other than these pesticide metabolites. Although initially ignored and dismissed as trivial by comparison with the DDT and dieldrin peaks, some tissue samples contained such high levels of the unidentified substances that it became difficult to measure the DDT and dieldrin peaks. Further investigations showed that those extra peaks were due to PCBs. Organochlorine levels reported before 1967 are likely to be erroneous, because the unknown presence of PCBs interfered with DDT and dieldrin signals (particularly the former), but from around

1967 most laboratories reported PCB levels separately. Initially PCB levels were reported as 'total PCBs', estimated by summation of peaks due to different isomers, but recent studies have tended to report individual isomer levels since these vary in toxicity and distribution. The GLC fingerprints vary among animals depending on the types of PCB ingested which in turn is affected by the length of time they have been in the food web, since some isomers are more readily catabolized than others.

Analysis of PCBs in bird tissues and eggs suffers from all the problems associated with the monitoring of DDT and dieldrin as well as the difficulties of quantifying PCBs themselves (Lambeck et al., 1991). PCB levels tend to correlate with those of other organchlorines in the same individual, probably because of individual differences in both lipid reserves and their metabolism related to the geographical and ecological distribution of the birds. Thus a seabird that winters in the Mediterranean will accumulate PCBs, DDE and HEOD to greater levels than a conspecific that winters in the Atlantic, for a bird with fluctuating fat reserves will metabolize more of the organchlorines in its body.

With the realization of widespread contamination of wildlife with PCBs, steps were taken to reduce PCB use, especially use that allows leakage into the environment. Monsanto, the main manufacturer, reduced production from 33 000 t in 1970 to 18 000 t in 1971, and restrictions have been progressively tightened since. This has produced measurable decreases in PCB levels in the wildlife of some areas (Figure 3.6), although PCB levels appear to have remained high in many ecosystems, particularly marine ones.

Although PCBs appear to be extremely toxic to seals and other marine mammals (Reijnders, 1986), an experimental study with puffins suggested little or no toxic impact (Harris and Osborn, 1981). A large wreck of seabirds in the Irish Sea in 1969 led to the discovery that many of these birds contained high levels of PCBs in their livers. However, it is now thought that these birds died of starvation and that the high PCB levels were a consequence of lipid and protein depletion accompanied by mobilization of PCBs from lipid stores to the liver. Although the PCB concentrations were high, the amounts of PCBs in these birds were no more than can be found in many healthy seabirds. Unfortunately, studies very rarely present pollutant burdens (amounts), in addition to concentrations, in tissues.

Studies of population trends and breeding performance tend to implicate DDE and dieldrin but find no additional toxic effects of PCBs (Newton et al., 1986, 1988, 1989), although Kubiak et al. (1989) argued that reproductive impairment of Forster's terns *Sterna forsteri* at a colony on Lake Michigan seemed to be associated with high PCB levels. I think it is fair to say that, at present, the toxic effects of PCBs on birds are unclear, and possibly minor. While surveillance of levels in eggs and tissues may provide useful evidence for reduced contamination of food chains after reductions in PCB usage, further toxicity studies would probably be required to indicate whether any bird populations have been reduced by PCBs.

3.4 HEAVY METALS

3.4.1 Introduction

Heavy metals, particularly mercury, cadmium and lead, have prompted many investigations following the documentation of pollution incidents in Japan, Sweden and Iraq (Borg et al., 1969: Fujiki et al., 1972; Bakir et al., 1973; Friberg et al., 1974). This is mainly because they are extremely toxic (Bryan, 1979). Most have no known biological function, and inputs into the environment, particularly of lead and mercury, largely result from human activities (Lantzy and Mackenzie, 1979). This gives rise to anxiety both with respect to the possible detrimental effects of heavy metals on resources and wildlife, and in terms of the impact of metals on human health.

In recent years concern about the long-term effects of such environmental contaminants has increased (Hutchinson and Meema, 1987) and biological monitoring is thought to be a satisfactory way to quantify heavy metal abundance and bioavailability (Phillips, 1980; Schubert, 1985; Thompson, 1990). Metals may accumulate in animals to levels much higher than are found in water or air and so are much easier to measure than in the physical environment. Fish can be used as monitors of metals (Leah et al., 1991), while diseases and recent declines in population of marine mammals have stimulated research into their metal burdens as it is thought that metals might influence their susceptibility to disease (Law et al., 1991; Marcovecchio et al., 1990). Birds offer a number of particular advantages as indicators of heavy metal pollution. The ecology of most bird species is quite well known. They feed at the upper trophic levels of ecosystems and so can provide information on the extent of contamination in the whole food chain (though most metals do not show bioamplification and so in some cases there are reasons to prefer the use of sessile invertebrates as monitors). Metal levels in birds may give a better picture of hazards to man than measurements in the physical environment, plants or invertebrates. Benefits of using seabirds to monitor metals are discussed by Brothers and Brown (1987) and Walsh (1990). However, it has proved difficult to assess toxic effects of metals on bird populations (Custer et al., 1986; Nicholson and Osborn, 1983) and some birds may have high metal burdens for reasons of natural accumulation or detoxification processes unrelated to pollution; thus high metal levels in birds do not necessarily indicate pollution (Muirhead and Furness, 1988; Murton *et al.*, 1978).

3.4.2 Problems of sampling from bird populations

For most metals the levels in juvenile birds are less than those found in adults, though copper tends to be higher in the livers of juveniles than of adults (Lock *et al.*, 1992). It is therefore important to sample either from juveniles or adults but not from a mixture of these classes. Recent studies of mercury levels in feathers of adults of known age have shown that they do not vary with age in adult red-billed gulls *Larus novaehollandiae scopulinus* (Furness *et al.*, 1990a) or in great skuas *Catharacta skua* (Thompson *et al.*, 1991). Both these studies made use of the unusual opportunities afforded by populations of birds in which a large proportion of chicks had been ringed, and subsequent capture of adults at the nest allowed feather samples to be taken from adults that had been ringed as chicks and whose age was therefore known. In both cases the variation in mercury levels among individuals was considerable but it was unrelated to age (Figure 3.7). These results are consistent with the fact that plumage mercury is all methylmercury (Thompson and Furness, 1989) and that birds seem normally to put all of their annual accumulation of methylmercury into their plumage during the moult in autumn (Braune and Gaskin, 1987a, b; Furness *et al.*, 1986a). By contrast, levels of cadmium (in the kidney) do seem to increase through adult life in some species, though this may not be true in others (Furness and Hutton, 1979; Hutton, 1981).

Although females can excrete mercury into eggs, the amount that they shed in this way is usually small compared to the amount put into feathers during moult, so there is little difference in mercury burdens between males and females (Furness *et al.*, 1990b; Honda *et al.*, 1986a, b; Thompson *et al.*, 1991). Indeed, little difference has been found between males and females in metal levels in any tissues, although Evans and Moon (1981) found higher levels of cadmium in female bar-tailed godwits *Limosa lapponica*; Gochfeld and Burger (1987) found significant differences in levels of several metals between male and female ducks; Hutton (1981) found differences in metal levels between male and female oystercatchers *Haematopus ostralegus*; Scharenberg (1989) found higher levels of lead in feathers of male grey herons than in females; and Stock *et al.* (1989) found higher levels of cadmium in soft tissues of male oystercatchers than females. Nevertheless, compared with the high levels of individual variation, differences in metal levels between sexes are small, so it is not

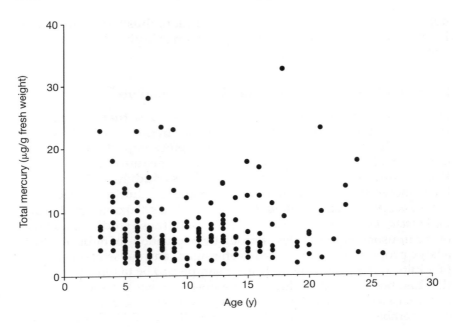

Figure 3.7 Concentrations of mercury in body feathers in relation to age of adult great skuas breeding at Foula, Shetland. (After Thompson *et al.*, 1991).

important to sample the sexes separately. Thus the sampling of adult birds can be done without concern that bias due to the age and sex of those birds sampled would substantially affect the results.

Unlike organochlorines, which are distributed widely among body tissues, largely in relation to their lipid content, heavy metals tend to be held in one particular tissue at much higher levels than in others. The site-specificity of metals has an important influence on the choice of tissue for monitoring studies.

(a) Internal tissues as monitoring units

Most studies of heavy metals in birds have used internal tissues, as the highest concentrations of particular metals tend to occur in these: cadmium is concentrated in the kidney, lead in bone, mercury in liver and kidney, zinc, copper and iron in liver. However, soft tissues have a number of drawbacks, including seasonal changes in mass, which can alter perceived metal concentrations even though the total amount in the body is unchanged. For practical and ethical reasons it is important to develop sampling methods that avoid the need to kill large numbers of birds in order to provide tissues for analysis.

(b) Blood as a monitoring unit

If birds can be caught safely, sampling of blood is comparatively easy and does not require that the bird is killed; but it has not been developed far except for studies of lead contamination (Scheuhammer, 1989; Tansy and Roth, 1970). Blood samples provide a picture of metal levels that reflect short-term exposure (immediate dietary intake) and physiological factors (such as mobilization of reserves for egg production, or increased blood volume at the start of moult).

(c) The egg as a monitoring unit

The avian egg has been used in many studies to monitor contaminants as it has several advantages over internal tissues (Barrett *et al.*, 1985; Becker, 1989; Burger and Gochfeld, 1988; Ohlendorf and Harrison, 1986; Parslow and Jefferies, 1977). Eggs have a highly consistent composition (unlike the other traditionally sampled tissue, the liver which changes in size and composition during both the day and year). They are produced by a clearly identified segment of the population, adult females, although this can be a disadvantage as it precludes sampling of other members. Sampling eggs takes little time, they are easy to handle and they can be readily sampled from the same location each year. They can be collected with little interference and their removal places less of a drain on the population than the sampling of adults, especially if only one egg is removed from each clutch. This is generally satisfactory because eggs from the same clutch tend to have similar pollutant levels (Potts, 1968). Finally, eggs have been shown to reflect metal uptake from local foraging more closely than do the tissues from adult birds (Barrett *et al.*, 1985; Parslow and Jefferies, 1975, 1977).

Eggs do have some drawbacks. Pollutants in eggs usually represent pollutant uptake in a short period before the egg is laid and so cannot be used to investigate pollutant burdens acquired at other times of year. In addition, levels in the egg contents do not adequately reflect body burdens or dietary intakes of some heavy metals (Becker, 1989). Unlike tissue samples, egg contents are not available from before about 1980, so that historical changes in pollutant levels in ecosystems cannot be traced from egg contents, though the possible use of eggshells in museums to do this has not yet been explored.

(d) Feathers as monitoring units

Although many pollutants do not enter feathers, many heavy metals become incorporated into the keratin structure. Sampling of feathers is an attractive alternative to collecting livers and kidneys. Feathers can be removed from live birds with virtually no effect on the birds sampled, especially if body feathers are taken rather than flight feathers, enabling a much larger sample size to be taken. Feathers can be stored without being frozen, so the logistics of sampling from remote populations are much simpler than with tissue collections. Museums contain large numbers of study skins of birds with data on the date and place of collection. Thus feathers provide an attractive means of studying historical changes and synoptic geographical patterns in pollutant burdens.

Detailed research has been done to assess the use of feathers to monitor levels of mercury pollution in particular. It has been demonstrated experimentally that mercury in feathers is strongly bonded and levels are not affected by storage or by vigorous treatments (Appelquist *et al.*, 1984). The same may be true of other heavy metals that enter feathers from the blood stream during feather growth but this has not been studied. However, heavy metals may be deposited onto the surface of feathers in quantities that can mask patterns due to incorporation of dietary or stored metals into feathers during moult. Indeed, at heavily polluted industrial sites bird feathers may provide a means of measuring rates of atmospheric deposition of heavy metals (Hahn, 1991). Research is needed to see if it is possible to discriminate between the quantities of metal in feathers from the diet and from atmospheric deposition (Ellenberg and Dietrich, 1982; Rose and Parker, 1982). Mercury is a special case in that atmospheric deposition (as inorganic mercury) appears to be slight and all of that incorporated into feathers from dietary sources is methylmercury. By extracting organic mercury it is possible to measure inputs from food and obtain a reliable indication of food chain contamination by mercury (see below).

(e) Chicks as monitoring units

Many of the limitations in using eggs or adult birds' tissues as monitoring units are less of a problem if chicks are sampled. Chick body burdens of metals reflect the amounts in the food they are fed during their development and are thus attributable to intake from a clearly defined time period and limited parental foraging area. There may be a slight complication in accounting for the dose present in the egg but this is likely to be a negligible proportion of the total burden in well grown chicks in most populations, since egg mass represents only about 2–8% of the body mass of fully grown chicks. The serious problem with egg sampling, namely the difficulty of assigning metal burdens to immediate dietary intake rather than to mobilized tissue stores, is not present when chicks are sampled. But like eggs, removing chicks from populations has less impact than removal of adults, particularly in long-lived species, so sample sizes can be greater with less damage to the population. Pollutant levels in chicks are likely to be less variable than in adults, because chicks tend to be fed on a selected diet of energy-rich foods; because dietary specializations of adults are averaged between two parents for species where both parents feed the young; and because the metal burden is obtained only from the narrow period of chick growth and from foods from near the breeding site. As a result, with chicks sample size need not be particularly large in order to obtain reliable results. Furthermore, the chick may be the stage of development at which toxicological effects are particularly evident. These arguments have been developed by a number of research and monitoring organizations, especially in North America. For example, the Canadian Wildlife Service considers the sampling of chicks as providing the best way of assessing the amelioration of pollution in the St Lawrence as a result of recent limitations on the discharges of pollutants into that system (J.L. DesGranges, pers. comm.). In some countries public opinion might not permit routine sampling of chicks, even were the case that it provides the best monitoring tool to be incontrovertable.

3.4.3 Mercury

Feathers have been shown experimentally to incorporate mercury in a dose-dependent fashion (Lewis and Furness, 1991; Scheuhammer, 1987). However the relationship between levels in feathers and levels in other tissues is poorly documented (Thompson *et al.*, 1990). In part this reflects the lack of multi-tissue studies but it also reflects the great variation of mercury concentrations among feathers of individual birds, largely due to moult.

Feathers replaced early in the moult cycle incorporate the highest concentrations of mercury, newly mobilized after accumulation in soft tissues between moults. This is most strikingly shown by the tendency for mercury concentrations to decrease linearly along such feather sequences as the primaries, corresponding to the order in which feathers have been renewed (Braune, 1987; Braune and Gaskin, 1987a, b; Furness *et al.*, 1986a; Lewis and Furness, 1991). Thus comparisons between studies are greatly complicated unless feathers from a similar position in the moult sequence are analysed. For environmental monitoring using feathers of adults, the effects of moulting on mercury levels of individual feathers can be minimized by pooling several small body feathers from a defined plumage area (Furness et al., 1986a). If nestlings are used little variation in mercury levels is found between feathers (Lewis, 1991).

In one study of seabird feathers exposed for ten months in Scotland, mercury was not deposited onto feathers in measurable quantities by atmospheric deposition (Lewis, 1991). Furthermore all mercury measured in feathers has been found to be methylmercury, so organic extraction of methylmercury from feathers ensures that complications caused by any contamination due to atmospheric deposition or storage in dusty museum drawers can be avoided (Thompson and Furness, 1989). This useful attribute of mercury is not, unfortunately, shared by other heavy metals (see below).

Mercury levels in feathers of birds vary from below detection limits up to about 50 $\mu g/g$. The highest levels found in feathers occur in the wandering albatross *Diomedea exulans* and other large biennially-breeding albatrosses (Lock et al., 1992; Thompson and Furness, 1989). These birds appear to live far from any source of pollution and there is evidence from analysis of feathers from nineteenth-century study skins in museum collections that the levels of mercury in these species have not increased much over the last 150 years, which suggests that these high levels are natural. Thus high levels of mercury in feathers do not necessarily imply pollution but may simply reflect natural dietary intakes or constraints on the ability to excrete mercury. Clearer evidence for pollution can be obtained by making geographical comparisons or determining time-series trends for particular species.

Levels of mercury in feathers of golden eagles *Aquila chrysaetos* in Scotland vary between west-coast birds, west-mainland birds and east-highland birds in a way that is correlated with overall breeding success and also reflects the amount of marine prey in the birds' diets (Furness et al., 1989). Birds breeding on coastal sites that fed extensively on colonial seabirds had elevated mercury levels. These might have caused the reduction in breeding performance, though this could have other causes such as a scarcity of the preferred terrestrial prey in the western areas.

Appelquist et al. (1985) showed that total mercury levels in feathers from guillemots *Uria aalge* and black guillemots *Cepphus grylle* during the 1960s and 1970s were much higher in populations in the Baltic than in the North Atlantic as a consequence of greater pollution in the Baltic.

(a) Historical trends

The first studies to examine long-term trends in mercury contamination of birds by analysing mercury levels of feathers were carried out in Scandinavia after the discovery that the use of alkylmercury-treated seed was resulting in serious contamination of wildlife and mortality of top predators (Berg *et al.*, 1966). These studies showed that mercury levels in predatory birds, especially fish-eaters, had increased several-fold between the 1950s and 1960s. Many of these early studies analysed total mercury levels in feathers and then retrospectively excluded from their data sets individual measurements that appeared to be elevated as a result of plumage contamination with inorganic mercury during storage. Such a procedure carries with it the possibility of generating desired patterns through selection of data, although in this case the changes were dramatic and unambiguous.

Studies of total mercury levels in feathers from guillemots and black guillemots (Appelquist *et al.*, 1985) showed that mercury levels had increased, especially in feathers from populations in the Baltic Sea, from 1–2 µg/g around 1900 to 2–8 µg/g in 1960–1975. This increase was attributed to the increased use of mercury seed dressings in Scandinavian countries in the 1960s, because levels in populations of these species from Atlantic areas showed smaller increases over the same period. Appelquist *et al.* (1985) measured total mercury levels of individual feathers by Neutron Activation Analysis. Many museum collections are contaminated with inorganic mercury, which was often used as a preservative on nineteenth-century study skins, and so it is preferable to use methods that measure only the organic mercury level. Such a procedure avoids the problem of selective exclusion of apparently 'aberrant' data points, which in fact are difficult to identify because the distribution of mercury levels in bird populations is highly skewed with a small number of individuals having much higher levels than most (Walsh, 1990).

Analysis of organic mercury levels in samples of body feathers from puffins *Fratercula arctica*, Manx shearwaters *Puffinus puffinus* and great skuas collected in summer from adults in breeding colonies in defined regions of the British Isles and dating from years between 1830 and 1990, has shown that there have been progressive increases in mercury levels in these populations (Thompson *et al.*, 1991; Thompson *et al.*, 1992). Mercury levels in body feathers were 2–4 times higher in the period 1960–90 than they were before 1900 (Figure 3.8). A similar analysis of organic mercury levels in body feathers of herring gulls from the German North Sea coast (D.R. Thompson *et al.*, in prep.) shows a different pattern in that region, with low levels in the period 1830–1930 (3.8 µg/g), a pronounced and short-lived peak of levels in the 1940s (12.1 µg/g), a drop in the 1950s to a level somewhat higher than pre-1940 (6.9 µg/g), followed by an increase from the 1950s to the 1970s (10.1 µg/g) and a decrease in the 1980s (5.4 µg/g). These changes

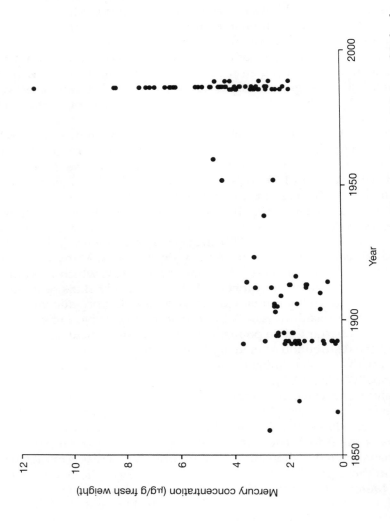

Figure 3.8 Concentrations of mercury in body feathers of puffins from south-west and west Britain and Ireland collected during the breeding season at colonies; samples from before 1970 are from museum collections. (After Thompson *et al.*, 1992.)

best be explained by a lack of discharge controls in Germany during the 1939–45 war, increased industrial production through the 1950s to 1970s and then greater control of pollution discharges in the 1980s, resulting in changes in the amount of mercury entering the North Sea estuaries from the major European rivers.

These long-term historical comparisons of mercury levels using feathers from museum skins are only useful if the observed changes reflect changes in mercury pollution of the ecosystem. This is likely to be true in most cases but it does rely upon the assumption that the diet of the bird has remained the same over the time period. If a population changes diet then it is quite likely that the mercury level in the birds will alter because different prey species may contain different mercury levels. For example, large fish such as adult herring *Clupea harengus* and mackerel *Scomber scombrus* contain higher concentrations of mercury than small short-lived fish such as the sandeel *Ammodytes marinus*. It is known that gannets in Shetland have recently switched diet from predominantly sandeels in the 1970s to predominantly herring and mackerel in the 1980s as a result of changes in abundance of these fish (Martin, 1990). In order to interpret long-term changes in mercury levels in feathers of birds, it is preferable to select species that have narrow and inflexible diets, rather than generalist feeders, and best to consider trends in a variety of species feeding on different prey within the same environment. A further test of the assumption of no change in diet over time could be made by measuring stable isotope ratios of carbon, nitrogen and possibly of other elements. The ratio of isotopes changes through the food chain, with slight isotope fractionation at each step in the food chain, so that changes in the trophic level are reflected by changes in the ratios of isotopes. This can be used to investigate the feeding relationships of birds, even of extinct ones (Hobson and Montevecchi, 1991). Since feathers can be used for analysis of stable isotope ratios as well as for mercury analysis, it would be possible to perform isotope ratio studies on the same museum samples used for mercury determinations, although no such study has yet been undertaken.

(b) Mercury and selenium in eggs

In cases of environmental contamination by mercury there is often a proportional accumulation of selenium on a 1:1 molar basis (Hutton, 1981). Mercury and selenium are appreciably transferred to eggs (Focardi *et al.*, 1988; Sell, 1977). Mercury in particular has been shown to enter eggs, even when the level in the diet is very low (Fimreite, 1971), in a dose-dependent fashion (Heinz, 1974; March *et al.*, 1983;

Tejning, 1967). Mercury levels in eggs rarely exceed 0.5 µg/g fresh weight (Barrett et al., 1985; Jensen et al., 1972; Ohlendorf and Harrison, 1986), although higher levels have been reported in those of royal terns *Sterna maxima* from the Texas coast (King et al., 1983) and of gannet from west Scotland, which have higher levels than found in samples from gannet populations in Norway (Fimreite et al., 1974, 1980; Parslow and Jefferies, 1977).

Selenium levels tend to be slightly higher than mercury levels, with concentrations reaching 2.9 µg/g dry weight in yellow-legged herring gulls *Larus michahellis* from the Mediterranean (Leonzio et al., 1986).

In some species the amount of mercury transferred to the eggs is small compared to the female's body-burden, so the removal of mercury through egg laying has been thought to be negligible (Helaner et al., 1982; Honda et al., 1986a, b). However, recent work has shown that excretion of mercury into the eggs can be fairly substantial in some species. In domesticated quails *Coturnix coturnix*, which lay one egg each day, over 40% of the female's body-burden of mercury was lost in this way (Lewis, 1991) and female herring gulls eliminated nearly 25% more of their body-burden of mercury than did males (Lewis, 1991). The ratio of mercury levels in various tissues to those in eggs has recently been estimated in herring gulls (Table 3.2). These figures allow internal tissue mercury contamination to be predicted on the basis of easily collectable samples such as eggs, and this enhances their value as a means of assessing mercury burdens of birds. Mercury levels in eggs have been reported to be 10–20% of those found in the female's liver in three Hawaiian seabirds (Ohlendorf and Harrison, 1986).

Table 3.2 Ratios of mercury levels in body tissues (dry weight) to those in eggs, in female herring gulls sampled from a colony on the German North Sea coast soon after egg laying (from Lewis et al., 1992)

Tissue	Mean ratio	S.D.	n[a]
Body feather	3.7	2.2	21
Primary feather	5.5	4.0	12
Liver	3.3	1.7	25
Pectoral muscle	1.5	0.7	25
Ovary	1.5	0.9	25

[a]Number of individuals sampled.

(c) Factors affecting mercury contamination in eggs

Eggs in the same clutch show similar levels of mercury and thus each egg reflects the contamination of the entire clutch and the female. However, in some birds the first egg laid contains significantly more mercury than subsequent eggs (Becker, 1989). When sampling for monitoring studies eggs should therefore be taken at the same position in the order of laying from each of the clutches.

Concentrations of mercury in eggs are affected by conditions during storage (even freezing) since much water can be lost (Stickel et al., 1973). This problem would be circumvented if dry weights were used when presenting contamination levels. This would also allow published egg data to be standardized and therefore avoid confusion and the need to use conversion factors.

Mercury transfer to eggs and the resultant toxic effects may be species-specific (Scheuhammer, 1987). In particular, eggs from fish-eating birds may be more tolerant of mercury pollution than those of non-piscivorous species. For example, in mallard reproductive dysfunctions were seen at levels of 6–9 μg/g of mercury in the egg (Heinz, 1974) whereas levels of 2–16 μg/g found in the eggs of herring gulls had no effect on hatching and fledging (Vermeer and Peakall, 1977).

Mercury accumulates particularly in the egg white proteins (Backstrom, 1969) and since these are derived from serum proteins (Romanoff and Romanoff, 1949) this suggests that egg concentrations more closely reflect mercury from recent dietary uptake than from accumulated storage tissues. This seems generally to be the case (Becker et al., 1985b; Becker, 1989). There is also evidence that the ovalbumin fraction of egg white has a specific affinity for dietary mercury while globulin fractions tend to accumulate low levels of 'non-dietary mercury' (Magat and Sell, 1979). A recent study found feather mercury levels of female herring gulls to be unrelated to levels found in the eggs (Lewis, 1991). This implied that the sources of mercury contamination of the eggs and feathers were rather different. Almost all of the accumulated body-burden of methylmercury enters the plumage during the autumn moult of the bird and therefore the feather concentration represents mercury intake during the inter-moult period (Furness et al., 1986a), whereas egg levels probably indicate mercury ingested one or two weeks prior to egg-laying and therefore reflect local contamination at the breeding site. This depends, however, on the physiology of egg formation of the monitor species. In many species, it is true that considerable quantities of food are required to produce eggs (a reason for courtship-feeding), and here egg residues do reflect recent feeding and thus the local contamination of the food web. By contrast, lesser snow geese *Chen caerulescens caerulescens* do not feed immediately prior to or during egg laying (Ankey and MacInnes, 1978), and here the mercury levels in these eggs would presumably reflect amounts of mercury accumulated on their wintering grounds, though this has not been investigated.

(d) Current evidence for geographical variation of mercury pollution from analysis of bird eggs

Eggs have been widely used to assess trends and regional variations in pollutant burdens, together with any effects these pollutants have on breeding (Barrett *et al.*, 1985; Fimreite *et al.*, 1980; Newton and Galbraith, 1991; Newton and Haas, 1988; Newton *et al.*, 1993; Renzoni *et al.*, 1986). Particularly striking patterns of geographical variation are evident from studies of mercury in European and Atlantic waters and common guillemots provide a good example. They are a fairly sedentary species (adult guillemots breeding in Britain disperse over only relatively short distances in winter and occupy the same nesting colony throughout their lives) and they have a relatively narrow food range (predominantly small shoaling-fish such as sandeels and sprats *Sprattus sprattus*). Eggs collected from 16 colonies around Great Britain and Ireland in the early 1970s showed a 20-fold variation in the mean mercury concentration with the lowest levels (0.8–1.9 µg/g dry weight) from north-west Scotland around to north-east England but markedly higher levels (1.9–4.9 µg/g fresh weight) at five Irish Sea colonies (Parslow and Jefferies, 1975). This is presumably connected with the slow rate of water exchange in the shallow Irish Sea as well as the larger amounts of industrial waste and effluent it receives.

Geographical variations in mercury levels of eggs have also been reported around British and Irish gannet and herring gull colonies. A fairly even trend was seen, with egg concentrations increasing from 0.09 µg/g in northern Norway to 1.28 in south-west Britain (Fimreite *et al.*, 1974; Holt *et al.*, 1979; Parslow and Jefferies, 1977).

Geographical variation in mercury levels along the German North Sea Coast have also been demonstrated using eggs. Herring gull and common tern eggs from seven regions of the German North Sea Coast averaged 0.2–0.5 µg/g and 0.3–0.7 µg/g respectively except for pronounced peaks of 1.55 and 4.58 µg/g at the Elbe estuary, where there is a particularly high industrial discharge (Becker, 1989).

Mercury levels in the Mediterranean, from both natural and manmade sources, are high and this is reflected in birds' eggs. Mercury concentrations in eggs of Cory's shearwaters *Calonectris diomedea* from three Mediterranean colonies were consistently higher than at Atlantic colonies (Renzoni *et al.*, 1986). Audouin's gulls *Larus audouini* and herring gulls in the Mediterranean also have higher mercury levels in their eggs than found in gulls from Atlantic or North Sea colonies (Bijleveld *et al.*, 1979; Lambertini, 1982; Lewis *et al.*, in press).

3.4.4 Cadmium

Cadmium accumulates in the kidneys of birds. Levels in other tissues tend to be very much lower, except that if a bird has recently been exposed to a high intake of cadmium the level in the liver may equal or exceed that in the kidney; the ratio of cadmium in the liver and kidney may give an indication of recent acute exposure (Scheuhammer, 1987). At present the only routine means of surveillance of cadmium in bird populations is to sample tissues from adults. Problems with sampling are similar to those already discussed for organochlorines except that cadmium is not lipid soluble, so levels are not necessarily higher in birds further up the food chain, and changes in concentrations in the kidney due to seasonal variation in body condition are less evident.

Cadmium levels reported for feathers range from zero to 27 µg/g. There is evidence in the literature of many erroneous measurements because it would appear that in most species feather cadmium levels are extremely low and so may be close to the limits of detection of analytical equipment. There are very few studies that suggest a correlation between measurements of cadmium levels in feathers and in soft tissues (but see Lee et al., 1989). One reason for somewhat inconsistent results in the literature is that cadmium levels measured may in many cases be due primarily to atmospheric deposition onto feathers rather than to elimination of the metal by the bird into its growing feathers.

Several studies have compared metal levels in feathers of birds from an industrially contaminated site with levels in a population from a 'control' site where pollution is assumed to be negligible. For example, Rose and Parker (1982) compared levels of several metals, including cadmium, in feathers of ruffed grouse *Bonasa umbellus* at a site of heavy industrial pollution with levels in a population of the same species 100 km away in an area considered to be free from metal contamination. They showed that grouse at the industrial site had higher cadmium levels in the plumage shortly before the annual moult but that levels were the same in feathers from the two sites immediately after the moult. They interpreted these patterns to mean that grouse assimilated similar amounts of cadmium from their diet at the two sites but that feathers accumulated cadmium on their surface during exposure to the contaminated atmosphere at the industrial site. Hahn (1991) found that cadmium levels of unwashed major flight feathers of goshawks *Accipiter gentilis* and magpies *Pica pica* tended to be highest at the tip, on the outer (exposed) side, and on the edges of the feather (Figure 3.9). By exposing plucked flight feathers he showed that cadmium levels increased

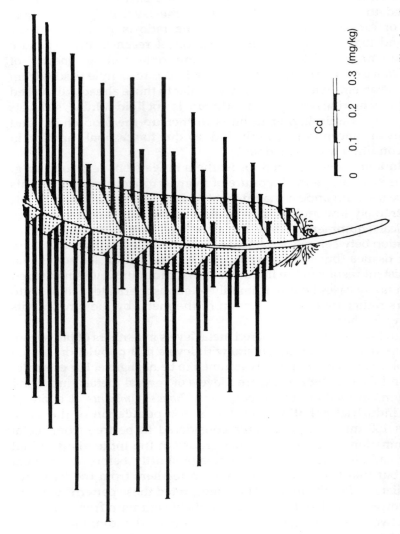

Figure 3.9 Cadmium levels from different parts of the first primary feather of a goshawk *Accipiter gentilis*, showing greater contamination of the parts of the feather more exposed to the atmosphere. (After Hahn, 1991.)

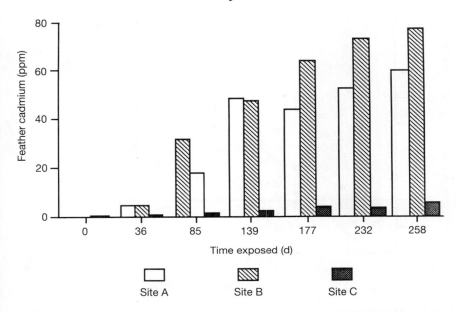

Figure 3.10 Cadmium levels in plucked feathers exposed to the atmosphere for different periods of time at three different sites. Site A and B are industrialized areas and site C is without local industrial contamination. (After Hahn, 1991.)

with exposure to the atmosphere (Figure 3.10) and that levels on feathers correlated closely with measured atmospheric deposition rates (Figure 3.11).

Laboratory experiments have shown that very little cadmium is transferred to eggs, regardless of the amount consumed (Sell, 1975). This is confirmed by seabird data, with concentrations usually less than 0.7 µg/g fresh weight in eggs despite high levels of cadmium in kidneys of many seabirds, especially pelagic species (Anderlini et al., 1972; Furness and Hutton, 1979; Honda et al., 1986c; Osborn et al., 1979; Renzoni et al., 1986), so low levels of cadmium in eggs of wild birds need not reflect a low dietary intake. It has been suggested that the barrier to cadmium transfer might break down if other tissues become overloaded (Hutton, 1981). There is limited evidence that cadmium concentrations in eggshells may be high, with means of 1.39 and 1.75 µg/g recorded for two laughing gull *Larus atricilla* populations on the Texas coast (Reid and Hackner, 1982). The possible use of eggshells to monitor cadmium levels has not yet been explored.

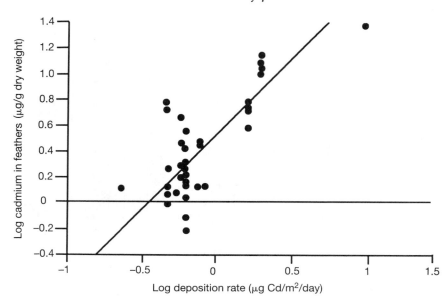

Figure 3.11 Correlation between levels of cadmium in magpie feathers and the rate of cadmium deposition measured by physical sampling of local atmospheric deposition of cadmium. (After Hahn, 1991.)

3.4.5 Lead

Decline in numbers and high mortality of adult mute swans *Cygnus olor* have been linked to ingestion of lead weights discarded or lost on riverbanks, canals and ponds by anglers, the lead being ground in the gizzard and assimilated (Birkhead, 1982; Birkhead and Perrins, 1985; Ogilvie, 1986). Blood lead levels, haematocrit and enzyme assays can be used to measure the exposure of individual birds to lead (Birkhead, 1983; O'Halloran et al., 1989). Such knowledge has resulted in the replacement of lead weights by other, non-toxic, materials in many places. In this case, studies of the birds provided the best means of assessing the impact of lead pollution by fishing weights and focussed attention on the quantities of lead being discarded in some waters. The popularity of swans no doubt assisted in shaping public opinion to encourage fishermen to find a less damaging alternative to lead. Similarly, studies of lead in waterfowl, including swans, have shown that lead shot from shooting can be ingested by birds and cause lead poisoning (Honda et al., 1990), leading to restrictions on the use of lead shot in some places.

Tansy and Roth (1970) showed that feral pigeons took up lead from the atmosphere and from ingested food and grit, and that, in samples that were known to be free of lead shot, pigeons from an urban area had significantly higher levels of lead in bone, feather, nail, liver and kidney than those from a rural area, though, surprisingly, they found no difference in lead levels in the blood. They suggested that feral pigeons would provide a useful means of assessing the hazards posed by atmospheric lead in urban environments. Ohi *et al.* (1974) showed that feral pigeons could be used to monitor atmospheric lead pollution in Tokyo. Not only did bone and kidney lead levels decrease along a transcect from central Tokyo to rural surroundings, but blood lead levels and the activity of the red blood cell enzyme ALA-D (delta-aminolevulinic acid dehydratase) also varied along the transect. The same research group later used blood lead levels and ALA-D activity to demonstrate a pronounced reduction in lead contamination in pigeons in central Tokyo after the banning of tetraethyl lead in petrol in Japan in 1975. There was a lag of a few years after the atmospheric lead concentration was reduced, indicating that pigeons were obtaining a substantial part of their body-burden from ingested grit coated with lead from earlier atmospheric deposition (Ohi *et al.*, 1981). This delayed response of lead in pigeons (relative to the reduction in atmospheric lead) makes monitoring in the birds different from and complementary to, rather than alternative to, atmospheric sampling and provides useful additional insights into the dynamics and hazards of lead in the urban environment.

Although lead concentrations can be measured in feathers, interpretation is difficult as there is evidence of heavy-metal contamination of feather surfaces by secretory products and exposure to the environment (Goede and de Bruin, 1984; Goede and de Voogt, 1985). However, this still allows some potential for indirect monitoring of lead exposure over known periods. As with cadmium, lead levels in feathers are often scarcely correlated with levels in soft tissues (but see Burger and Gochfeld, 1991; Lock *et al.*, 1992). This may be partly because feathers can take up metals from sources other than the diet. Nickel, copper, iron, cadmium, lead and cobalt from bird feathers near sites of high industrial outputs were reported to originate primarily from atmospheric deposition rather than by excretion of metals into the feathers from internal tissues (Hahn, 1991; Hahn *et al.*, 1989; Rose and Parker, 1982). As for cadmium, Hahn (1991) found that lead levels of major flight feathers tended to be highest where exposure to the atmosphere was greatest and that lead levels on exposed feathers increased over time, closely correlated with atmospheric deposition rates. Where atmospheric deposition of heavy metals is intense, it may be impossible to assess levels of metals entering feathers from dietary sources. Studies are required of the distribution of lead and other metals within feathers, by techniques such as X-ray microanalysis, to assess whether dietary and atmospheric deposition can be distinguished.

Lead transfer to eggs is invariably low. Concentrations in seabird eggs (usually less than 0.4 µg/g) are often below detection limits (Anderlini *et al.*, 1972; Munoz *et al.*, 1976; Reid and Hacker, 1982; Renzoni *et al.*, 1986). However, Burger and Gochfeld (1991) found correlations between the levels of lead in egg contents and body feathers of the females that produced them in a sample of common terns. Bird eggs seem to provide a particularly useful means of monitoring mercury and selenium pollution in the vicinity of a breeding colony in the immediate pre-laying season. Their value in assessing pollution by other heavy metals is not yet clearly established, but, as suggested for cadmium, eggshells might provide a means of assessing levels of lead which cannot satisfactorily be measured using egg contents. Further research is required.

3.4.6 Other metals

Few authors have examined the levels of other metals in soft tissues, bones, eggs or feathers in detail. Data for zinc and arsenic in feathers are given by Goede (1985). Hahn (1991) indicates the extent of atmospheric deposition of various metals onto feathers at an industrial site in Germany. Nyholm (1981) and Nyholm and Myhrberg (1977) examined aluminium levels in eggs and their effects on eggshell structure.

3.5 PLASTIC

Plastic production in large quantities began in the 1940s and has increased considerably over the decades since then. It was not until the 1970s that the widespread pollution of oceans by plastic became evident.

3.5.1 User-plastic litter and entanglement

Waste disposal practices led to such 'user-plastics' as bottles, bags, packaging materials and toys being dumped on land but finding their way into rivers, and then into the sea. Other 'user-plastics' may be dumped at sea by ships, while fishing boats may lose nets and ropes. Large plastic items that enter the sea are often broken up into smaller fragments. These may cause problems because they float, and are largely non-biodegradable, and are only slowly broken down by photooxidation through exposure to ultraviolet radiation. Indeed, most plastics incorporate antioxidants and UV-stabilizers to prevent breakdown, so increasing their persistance. Much of the assessment of the extent of marine pollution by user-plastic waste has been carried out by beach observations of the amount, composition and origins of plastic litter (Dixon and Dixon, 1981; Vauk and Schrey, 1987), although assessments can be made by considering materials discarded from the world fleet of vessels, the littering habits of beachgoers in holiday resorts, or by visual surveys at sea and collections with research nets (Pruter, 1987). User-plastics affect birds mainly in two ways: they may become tangled in plastics, such as lost fragments of fishing nets or long lines (Laist, 1987), or may use plastic litter in construction of their nests (Podolsky and Kress, 1989) with some adults or chicks dying after becoming entangled in plastic built into the nest (Montevecchi, 1991).

The hazards of waste user-plastics are not very well known, though they seem to be minor for birds. Indeed, concern about plastic waste is more often for aesthetic reasons, so the beach litter surveys described by Dixon and Dixon (1981) provide an appropriate means of assessing user-plastic pollution. However, Montevecchi (1991) considered that surveys of gannet nests provided a useful indication of the amount of plastic pollution in the sea around colonies. Bourne (1976) reported that one fifth of the gannets and one third of the shags on the Bass Rock had plastic in their nests in 1972 and nearly half of the gannets at Grassholm had plastic in their nests in 1974. Nelson (1978) reported a figure of 50% for gannets at the Bass Rock. Montevecchi (1991) found that virtually all gannet nests in Canadian colonies contained plastic, suggesting an increase in the level of plastic pollution over the last 20 years or a higher level on the west side of the Atlantic. He suggested that quantifying the incidence and kinds of plastic in newly constructed nests could provide information on pollution levels in different regions and over time.

3.5.2 Ingested plastic

Although some seabirds will ingest small fragments of broken user-plastic or even whole plastic objects (Fry *et al.*, 1987), most of the plastic found in seabird stomachs consists of particles of industrial plastic (Azzarello and van Vleet, 1987; Fry *et al.*, 1987; Furness 1985a,b; Ryan, 1987a; van Franeker 1985). These are the bulk material in which plastics are manufactured and transported prior to being melted into user-plastic items. Harmful effects of ingested plastic are evident in cases where birds suffer ulceration and cuts to the intestine from plastic, especially from sharp fragments of user-plastic (Fry *et al.*, 1987), but several studies of the amounts of ingested plastic particles and indices of body condition among pelagic seabirds have provided only very weak evidence for any harmful effect on digestive ability (Connors and Smith, 1982; Furness, 1985a,b; Ryan, 1987b). Experiments with chickens showed that plastic in the gizzard reduced food consumption and energy assimilation (Ryan, 1988) and plastic in the gizzard may increase PCB uptake (Ryan et al., 1988). Thus plastic particles on the sea surface are probably undesirable and a hazard to marine animals likely to ingest them, such as fish and seabirds. Assessment of the level of plastic particle pollution of the seas and of trends over time may be possible by comparing catches of plastic particles in research surface trawls with neuston nets, as reported by Wong et al. (1974). Alternatively, collection of samples of the seabirds that tend to accumulate plastic particles, which means mostly the smaller and medium sized procellariiforms (Ryan, 1987a), could permit monitoring of the levels of plastic-particle pollution over a period of years. The use of birds as monitors does make a number of assumptions. Differences in diet may affect the likelihood of a bird ingesting plastic, species taking small surface prey probably being more likely to ingest plastic than those feeding on larger prey or prey below the sea surface. Hungry birds may be particularly likely to ingest plastic. Since plastic particles will tend to accumulate at some convergent fronts and at other oceanographic features, seabirds feeding in association with frontal systems may be especially likely to ingest plastic. Inexperienced young birds may be more likely to eat plastic than adults. Most of these potential biases have not been examined but Ryan (1987a) showed that ingested plastic particles differed in size among seabird species, with larger species tending to select larger particles. Pale coloured particles were under-represented in stomach contents, suggesting that birds tend to select darker coloured particles, though smaller petrels were less colour-selective and ingested more plastic than did larger species. Nevertheless, one could reasonably expect that a long-term increase in plastic particle pollution of the seas would be paralleled by an increase in the amount of plastic ingested by seabirds, and especially by procellariiforms.

3.6 NUTRIENTS

At first sight it may seem perverse to class nutrients as pollutants, even when they comprise the by-products of human activities (such as the inorganic nitrogen or phosphorus leaching into water courses from agricultural land) or human waste products (such as sewage, discarded agricultural products, or distillery waste). It seems especially perverse when we find that, at least for moderate increases in nutrient input, there is often increased productivity in the ecosystem, which may be reflected in greater numbers of birds. However, some oligotrophic systems may be valued or valuable because they are nutrient poor. More generally, increased nutrient loadings may lead to the reduction of species diversity in aquatic ecosystems, to communities becoming dominated by a few specialists, and to the loss of species that are aesthetically or economically more desirable. Cyanobacterial blooms, resulting from elevated nutrient inputs to aquatic ecosystems, can be toxic to other organisms and lower the productivity at higher trophic levels. And, at the extreme, levels of organic enrichment can be so high as to raise the biological oxygen demand to levels that eliminate all but a few specialist anaerobes and produce water bodies that are aesthetically extremely undesirable.

Because the responses of birds to changes in nutrient loadings in an environment are likely to be secondary and complex, depending on the interactions at lower trophic levels and on the ability of the bird population to respond to the changes in their food supply, it would be foolish to suggest that birds provide the best means of monitoring changes in nutrient pollution. However, nutrient inputs may be difficult to assess from water or sediment samples because they vary markedly over short time-periods, and biological monitoring that averages out such variations may be the best approach. It would seem logical to monitor effects on bacteria and algae, or if a longer period of integration were required, invertebrates, and if a new programme of monitoring of change in the level of nutrient pollution into, say, an estuary, were to be started now, it would probably be sensible to start with a programme of monitoring population densities and communities of invertebrates. However, for most ecosystems long-term historical data on invertebrates are scarce, while bird populations have been widely monitored. Furthermore, invertebrates tend to be patchy in distribution, so that large numbers of samples need to be collected to obtain sufficiently precise means to provide a reasonable prospect of demonstrating changes between years or between areas. Invertebrates of many species also show pronounced seasonal changes in abundance, so that it is inappropriate to make comparisons between years unless samples have been collected at the same time in the season. It may even be necessary to consider the relative timing of seasonal events in different years to allow for early or late seasonal population change in different years. Bird populations are less subject to such short-term variations.

Long time-series of high quality counts of birds already exist for many estuaries and freshwater bodies where nutrient pollution problems may be an issue and where suitable invertebrate data are generally lacking. This is perhaps best illustrated with examples, and two are given below.

3.6.1 The Wadden Sea

Although it tends to be assumed that bird populations will increase if pollution is reduced and decrease when pollution levels increase, van Impe (1985) pointed out that a long-term increase of estuarine birds need not always be a good indication of the improvement of the general environmental quality of the estuary. In part of the Scheldt estuary he noted that 'an incontestable ecological deterioration' due to increased inputs of organic matter, and consequent reductions in species richness of the macrobenthic fauna but increased biomass (dominated by the annelid *Nereis diversicolor* and the crustacean *Corophium volutator*), led to an increase in bird numbers by a factor of between three and five for most species. He suggested that long-term increases of intertidal bird numbers might even be used as an indicator of environmental alteration by nutrient pollution, since information on bird numbers is often available, whereas data on nutrient inputs and benthic faunal diversity and abundances are usually lacking.

3.6.2 The Clyde Estuary

Data from the British Trust for Ornithology Birds of Estuaries Enquiry (BoEE) for the Clyde Estuary, down-stream from Glasgow in west Scotland, show a dramatic fall in numbers of overwintering dunlins *Calidris alpina* (by 85%), redshanks *Tringa totanus* (by 70%) and lapwings *Vanellus vanellus* (by 60%) between the early 1970s and the late 1970s and 1980s. Furness *et al.* (1986b) showed that the decrease in numbers of these species occurred rather abruptly between 1977 and 1978, and that numbers of oystercatchers and curlews *Numenius arquata* showed little or no change. The decreases on the Clyde did not correspond to national trends: dunlin, redshank and lapwing decreased in numbers far more on the Clyde than on other British estuaries, while oystercatchers increased nationally but not on the Clyde. Studies of the breeding origins of the populations wintering on the Clyde showed that the lapwings came largely from rural areas close to the estuary, redshanks came from Iceland and from west Scotland, oystercatchers came predominantly from east Scotland, and dunlin from Scandinavia, the Baltic States and Russia. These diverse breeding origins, and the lack of any change in other British estuaries, make it unlikely that the decreases resulted from changed conditions in the breeding areas. Studies of roosting habits and other factors thought likely to influence numbers of waders left only food as a probable cause. Dunlin, redshank and lapwing feed predominantly on the burrowing amphipod *Corophium volutator*, and also on the burrowing polychaete *Nereis diversicolor* on the Clyde mudflats, and rarely feed other than on intertidal mudflats. Oystercatchers also feed on *Nereis* but take other prey such as mussels. Curlews feed rather little on *Nereis* and take mainly crabs and molluscs. The waders that declined in numbers seem to be the species dependent on *Corophium* and *Nereis*. Furness et al. (1986b) found evidence from research by others to suggest that *Corophium* and *Nereis* densities on the Clyde mudflats had decreased between the early 1970s and the 1980s and that this change coincided with an increase of invertebrate community diversity, recolonization of the estuary by fish and reduced discharge of sewage solids into the estuary after improvements to sewage-treatment plants. Although it is not clear whether the decrease in dunlin, redshank and lapwing numbers was primarily due to increased competition for invertebrate prey when the estuary was recolonized by fish, or whether it was due directly to their particular prey being provided with less organic matter and bacteria when sewage treatment facilities were improved, the detailed and long-term wader census data collected by volunteers for the BoEE provided an unequivocal picture of the response of the birds to the changed environmental conditions in the estuary, though not an explanation of the mechanism.

3.6.3 Potential for use of bird counts to assess nutrient change

Detailed count data exist for wader numbers on most estuaries in the UK and for many throughout Europe. Similar long-term count data also exist for many freshwater bodies, through the wildfowl counts of the Wildfowl and Wetlands Trust and the International Waterfowl Research Bureau. With an improved understanding resulting from analyses of the factors that determine community composition and sizes of overwintering populations on estuaries, long-term data-sets, such as those of the BoEE, should permit these bird counts from particular sites to be used to assess how changes in nutrient loading affect particular ecosystems. Wader numbers may be correlated with nutrient inputs (unless these are very high), and so ornithologists may have some reservations as to whether reduced organic input to estuaries represents a reduction of pollution or not, but it would be valuable to be able to demonstrate the effect of measures to reduce nutrient inputs in terms of the responses of populations of waders and other birds.

3.7 RADIONUCLIDES

The uptake of radionuclides by birds and the use of birds as monitors of radionuclide contamination of the environment is reviewed in detail in Chapter 4 by Brisbin. His review shows that birds can provide valuable information on the extent of radionuclide contamination over a wide range of spatial scales. However, apart from the detailed work of biologists at the Savannah River Laboratory, there have been few studies that have attempted to survey the hazards to wildlife of radionuclides discharged into the environment.

One particularly large discharge of radionuclides has been from Sellafield, Cumbria, on the east shore of the Irish Sea. Although most of the radionuclides become incorporated into sediments close to the discharge, caesium-137 is more mobile and is carried in the drifting seawater around the northern Irish Sea and then along the west of Scotland and into the northern North Sea. Numbers of black-headed gulls *Larus ridibundus* nesting near to this pollution source have declined in recent years, leading to some concern that radionuclide pollution could be harming the birds. Studies of the levels of caesium-134, caesium-137, plutonium-238, plutonium-239/240 in birds, their eggs and invertebrate prey (Lowe, 1991) indicated that the radiation dose to the whole body of the most contaminated birds was three orders of magnitude too low to have caused the population decline, and other factors such as predation are more likely to be the cause. However, the radiocaesium and plutonium concentrations in the body tissues of black-headed gulls were among the lowest found in any of the species of birds sampled by Lowe (1991), and so the hazard of radiation to most contaminated birds such as the mollusc-eating oystercatcher (with levels 50 times those in black-headed gulls) would have been of greater interest. The paper by Lowe (1991) provides a good example of a study initiated because of an environmental concern generated by ornithologists, although in this case their suspicion that gulls were being harmed by radionuclide pollution proved to be unfounded. Routine monitoring of radionuclide levels in animals concentrates on contamination of shellfish, with particular concern for the hazard to humans eating them. In this context, birds are not satisfactory as monitors. Most of the bird species inhabiting the shallow sea and intertidal area around Cumbria are semi-nomadic or long-distance migrants (waders, ducks, geese) and feed in a mixture of habitats, which makes interpretation of radionuclide levels very difficult, though it might be useful to focus on the cormorant *Phalacrocorax carbo*, a resident inshore fish-eater, and curlew or oystercatcher, intertidal feeders specializing on shellfish prey and present through a substantial part of the year. There is, to my mind, no compelling evidence that radionuclide levels in these species are too low to be of concern. Lowe (1991) cites the fact that oystercatcher numbers are increasing on the Cumbrian coast as evidence that there is no detrimental effect of radionuclides on that species but oystercatcher numbers have been increasing generally in the UK, and so the increase in numbers in Cumbria does not preclude the possibility of harmful effects to individuals with high radionuclide burdens.

3.8 ACIDIFICATION

Human activities have increased the input of sulphur dioxide, nitrogen oxides, and ammonia to the atmosphere, leading to increased acid deposition. This has been much debated and researched. The primary effects of acidification of freshwater bodies are to increase metal solubilities and to enhance metal toxicity to invertebrates and fish (Wren and Stephenson, 1991; Scheuhammer, 1991a,b). Nyholm (1981) and Nyholm and Myhrberg (1977) suggested that elevated aluminium levels due to acidification could impair eggshell formation, in which case studies of shell structure might indicate historical changes in the effect of acid deposition.

Acidification can affect birds in a number of ways but these are generally indirect, via other parts of the ecosystem. Invertebrates, fish or the physical environment, may be more appropriate subjects for monitoring than bird populations but large quantities of census and breeding surveillance data are available for birds, so there is reason to consider the indications of changes in acidification that birds can provide, especially where data may allow 'retrospective monitoring' of historical changes for which physical, invertebrate and fish data are lacking.

Ormerod and co-workers have shown that in riparian birds such as dippers *Cinclus cinclus* river pH affects breeding density, diet, breeding success, egg mass and shell thickness (Ormerod *et al.*, 1986, 1987, 1988; Vickery, 1991; Chapter 5). Such evidence may then allow historical data such as the British Trust for Ornithology's nest record cards and museum collections of eggshells to be used to monitor retrospectively the organochlorine histories of different rivers or regions. It is possible that eggshell chemistry and structure may also be useful tools, since metal solubilities, which are affected by pH, may be indicated by eggshell microanalysis, although such possibilities have yet to be tested.

Mild acidification of lakes can result in increased numbers of ducks that feed on benthic invertebrates (because fish have been killed off benthic invertebrate populations increase) (DesGranges and Rodrigue, 1986; Eadie and Keast, 1982; Eriksson, 1979, 1984, 1987; Eriksson *et al.*, 1980; Hunter *et al.*, 1986); in contrast, piscivorous ducks and divers are lost. Thus data on changes in waterfowl communities and numbers may be of some use in broad-brush assessment of acidification over regions of North America or Europe (McNicol *et al.*, 1987a,b; Rutschke, 1987).

3.9 OIL

Although oil spills at sea can kill very large numbers of seabirds (some 69 000 are thought to have been killed by oil in the eastern North Sea in winter 1980–81 (Baillie and Mead, 1982)), monitoring of seabird breeding populations, some of which has been carried out specifically in response to the perceived hazards of oil pollution from production close to major seabird breeding sites, has not shown any changes that can be attributed to oil pollution (Dunnet, 1987). It is now clearly evident that many other factors affect seabird populations, including fish stocks, weather, and entanglement in nets, so that the monitoring of breeding numbers can do no more than alert scientists to unidentified factors causing changes in population size. The monitoring of breeding seabirds continues, but emphasis is now being put on monitoring not only numbers but also breeding success, diet and other factors, in order to obtain some clues as to likely causes of detected changes.

Most seabird mortality due to oil occurs as a result of chronic oil pollution rather than catastrophic accidents (Clark, 1984), and probably accounts for only a small part of the total annual mortality in seabird populations (Dunnet, 1982). Ford *et al.* (1982) and Wiens *et al.* (1984) discussed the possibility of modelling the impact of oil spills on colonially breeding seabirds in order to be able to predict the impacts of spills on seabirds during the breeding season.

Evidence for specific oiling incidents and for the level of chronic oil pollution can be obtained from beached-bird surveys, where all dead birds found on stretches of beach are recorded during periodic surveys and the numbers contaminated with oil are noted. Beached-bird surveys have been organized in many parts of the world; particularly detailed sets of data exist for the coasts of Europe and for selected sites in South Africa and New Zealand. That numbers of dead birds on beaches, and the proportion of these that are oiled, reflect the level of oil pollution at sea in the vicinity, is an idea that has attracted much severe criticism. Numbers of birds washed up may depend most on wind direction (Stowe, 1982). Birds may die in one place, be carried by winds and currents, and be beached elsewhere, perhaps very great distances away. Birds may become oiled by contact with slicks after they have died (just as polythene bottles washed up on beaches are often covered with oil or tar). Less severe criticisms can be raised regarding identification of oil on dark plumage by the relatively untrained volunteers who participate in such surveys, and problems of corpses being removed by predators or people before the survey is done or of corpses being redeposited by later tidal actions. Stowe (1982) considered the limitations in detail and recommended that beached-bird surveys should be continued throughout Britain, but shortly afterwards the national survey was suspended. Beached-bird surveys have since continued in Orkney and Shetland and a mid-winter pan-European survey continues. Beached-bird surveys have been carried out in the Netherlands since 1915 and provide a picture of patterns and causes of seabird mortality and changes over the decades (Camphuysen, 1989). Extensive changes in the numbers and distributions of some seabirds in the southern North Sea are well shown by the beached-bird data, and there are some indications of changes in oiling rates, though it is difficult to attribute this to reductions in oil mortality rather than to increases in other causes of death. Beached-bird surveys do show large variations in proportions of dead seabirds that are oiled, among species and among regions. Oil on corpses is most frequent with diving seabirds such as auks and sea ducks, and is more frequent in the south-east North Sea and lower in the north-west North Sea.

New techniques now permit 'fingerprinting' of oil from seabird corpses to allow the source of pollution to be identified in some cases (Dahlmann and Timm, 1991), but whether the frequency of oiled seabirds on beaches provides an accurate measure of the amount of oil pollution at sea is still unclear. Studies currently underway in the Netherlands aim to answer this, by relating aerial observations of the numbers of oil slicks to numbers of oiled birds reported in beached-bird surveys.

3.10 AIR POLLUTION

The use of the presence of particular species of lichens as indicators of air pollution levels is well known, as is the use of moss bags as accumulators of airborne pollutants. Rates of deposition of some heavy metals may be indicated by metal levels of flight feathers of resident birds (sections 3.4.2 and 3.4.3). Newman (1979), Newman and Schreiber (1984), Newman et al. (1985, 1989) advocate the wider use of wildlife as monitors of air pollution levels. In particular, they suggest that house martins *Delichon urbica* are sensitive to variations in insect abundance caused by air pollution, so that low numbers or poor breeding success of house martins in particular areas within a region may provide an indication of areas of poor air quality. While the use of sessile indicators such as lichens or deposition traps would probably give a more reliable picture, background data on house martin populations make it worth

Table 3.3 Some examples of national pollutant monitoring schemes using birds

Country and organization		Principal species used in monitoring	Samples collected	Reference
a) oil				
Netherlands	NSO	Seabirds	–	Camphuysen (1989)
UK	RSPB	Seabirds	–	Stowe (1982)
b) Chemical pollutants (especially organochlorines mercury)				
UK	ITE	Birds of prey	Tissues, eggs	Newton (1986)
UK	ITE	Guillemots, gannets	Eggs	Newton et al. (1990)
Sweden	NEPB	Guillemot	Eggs	Olsson and Reutergardh (1986)
Germany	IV	Herring gulls	Eggs (banked)	
Norway	NINA	Seabirds	Eggs (banked)	
USA	USFWS	Herring gulls	Tissues, eggs	
USA	USFWS	Starlings, ducks, eagles	Tissues	Cain and Bunck (1983)
Canada	CWS	Seabirds	Tissues, eggs	Noble and Elliott (1986)
Canada	CWS	Gannets	Eggs	Chapdelaine et al. (1987)
c) Pesticides				
UK	MAFF	All wildlife	Dead birds	Hardy et al. (1987)
d) Population monitoring				
UK	BTO	Many land birds	Population data	Baillie (1991)

investigating whether these birds can also be of use in assessing air pollution.

3.11 NATIONAL POLLUTANT MONITORING SCHEMES USING BIRDS

Many countries have recently run, or are currently running, pollutant monitoring schemes based on biological sampling, but the different national schemes tend not to be co-ordinated internationally. Some examples of the schemes that include birds are given in Table 3.3.

In recent years schemes have been developed to store annual or five-yearly samples at very low temperatures ($-60°C$ to $-170°C$) so that these would be available in the future should some new pollution problem become evident, allowing retrospective studies such as those utilizing eggshell thickness to prove the impact of DDT (Elliott, 1985; Lewis and Lewis, 1979; Wise and Zeisler, 1985). Such a strategy has the powerful advantage of allowing new, more refined, analytical methods to be used on old as well as recent material, while avoiding costly annual analyses of residues that eventually turn out to be of no long-term interest.

REFERENCES

Anderlini, V.C., Connors, P.G., Risebrough, R.W. and Martin, J.H. (1972) Concentrations of heavy metals in some Antarctic and North American seabirds, in *Proc. Symp. Conservation Problems Antarctica*, Blackburg Virginia Polytechnic Institute and State University, pp. 49–62.

Anderson, D.W. and Hickey, J.J. (1976) Dynamics of storage of organochlorine pollutants in herring gulls. *Environ. Pollut.*, **10**, 183–200.

Anderson, D.W., Raveling, D.G., Risebrough, R.W. and Springer, A.M. (1984) Dynamics of low-level organochlorines in adult cackling geese over the annual cycle. *J. Wildl. Manage.*, **48**, 1112–27.

Ankney, C.D. and MacInnes, C.D. (1978) Nutrient reserves and reproductive performance of female lesser snow geese. *Auk*, **95**, 459–71.

Appelquist, H., Asbirk, S. and Drabaek, I. (1984) Mercury monitoring: mercury stability in bird feathers. *Mar. Pollut. Bull.*, **15**, 22–4.

Appelquist, H., Drabaek, I. and Asbirk, S. (1985) Variation in mercury content of guillemot feathers over 150 years. *Mar. Pollut. Bull.*, **16**, 244–8.

Azzarello, M.Y. and van Vleet, E.S. (1987) Marine birds and plastic pollution. *Mar. Ecol. Prog. Ser.*, **37**, 295–303.

Backstrom, J. (1969) Distribution studies of mercuric pesticides in quail and some freshwater fishes. *Acta Pharmacol. Toxicol.*, **27**, suppl. 3.

Baillie, S.R. (1991) Monitoring terrestrial bird populations, in *Monitoring for Conservation and Ecology* (ed. F.B. Goldsmith), Chapman & Hall, London, pp. 112–32.

Baillie, S.R. and Mead, C.J. (1982) The effect of severe oil pollution during the winter of 1980–81 on British and Irish auks. *Ring. & Migr.*, **4**, 33–44.

Bakir, F., Damiuji, S.F., Amin-Zaki, L. *et al.* (1973) Methyl mercury poisoning in Iraq. *Science*, **181**, 176–9.

Barrett, R.T., Skaare, J.U., Norheim, G. *et al.* (1985) Persistent organochlorines and mercury in eggs of Norwegian seabirds 1983. *Environ. Pollut.*, **39**, 79–93.

Becker, P.H. (1989) Seabirds as monitor organisms of contaminants along the German North Sea coast. *Helgolander Meeresuntersuchungen*, **43**, 395–405.

Becker, P.H. (1991) Population and contamination studies in coastal birds: the Common Tern *Sterna hirundo*, in *Bird Population Studies: Relevance to Conservation and Management* (eds C.M. Perrins, J.D. Lebreton and G.J.M. Hirons), Oxford University Press, Oxford, pp. 433–60.

Becker, P.H., Buthe, A. and Heidmann, W. (1985a) Schadstoffe in Gelegen von Brutvogeln der deutschen Nordseekuste. I. Chlororganische Verbindungen. *J. Orn.*, **126**, 29–51.

Becker, P.H., Ternes, W. and Russel, H.A. (1985b) Schadstoffe in Gelegen von Brutvogeln der deutschen Nordseekuste. II. Quecksilber. *J. Orn.*, **126**, 253–62.

Becker, P.H., Buthe, A. and Heidmann, W. (1988) Decreasing contaminants in shorebird eggs on the German North Sea coast? *J. Orn.*, **129**, 104–5.

Berg, W., Johnels, A., Sjostrand, B. and Westermark, T. (1966) Mercury content in feathers of Swedish birds from the past 100 years. *Oikos*, **17**, 71–83.

Bijleveld, M.F.I.J., Goeldlin, P. and Mayol, J. (1979) Persistent pollutants in Audouin's gull (*Larus audouinii*) in the western Mediterranean: a case study with wide implications? *Environ. Conserv.*, **6**, 139–43.

Birkhead, M.E. (1982) Causes of mortality in the mute swan, *Cygnus olor* on the River Thames. *J. Zool., Lond.*, **198**, 15–25.

Birkhead, M.E. (1983) Lead levels in the blood of mute swans, *Cygnus olor* on the River Thames. *J. Zool., Lond.*, **199**, 59–73.

Birkhead, M.E. and Perrins, C.M. (1985) The breeding biology of the mute swan, *Cygnus olor* on the River Thames with special reference to lead poisoning. *Biol. Conserv.*, **32**, 1–11.

Boersma, D.C., Ellenton, J.A. and Yagminas, A. (1986) Investigation of the hepatic mixed-function oxidase system in herring gull embryos in relation to environmental contaminants. *Environ. Toxicol. Chem.*, **5**, 309–18.

Bogan, J.A. and Newton, I. (1977) Redistribution of DDE in sparrowhawks during starvation. *Bull. Environ. Contam. Toxicol.*, **18**, 317–21.

Borg, K., Wanntorp, H., Erne, K. and Hanko, E. (1969) Alkyl mercury poisoning in terrestrial Swedish wildlife. *Viltrevy*, **6**, 301–79.

Bourne, W.R.P. (1976) Seabirds and pollution, in *Marine Pollution* (ed. R. Johnston), Academic Press, London, pp. 403–502.

Bourne, W.R.P. and Bogan, J.A. (1972) Polychlorinated biphenyls in North Atlantic seabirds. *Mar. Pollut. Bull.*, **11**, 171–5.

Braune, B.M. (1987) Comparison of total mercury levels in relation to diet and molt for nine species of marine birds. *Arch. Environ. Contam. Toxicol.*, **16**, 217–24.

Braune, B.M. and Gaskin, D.E. (1987a) Mercury levels in Bonaparte's gulls (*Larus philadelphia*) during autumn molt in the Quoddy Region, New Brunswick, Canada. *Arch. Environ. Contam. Toxicol.*, **16**, 539–49.

Braune, B.M. and Gaskin, D.E. (1987b) A mercury budget for the Bonaparte's gull during autumn moult. *Ornis Scand.*, **18**, 244–50.

Brothers, N.P. and Brown, M.J. (1987) The potential use of fairy prions (*Pachyptila turtur*) as monitors of heavy metal levels in Tasmanian waters. *Mar. Pollut. Bull.*, **18**, 132–4.

Bryan, G.W. (1979) Bioaccumulation of marine pollutants. *Phil. Trans. Roy. Soc. Lond.*, B, **286**, 483–505.

Bunck, C.M., Spann, J.W., Pattee, O.H. and Fleming, W.J. (1985) Changes in eggshell thickness during incubation: implications for evaluating the impact of organochlorine contaminants on productivity. *Bull. Environ. Contam. Toxicol.*, **37**, 173–82.

Burger, J. and Gochfeld, M. (1988) Metals in tern eggs in a New Jersey estuary; a decade of change. *Environ. Monitor. Assess.*, **11**, 127–35.

Burger, J. and Gochfeld, M. (1991) Cadmium and lead in common terns (Aves: *Sterna hirundo*): relationships between levels in parents and eggs. *Environ. Monitor. Assess.*, **16**, 253–8.

Busby, D.G., Pearce, P.A., Garrity, N.R. and Reynolds, L.M. (1983) Effect of an organophosphorus insecticide on brain cholinesterase activity in white-throated sparrows exposed to aerial forest spraying. *J. Anim. Ecol.*, **20**, 255–63.

Cade, T.J., Lincer, J.L., White, C.M. et al. (1971) DDE residue and eggshell changes in Alaskan falcons and hawks. *Science*, **172**, 955–7.

Cain, B.W. and Bunck, C.M. (1983) Residues of organochlorine compounds in starlings (*Sturnus vulgaris*). *Environ. Monit. Assess.*, **3**, 161–72.

Camphuysen, C.J. (1989) *Beached bird surveys in the Netherlands 1915–1988*, Nederlands Stookolieslachtoffer-Onderzzoek, Amsterdam.

Chapdelaine, G., Laporte, P. and Nettleship, D.N. (1987) Population, productivity, and DDT contamination trends of Northern Gannets (*Sula bassanus*) at Bonaventure Island, Quebec, 1967–1984. *Can. J. Zool.*, **65**, 2992–6.

Clark, R.B. (1984) Impact of oil pollution on seabirds. *Environ. Pollut. Ser. A*, **33**, 1–22.

Clark, T.P., Norstrom, R.J., Fox, G.A. and Won, H.T. (1987) Dynamics of organochlorines in herring gulls (*Larus argentatus*): II. A two compartment model and data for ten compounds. *Environ. Toxicol. Chem.*, **6**, 547–59.

Connors, P.G. and Smith, K.G. (1982) Oceanic plastic particle pollution: suspected effect on fat deposition in red phalaropes. *Mar. Pollut. Bull.*, **13**, 18–20.

Coulson, J.C., Deans, I.R., Potts, G.R. et al. (1972) Changes in organochlorine contamination of the marine environment of eastern Britain monitored by shag eggs. *Nature*, **236**, 454–6.

Custer, T.W., Franson, J.C., Moore, J.F. and Myers, J.E. (1986) Reproductive success and heavy metal contamination in Rhode Island common terns. *Environ. Pollut*, **41**, 33–52.

Dahlmann, G. and Timm, D. (1991) First analytical results of the EC-project 'Oiled Seabirds': comparative investigations on oiled seabirds and oiled beaches in the Netherlands, Denmark, and the Federal Republic of Germany. *Sula*, **5** (special issue), 12–14.

DesGranges, J-L. and Rodrigue, J. (1986) Influence of acidity and competition with fish on the development of ducklings in Quebec. *Water, Air and Soil Pollut.*, **30**, 743–50.

Diamond, A.W. and Filion, F. (eds) (1987) *The Value of Birds*, ICBP Tech. Publ. 6, International Council for Bird Preservation, Cambridge.

Dixon, T.R. and Dixon, T.J. (1981) Marine litter surveillance. *Mar. Pollut. Bull.*, **12**, 289–95.

Dunnet, G.M. (1982) Oil pollution and seabird populations. *Phil. Trans. R. Soc. Lond. B.*, **297**, 413–27.

Dunnet, G.M. (1987) Seabirds and North Sea oil. *Phil. Trans. R. Soc. Lond. B.*, **316**, 513–24.

Dustman, E.H., Martin, W.E., Heath, R.G. and Reichel, W.L. (1971) Monitoring pesticides in wildlife. *Pestic. Monit. J.*, **5**, 50–2.

Eadie, J.M. and Keast, A. (1982) Do goldeneye and perch compete for food? *Oecologia*, **55**, 225–30.

Ellenberg, H. and Dietrich, J. (1982) The goshawk as a bioindicator, in *Understanding the Goshawk* (eds R.E. Kenward and D. Lindsay), Int. Ass. Falconry, Oxford.

Ellenton, J.A. and McPherson, M.F. (1983) Mutagenicity studies on herring gulls from different locations on the Great Lakes. I. Sister chromatid exchange rates in herring gull embryos. *J. Toxicol. Environ. Health*, **12**, 317–24.

Ellenton, J.A., Brownlee, L.J. and Hollebone, B.R. (1985) Aryl hydrocarbon hydroxylase levels in the herring gull embryos from different locations on the Great Lakes. *Environ. Toxicol. Chem.*, **4**, 615–22.

Ellenton, J.A., McPherson, M.F. and Maus, K.L. (1983) Mutagenicity studies on herring gulls from different colonies on the Great Lakes. II. Mutagenic evaluation of extracts of herring gull eggs in a battery of in vitro mammalian and microbial tests. *J. Toxicol. Environ. Health*, **12**, 325–36.

Elliott, J.E. (1985) Specimen banking in support of monitoring for toxic contaminants in Canadian wildlife, in *International Review of Environmental Specimen Banking* (eds S.A. Wise and R. Zeisler), *National Bureau of Standards, Spec. Publ. 706.*, National Bureau of Standards, Washington, D.C.

Eriksson, M.O.G. (1979) Competition between freshwater fish and golden eyes (*Bucephala clangula*) for common prey. *Oecologia*, **41**, 99–107.

Eriksson, M.O.G. (1984) Acidification of lakes: effects on waterbirds in Sweden. *Ambio*, **13**, 260–2.

Eriksson, M.O.G. (1987) Some effects of freshwater acidification on birds in Sweden, in *The Value of Birds* (eds A.W. Diamond and F. Filion), *ICBP Tech. Publ. 6*, International Council for Bird Preservation, Cambridge, pp. 183–90.

Eriksson, M.O.G., Henrikson, L., Nilsson, B.L. *et al.* (1980) Predator-prey relations important for the biotic changes in acidified lakes. *Ambio*, **9**, 248–9.

Evans, P.R. and Moon, S.J. (1981) Heavy metals in shore birds and their prey, in *Heavy Metals in Northern England: Environmental and biological aspects* (eds P.J. Say and B.A Whitton), University of Durham, Durham, pp. 181–90.

Fimreite, N. (1971) Effects of dietary methyl mercury on ring-necked pheasants. *Can. Wildl. Serv. Occ. Pap. 9*, Canadian Wildlife Service, Ottawa.

Fimreite, N., Brun, E., Froslie, A. *et al.* (1974) Mercury in eggs of Norwegian seabirds. *Astarte*, **1**, 71–5.

Fimreite, N., Kveseth, N. and Brevik, E.M. (1980) Mercury, DDE and PCBs in eggs from a Norwegian gannet colony. *Bull. Environ. Contam. Toxicol.*, **24**, 142–4.

Focardi, S., Fossi, C., Lambertini, M. et al. (1988) Long term monitoring of pollutants in eggs of yellow-legged herring gulls from Capraia Island (Tuscan Archipelago). *Environ. Monit. Assess.*, **10**, 43–50.

Ford, R.G., Wiens, J.A., Heinemann, D. and Hunt, G. (1982) Modelling the sensitivity of colonial breeding marine birds to oil spills: guillemot and kittiwake populations on the Pribilof Islands, Bering Sea. *J. Appl. Ecol.*, **19**, 1–31.

Fox, G.A. (1976) Eggshell qality: its ecological and physiological significance in a DDE-contaminated common tern population. *Wilson Bull.*, **88**, 459–77.

Fox, G.A. and Weseloh, D.V. (1987) Colonial waterbirds as bio-indicators of environmental contamination in the great lakes, in *The Value of Birds* (eds A.W. Diamond and F. Filion), *ICBP Tech. Publ. 6*, International Council for Bird Preservation, Cambridge, pp. 209–16.

Franklin, A. (1990) Monitoring and surveillance of non-radioactive contaminants in the aquatic environment, 1984–1987. *Aquat. Environ. Monit. Rep. 22*, MAFF Direct. Fish. Res., Lowestoft.

Franklin, A. (1991) Monitoring and surveillance of non-radioactive contaminants in the aquatic environment and activities regulating the disposal of wastes at sea, 1988–1989. *Aquat. Environ. Monit. Rep. 26*, MAFF Direct. Fish. Res., Lowestoft.

Frenzel, R.W. and Anthony, R.G. (1989) Relationship of diets and environmental contaminants in wintering bald eagles. *J. Wildl. Manage.*, **53**, 792–802.

Friberg, L., Iscator, M., Nordberg, G. and Kjellstrom, T. (1974) *Cadmium in the environment.*, 2nd end, CRC Press, New York.

Fry, D.M. and Toone, C.K. (1981) DDT-induced feminization of gull embryos. *Science*, **213**, 922–4.

Fry, D.M., Fefer, S.I. and Sileo, L. (1987) Ingestion of plastic debris by Laysan albatrosses and wedge-tailed shearwaters in the Hawaiian Islands. *Mar. Pollut. Bull.*, **18**, 339–43.

Fujiki, M., Tajima, S. and Omori, A. (1972) The transition of the mercury contamination in Mimimata District. *Japan J. Hyg.*, **27**, 115–17.

Furness, R.W. (1985a) Plastic particle pollution: accumulation by Procellariiform seabirds at Scottish colonies. *Mar. Pollut. Bull.*, **16**, 103–6.

Furness, R.W. (1985b) Ingestion of plastic particles by seabirds at Gough Island, South Atlantic Ocean. *Environ. Pollut.*, **38**, 261–72.

Furness, R.W. and Hutton, M. (1979) Pollutant levels in the great skua *Catharacta skua*. *Environ. Pollut.*, **19**, 261–8.

Furness, R.W., Lewis, S.A. and Mills, J.A. (1990a) Mercury levels in the plumage of adult red-billed gulls *Larus novaehollandiae scopulinus* of known sex and age. *Environ. Pollut.*, **63**, 33–9.

Furness, R.W., Thompson, D.R. and Walsh, P.M. (1990b) Evidence from biological samples for historical changes in global metal pollution, in *Heavy Metals in the Marine Environment* (eds R.W. Furness and P.S. Rainbow), CRC Press, New York, pp. 219–25.

Furness, R.W., Muirhead, S.J. and Woodburn, M. (1986a) Using bird feathers to measure mercury in the environment: relationships between mercury content and moult. *Mar. Pollut. Bull.*, **17**, 27–30.

Furness, R.W., Galbraith, H., Gibson, I.P. and Metcalfe, N.B. (1986b) Recent changes in numbers of waders on the Clyde Estuary, and their significance for conservation. *Proc. Roy. Soc. Edin.*, **90B**, 171–84.

Furness, R.W., Johnson, J.L., Love, J.A. and Thompson, D.R. (1989)

Pollutant burdens and reproductive success of golden eagles *Aquila chrysaetos* exploiting marine and terrestrial food webs in Scotland, in *Raptors in the Modern World* (eds B-U. Meyburg and R.D. Chancellor), World Working Group on Birds of Prey and Owls, Berlin, pp. 495–500.

Gilbertson, M., Eliott, J.E. and Peakall, D.B. (1987) Seabirds as indicators of marine pollution, in *The Value of Birds* (eds A.W. Diamond and F. Filion), ICBP Tech. Publ. 6, International Council for Bird Preservation, Cambridge, pp. 231–48.

Gochfeld, M. and Burger, J. (1987) Heavy metal concentrations in the liver of three duck species: influences of species and sex. *Environ. Pollut. Ser. A*, **45**, 1–15.

Goede, A.A. (1985) Mercury, selenium, arsenic and zinc in waders from the Dutch Wadden Sea. *Environ. Pollut.*, **37**, 287–309.

Goede, A.A. and de Bruin, M. (1984) The use of bird feather parts as a monitor for metal pollution. *Environ. Pollut.*, **8**, 281–98.

Goede, A.A. and de Voogt, P. (1985) Lead and cadmium in waders from the Dutch Wadden Sea. *Environ. Pollut. Ser. A*, **37**, 311–22.

Grue, C.E., Fleming, W.J., Busby, D.G. and Hill, E.F. (1983) Assessing hazards of organophosphate pesticides to wildlife. *Trans. N. Am. Wildl. Nat. Res. Conf.*, **48**, 200–20.

Hahn, E. (1991) *Schwermetallgehalte in Vogelfedern – ihre Ursache und der Einsatz von Federn standorttreuer Vogelarten im Rahmen von Bioindikationsverfahren*. Berichte des Forschungszentrums, Julich.

Hahn, E., Hahn, K. and Stoeppler, M. (1989) Schwermetalle in Federn von Habichten (*Accipiter gentilis*) aus unterschiedlich belasteten Gebieten. *J. Orn.*, **130**, 303–9.

Hardy, A.R., Stanley, P.I. and Greig-Smith, P.W. (1987) Birds as indicators of the intensity of use of agricultural pesticides in the UK, in *The Value of Birds* (eds A.W. Diamond and F. Filion), ICBP Tech. Publ. 6, International Council for Bird Preservation, Cambridge, pp. 119–32.

Harris, M.P. and Osborn, D. (1981) Effect of a polychlorinated biphenyl on the survival and breeding of puffins. *J. Appl. Ecol.*, **18**, 471–9.

Heinz, G. (1974) Effects of low dietary levels of methyl mercury on mallard reproduction. *Bull. Environ. Contam. Toxicol.*, **11**, 386–92.

Helander, B., Olsson, M. and Reutergardh, L. (1982) Residue levels of organochlorine and mercury compounds in unhatched eggs and the relationships to breeding success in white-tailed sea eagles *Haliaeetus albicilla* in Sweden. *Holarctic Ecol.*, **5**, 346–66.

Hill, E.F. and Fleming, W.J. (1982) Anticholinesterase poisoning of birds: field monitoring and diagnosis of acute poisoning. *Environ. Toxicol. Chem*, **1**, 27–38.

Hill, E.F. and Hoffman, D.J. (1984) Avian models for toxicity testing. *J. Am. Coll. Toxicol.*, **3**, 357–76.

Hobson, K.A. and Montevecchi, W.A. (1991) Stable isotopic determinations of trophic relationships of great auks. *Oecologia*, **87**, 528–31.

Hoffman, D.J., Rattner, B.A., Sileo, L. *et al.* (1987) Embryotoxicity, teratogenicity, and aryl hydrocarbon hydroxylase activity in Forster's terns on Green Bay, Lake Michigan. *Environ. Res.*, **42**, 176–84.

Holt, G., Froslie, A. and Norheim, G. (1979) Mercury, DDE, and PCB in the avian fauna in Norway (1965–1976), *Acta Vet. Scand. Suppl.*, **70**, 1–28.

Honda, K., Nasu, T. and Tatsukawa, R. (1986a) Seasonal changes in mercury accumulation in the black-eared kite, *Milvus migrans lineatus*. *Environ. Pollut.*, **42**, 325–34.

Honda, K., Min, B.Y. and Tatsukawa, R. (1986b) Distribution of heavy metals and their age-related changes in the eastern great white egret, *Egretta alba modesta*, in Korea. *Arch. Environ. Contam. Toxicol.*, **15**, 185–97.

Honda, K., Yamamoto, Y., Hidaka, H. and Tatsukawa, R. (1986c) Heavy metal accumulations in Adelie penguin, *Pygoscelis adeliae*, and their variations with the reproductive process. *Mem. Nat. Inst. Polar Res.*, Special Issue, **40**, 443–53.

Honda, K., Lee, D.P. and Tatsukawa, R. (1990) Lead poisoning in swans in Japan. *Environ. Pollut.*, **65**, 209–18.

Hudson, P.J. and Rands, M.R.W. (1988) *Ecology and Management of Gamebirds*, BSP Professional Books, Oxford.

Hunter, M.L., Jones, J.J., Gibbs, K.E. and Morning, J.R. (1986) Duckling responses to lake acidification: do black ducks and fish compete? *Oikos*, **47**, 26–32.

Hutchinson, T.W. and Meema, K.M. (1987) *Lead, Mercury, Cadmium and Arsenic in the Environment*, Wiley, New York.

Hutton, M. (1981) Accumulation of heavy metals and selenium in three seabird species from the United Kingdom. *Environ. Pollut. Ser. A*, **26**, 129–45.

Jensen, S., Johnels, A.G., Olsson, M. and Westermark, T. (1972) The avifauna of Sweden as indicators of environmental contamination with mercury and chlorinated hydrocarbons. *Proc. XVth Int. Ornithol. Congr., Leiden.*, pp. 455–6.

Jorgensen, O.H. and Kraul, I. (1974) Eggshell parameters, and residues of PCB and DDE in eggs from Danish herring gulls *Larus a. argentatus*. *Ornis Scand.*, **5**, 173–9.

King, K.A., Lefebvre, C.A. and Mulhern, B.M. (1983) Organochlorine and metal residues in royal terns nesting on the central Texas coast. *J. Field Ornithol.*, **54**, 295–303.

Koeman, J.H., de Brauw, M.C.T.N. and de Vos, R.H. (1969) Chlorinated biphenyls in fish, mussels and birds from the river Rhine and the Netherlands coastal area. *Nature*, **221**, 1126–8.

Kubiak, T.J., Harris, H.J., Smith, L.M. *et al.* (1989) Microcontaminants and reproductive impairment of the Forster's tern on Green Bay, Lake Michigan – 1983. *Arch. Environ. Contam. Toxicol.*, **18**, 706–27.

Laist, D.W. (1987) Overview of the biological effects of lost and discarded plastic debris in the marine environment. *Mar. Pollut. Bull.*, **18**, 319–26.

Lambeck, R.H.D., Nieuwenhuize, J. and van Liere, J.M. (1991) Polychlorinated biphenyls in oystercatchers (*Haematopus ostralegus*) from the Oosterschelde (Dutch delta area) and the Western Wadden Sea, that died from starvation during severe winter weather. *Environ. Pollut.*, **71**, 1–16.

Lambertini, M. (1982) Mercury levels in *Larus audouinii* and *Larus argentatus michahellis* breeding in Capraia Island (Tyrrhenian Sea). *Riv. Ital. Ornithol. Milano*, **52**, 75–99.

Lantzy, R.J. and Mackenzie, F.T. (1979) Atmospheric trace metals: global assessment of man's impact. *Geochimica et Cosmochimica Acta*, **43**, 511–25.

Law, R.J., Fileman, C.F., Hopkins, A.D. *et al.* (1991) Concentrations of trace metals in the livers of marine mammals (seals, dolphins and porpoises) from waters around the British Isles. *Mar. Pollut. Bull.*, **22**, 183–91.

Leah, R.T., Evans, S.J., Johnson, M.S. and Collings, S. (1991) Spatial patterns in accumulation of mercury by fish from the NE Irish Sea. *Mar. Pollut. Bull.*, **22**, 172–5.

Lee, D.P., Honda, K., Tatsukawa, R. and Won, P. (1989) Distribution and residue levels of mercury, cadmium and lead in Korean birds. *Bull. Environ. Contam. Toxicol.*, **43**, 550–5.

Leonzio, C., Fossi, C. and Focardi, S. (1986) Heavy metal and selenium in a migratory bird wintering in a mercury polluted lagoon. *Bull. Environ. Contam. Toxicol.*, **37**, 219–25.

Lewis, R.A. and Lewis, C.W. (1979) Terrestrial vertebrate animals as biological monitors of pollution, in *Monitoring Environmental Materials and Specimen Banking* (ed. N.P. Luepke), Martinus Nijhoff Publ., Den Haag.

Lewis, S.A. (1991) Studies of mercury dynamics in birds. PhD thesis, University of Glasgow.

Lewis, S.A. and Furness, R.W. (1991) Mercury accumulation and excretion in laboratory reared black-headed gull *Larus ridibundus* chicks. *Arch. Environ. Contam. Toxicol,.* **21**, 316–20.

Lewis, S.A., Becker, P.H. and Furness, R.W. (in press) Mercury levels in eggs, internal tissues and feathers of herring gulls *Larus argentatus* from the German Wadden Sea coast. *Environ. Pollut.*

Lock, J.W. Thompson, D.R., Furness, R.W. and Bartle, J.A. (1992) Metal concentrations in seabirds of the New Zealand region. *Environ. Pollut*, **75**, 289–300.

Lowe, V.P.W. (1991) Radionuclides and the birds at Ravenglass. *Environ. Pollut.*, **70**, 1–26.

McNicol, D.K., Bendell, B.E. and Ross, R.K. (1987a), Studies of the effects of acidification on aquatic wildlife in Canada: waterfowl and trophic relationships in small lakes in northern Ontario. *Can. Wildl. Serv. Occ. Pap.* **62.**, Canadian Wildlife Service, Ottawa.

McNicol, D.K., Blancher, P.J. and Bendell, B.E. (1987b), Waterfowl as indicators of wetland acidification in Ontario, in *The Value of Birds* (eds A.W. Diamond and F. Filion), *ICBP Tech. Publ.* 6, International Council for Bird Preservation, Cambridge, pp. 149–66.

Magat, W. and Sell, J.L. (1979) Distribution of mercury and selenium in egg components and egg-white proteins. *Proc. Soc. Exp. Biol. Med.*, **161**, 458–61.

March, B.E., Poon, R. and Chu, S. (1983) Metabolism and nutrition: The dynamics of ingested methyl mercury in growing and laying chickens. *Poultry Sci.*, **62**, 1000–9.

Marcovecchio, J.E., Moreno, V.J., Bastida, R.O. *et al.* (1990) Tissue distribution of heavy metals in small cetaceans from the southwestern Atlantic Ocean. *Mar. Pollut. Bull.*, **21**, 299–304.

Martin, A.R. (1990) The diet of Atlantic puffin *Fratercula arctica* and northern gannet *Sula bassana* chicks at a Shetland colony during a period of changing prey availability. *Bird Study.*, **36**, 170–80.

Mejstrik, V. and Pospisil, J. (1989) Theoretical principles of biomonitoring, in *Proc. Vth Int. Conf. Bioindicatores Deteriorisationis Regionis* (eds J. Bohac and V. Ruzicka), Inst. Landscape Ecology CAS, Ceske Budejovice, pp. 10–14.

Mineau, P. (1982) Levels of major organochlorine contaminants in sequentially-laid herring gull eggs. *Chemosphere*, **11**, 679–85.

Mineau, P., Sheehan, P.J. and Barl, A. (1987) Pesticides and waterfowl on the Canadian prairies: a pressing need for research and monitoring, in *The Value of Birds* (eds A.W. Diamond and F. Filion), *ICBP Tech. Publ. 6*, International Council for Bird Preservation, Cambridge, pp. 133–48.

Montevecchi, W.A. (1991) Incidence and types of plastic in gannets' nests in the northwest Atlantic. *Can. J. Zool.*, **69**, 295–7.

Moore, N.W. and Ratcliffe, D.A. (1962) Chlorinated hydrocarbon residues in the egg of a peregrine falcon (*Falco peregrinus*) from Perthshire. *Bird Study*, **9**, 242–4.

Moriarty, F. (1988) *Ecotoxicology: The Study of Pollutants in Ecosystems*, Academic Press, London.

Munoz, R.V. Jr., Hacker, C.S. and Gesell, T.F. (1976) Environmentally acquired lead in the laughing gull, *Larus atricilla*. *J. Wildl. Dis.*, **12**, 139–43.

Muirhead, S.J. and Furness, R.W. (1988) Heavy metal concentrations in the tissues of seabirds from Gough Island, South Atlantic Ocean. *Mar. Pollut. Bull.*, **19**, 278–83.

Murton, R.K., Osborn, D. and Ward, P. (1978) Are heavy metals pollutants in Atlantic seabirds? *Ibis*, **120**, 106–7.

Nelson, J.B. (1978) *The Sulidae*, Oxford University Press, Oxford.

Nettleship, D.N. and Peakall, D.B. (1987) Organochlorine residue levels in three high Arctic species of colonially-breeding seabirds from Prince Leopold Island. *Mar. Pollut. Bull.*, **18**, 434–8.

Newman, J.R. (1979) The effects of air pollution on wildlife and their use as biological indicators, in *Animals as Monitors of Environmental Pollutants*, Nat. Acad. Sci., Washington D.C.

Newman, J.R. and Schreiber, R.K. (1984) Animals as indicators of ecosystem responses to air emissions. *Environ. Manage.*, **8**, 309–24.

Newman, J.R., Novakova, E. and McClave, J.T. (1985) The influence of industrial air emissions on the nesting ecology of the house martin (*Delichon urbica*) in Czechoslovakia. *Biol. Conserv.*, **31**, 229–48.

Newman, J.R., Novakova, E. and McClave, J.T. (1989) The use of the house martin, *Delichon urbica*, as a bioindicator of air pollution, in *Proc. Vth Int. Conf. Bioindicatores Deteriorisationis Regionis* (eds J. Bohac and V. Ruzicka). Inst. Landscape Ecology CAS, Ceske Budejovice, pp. 179–87.

Newton, I. (1979) *Population Ecology of Raptors*, T. & A.D. Poyser, Berkhamsted.

Newton, I.(1986) *The Sparrowhawk*, T. & A.D. Poyser, Calton.

Newton, I. (1988) Determination of critical pollutant levels in wild populations, with examples from organochlorine insecticides in birds of prey. *Environ. Pollut.*, **55**, 29–40.

Newton, I. and Galbraith, E.A. (1991) Organochlorines and mercury in the eggs of golden eagles *Aquila chrysaetos* from Scotland. *Ibis*, **133**, 115–20.

Newton, I. and Haas, M.B. (1988) Pollutants in merlin eggs and their effects on breeding. *Brit. Birds*, **81**, 258–69.

Newton, I. and Wyllie, I. (1992) Recovery of a sparrowhawk population in relation to declining pesticide contamination. *J. Appl. Ecol.*, **29**, 476–84.

Newton, I., Bogan, J.A. and Rothery, P. (1986) Trends and effects of organochlorine compounds in sparrowhawk eggs. *J. Appl. Ecol.*, **23**, 461–78.

Newton, I., Bogan, J.A. and Haas, M.B. (1989) Organochlorines and mercury in the eggs of British peregrines *Falco peregrinus*. *Ibis*, **131**, 355–76.

Newton, I., Wyllie, I. and Freestone, P. (1990) Rodenticides in British barn owls. *Environ. Pollut.*, **68**, 101–17.

Newton, I., Wyllie, I. and Asher, A. (1993) Long-term trends in organochlorine and mercury residues in some predatory birds in Britain. *Environ. Pollut.*, **79**, 143–52.

Nicholson, J.K. and Osborn, D. (1983) Kidney lesions in pelagic seabirds with high tissue levels of cadmium and mercury. *J. Zool., Lond.*, **200**, 99–118.

Noble, D.G. and Elliott, J.E. (1986) Environmental contaminants in Canadian seabirds, 1968–1985: trends and effects. *Tech. Rep. Ser. 13*, Canadian Wildlife Service, Ottawa.

Nyholm, N.E.I. (1981) Evidence of involvement of aluminium in causation of defective formation of eggshells and impaired breeding in wild passerine birds. *Environ. Res.*, **26**, 363–71.

Nyholm, N.E.I. and Myhrberg, H.E. (1977) Severe eggshell defects and impaired reproductive capacity in small passerines in Swedish Lapland. *Oikos*, **29**, 336–41.

Ogilvie, M.A. (1986) The mute swan *Cygnus olor* in Britain 1983. *Bird Study*, **33**, 121–37.

O'Halloran, J., Myers, A.A. and Finbarr Duggan, P. (1989) Some sub-lethal effects of lead on mute swan *Cygnus olor*. *J. Zool., Lond.*, **218**, 627–32.

Ohi, G., Seki, H., Akiyama, K. and Yagyo, H. (1974) The pigeon, a sensor of lead pollution. *Bull Environ. Contam. Toxicol.*, **12**, 92–8.

Ohi, G., Seki, H., Minowa, K. *et al.* (1981) Lead pollution in Tokyo – the pigeon reflects its amelioration. *Environ. Res.*, **26**, 125–9.

Ohlendorf, H.M. and Harrison, C.S. (1986) Mercury, selenium, cadmium and organochlorines in eggs of three species of Hawaiian seabird species. *Environ. Pollut.*, **11**, 169–91.

Olsson, M. and Reutergardh, L. (1986) DDT and PCB pollution trends in the Swedish aquatic environment. *Ambio*, **15**, 103–9.

Ormerod, S.J. and Tyler, S.J. (1987), Dippers (*Cinclus cinclus*) and grey wagtails (*Motacilla cinerea*) as indicators of stream acidity in upland Wales in *The Value of Birds* (eds A.W. Diamond and F. Filion), *ICBP Tech. Publ. 6*, International Council for Bird Preservation, Cambridge, pp. 191–208.

Ormerod, S.J., Allinson, N., Hudson, D. and Tyler, S.J. (1986) The distribution of breeding dippers (*Cinclus cinclus* L., Aves) in relation to stream acidity in upland Wales. *Freshwat. Biol.*, **16**, 501–9.

Ormerod, S.J., Bull, K.R., Cummins, C.P., Tyler, S.J. and Vickery, J.A. (1988) Egg mass and shell thickness in dippers *Cinclus cinclus* in relation to stream acidity in Wales and Scotland. *Environ. Pollut*, **55**, 107–21.

Osborn, D., Harris, M.P. and Nicholson, J.K. (1979) Comparative tissue distribution of mercury, cadmium and zinc in three species of pelagic seabirds. *Comp. Biochem. Physiol.*, **64C**, 61–7.

Parslow, J.L.F. and Jefferies, D.J. (1975) Geographical variation in pollutants in guillemot eggs. *Ann. Rep. Inst. Terr. Ecol. 28.*, Institute of Terrestrial Ecology, Cambridge.

Parslow, J.L.F. and Jefferies, D.J. (1977) Gannets and toxic chemicals. *Brit. Birds*, **70**, 366–72.

Peakall, D.B. (1992) *Animal Biomarkers as Pollution Indicators*, Chapman & Hall, London.

Peakall, D.B., Cade, T.J., White, C.M. and Haugh, J.R. (1975) Organochlorine residues in Alaskan peregrines. *Pesticide Monitoring Journal*, **8**, 255–60.

Peakall, D.B. and Boyd, H. (1987) Birds as bio-indicators of environmental conditions, in *The Value of Birds* (eds A.W. Diamond and F. Filion). *ICBP Tech. Publ. 6*, International Council for Bird Preservation, Cambridge, pp. 113–18.

Peakall, D.B., Fox, G.A., Gilman, A.P. et al. (1980) Reproductive success of herring gulls as an indicator of Great Lakes water quality, in *Hydrocarbons and Halogenated Hydrocarbons in the Aquatic Environment* (eds B.K. Afghan and D. MacKay). Plenum Press, London, pp. 337–44.

Phillips, D.J.H. (1980) *Quantitative Aquatic Biological Indicators: Their Use to Monitor Trace Metal and Organochlorine Pollution*, Applied Sci. Publ., London.

Podolsky, R.H. and Kress, S.W. (1989) Plastic debris incorporated into double-crested cormorant nests in the Gulf of Maine. *J. Field Ornithol.*, **60**, 248–50.

Potts, G.R. (1968) Success of eggs of shags on the Farne Islands, Northumberland, in relation to their content of dieldrin and pp'DDE. *Nature*, **217**, 1282–4.

Potts, G.R. (1986) *The Partridge: Pesticides, Predation and Conservation*, Collins, London.

Powell, G.N. (1984) Reproduction by an altricial songbird, the red-winged blackbird, in fields treated with the organophosphate insecticide fenthion. *J. Appl. Ecol.*, **21**, 83–95.

Pruter, A.T. (1987) Sources, quantities and distribution of persistent plastics in the marine environment. *Mar. Pollut. Bull.*, **18**, 305–10.

Ratcliffe, D.A. (1967) Decrease in eggshell weight in certain birds of prey. *Nature*, **215**, 208–10.

Ratcliffe, D.A. (1970) Changes attributed to pesticides in egg breakage frequency and eggshell thickness in some British birds. *J. Appl. Ecol.*, **7**, 67–115.

Ratcliffe, D.A. (1980) *The Peregrine Falcon*, T. & A.D. Poyser, Calton.

Reid, M. and Hacker, C.S. (1982) Spatial and temporal variation in lead and cadmium in the laughing gull, *Larus atricilla*. *Mar. Pollut. Bull.*, **13**, 387–9.

Reijnders, P.J.H. (1986) Reproductive failure in common seals feeding on fish from polluted coastal waters. *Nature*, **324**, 456–7.

Renzoni, A., Focardi, S., Fossi, C. et al. (1986) Comparison between concentrations of mercury and other contaminants in eggs and tissues of Cory's shearwater *Calonectris diomedea* collected on Atlantic and Mediterranean islands. *Environ. Pollut.*, **40**, 17–35.

Romanoff, A.L. and Romanoff, A.J. (1949) *The Avian Egg.*, J. Wiley and Sons, New York.

Rose, G.A. and Parker, G.H. (1982) Effects of smelter emissions on metal levels in the plumage of ruffed grouse near Sudbury Ontario, Canada. *Can. J. Zool.*, **60**, 2659–67.

Rutschke, E. (1987) Waterfowl as bio-indicators in *The Value of Birds* (eds. A.W. Diamond and F. Filion), *ICBP Tech. Publ. 6*, International Council for Bird Preservation, Cambridge, pp. 167–72.

Ryan, P.G. (1987a) The incidence and characteristics of plastic particles ingested by seabirds. *Mar. Environ. Res.*, **23**, 175–206.

Ryan, P.G. (1987b) The effects of ingested plastic on seabirds: correlations between plastic load and body condition. *Environ. Pollut.*, **46**, 119–25.

Ryan, P.G. (1988) Effects of ingested plastic on seabird feeding: evidence from chickens. *Mar. Pollut. Bull.*, **19**, 125–8.

Ryan, P.G., Connell, A.D. and Gardner, B.D. (1988) Plastic ingestion and PCBs in seabirds: Is there a relationship? *Mar. Pollut. Bull.*, **19**, 174–76.

Scharenberg, W. (1989) Heavy metals in tissue and feathers of grey herons, *Ardea cinerea* and cormorants *Phalacrocorax carbo sinensis*. *J. Orn.*, **130**, 25–33.

Scheuhammer, A.M. (1987) The chronic toxicity of aluminium, cadmium, mercury, and lead in birds: a review. *Environ. Pollut.*, **46**, 263–95.

Scheuhammer, A.M. (1989) Monitoring wild bird populations for lead exposure. *J. Wildl. Manage.*, **53**, 759–65.

Scheuhammer, A.M. (1991a) Acidification-related changes in the biogeochemistry and ecotoxicology of mercury, cadmium, lead and aluminium: overview. *Environ. Pollut.*, **71**, 87–90.

Scheuhammer, A.M. (1991b) Effects of acidification on the availability of toxic metals and calcium to wild birds and mammals. *Environ. pollut.*, **71**, 329–75.

Schubert, R. (1985) *Bioindikation in Terrestrischen Okosystemen*, G. Fischer Verlag, Stuttgart.

Sell, J.L. (1975) Cadmium and the laying hen: Apparent absorption, tissue distribution and virtual absence of transfer into eggs. *Poult. Sci.*, **54**, 1674–8.

Sell, J.L. (1977) Comparative effects of selenium on metabolism of methyl mercury by chickens and quail: tissue distribution and transfer into eggs. *Poult. Sci.*, **56**, 939–48.

Springer, A.M., Walker, W.II., Risebrough, R.E. *et al.* (1984) Origins of organochlorines accumulated by peregrine falcons, *Falco peregrinus*, breeding in Alaska and Greenland. *Can. Field Nat.*, **98**, 159–66.

Stickel, L.F., Wieneyer, S.N. and Blus, L.J. (1973) Pesticide residues in eggs of wild birds: adjustment for loss of moisture and lipid. *Bull. Environ. Contam. Toxicol.*, **9**, 193–6.

Stock, M., Herber, R.F.M. and Geron, H.M.A. (1989) Cadmium levels in oystercatcher *Haemaetopus ostralegus* from the German Wadden Sea. *Mar. Ecol. Prog. Ser.*, **53**, 227–34.

Stowe, T.J. (1982) *Beached Bird Surveys and Surveillance of Cliff-breeding Seabirds*, RSPB, Sandy, Bedfordshire.

Svecova, Z. (1989) Decrease in the clutch size of the black-headed gull (*Larus ridibundus*) as an indicator of the environmental reproduction toxic potential, in *Proc. Vth Int. Conf. Bioindicatores Deteriorisationis Regionis* (eds J. Bohac and V. Ruzicka), Inst. Landscape Ecology CAS, Ceske Budejovice, pp. 188–93.

Tansy, M.F. and Roth, R.P. (1970) Pigeons: a new role in air pollution. *Air Pollut. Control Assoc. J.*, **20**, 307–9.

Tejning, S. (1967) Mercury in pheasants (*Phasianus colchicus*) deriving from seed grain dressed with methyl and ethyl mercury compounds. *Oikos*, **18**, 334–44.

Thompson, D.R. (1990), Metal levels in marine vertebrates, in *Heavy Metals in the Marine Environment* (eds R.W. Furness and P.S. Rainbow), CRC Press, New York, pp. 143–82.

Thompson, D.R. and Furness, R.W. (1989) Comparison of the levels of total and organic mercury in seabird feathers. *Mar. Pollut. Bull.*, **20**, 577–9.

Thompson, D.R., Stewart, F.M. and Furness, R.W. (1990) Using seabirds to monitor mercury in marine environments: the validity of conversion ratios for tissue comparisons. *Mar. Pollut. Bull.*, **21**, 339–42.

Thompson, D.R., Hamer, K.C. and Furness, R.W. (1991) Mercury accumulation in great skuas *Catharacta skua* of known age and sex, and its effects upon breeding and survival. *J. Appl. Ecol.*, **28**, 672–84.

References

Thompson, D.R., Furness, R.W. and Walsh, P.M. (1992) Historical changes in mercury concentrations in the marine ecosystem of the north and northeast Atlantic Ocean as indicated by seabird feathers. *J. Appl. Ecol.*, **29**, 79–84.

Townsend, M.G., Odam, E.M,. Stanley, P.I. and Wardall, H.P. (1981) Assessment of secondary poisoning hazard of warfarin to tawny owls. *J. Wildl. Manage.*, **45**, 242–7.

van Franeker, J.A. (1985) Plastic ingestion in the North Atlantic fulmar. *Mar. Pollut. Bull.*, **16**, 367–9.

van Impe, J. (1985), Estuarine pollution as a probable cause of increase of estuarine birds. *Mar. Pollut. Bull.*, **16**, 271–6.

Vauk, G.J.M. and Schrey, E. (1987) Litter pollution from ships in the German Bight. *Mar. Pollut. Bull.*, **18**, 316–19.

Vermeer, K. and Peakall, D.B. (1977) Toxic chemicals in Canadian fish-eating birds. *Mar. Pollut. Bull.*, **8**, 205–10.

Vickery, J. (1991) Breeding density of dippers *Cinclus cinclus*, grey wagtails *Motacilla cinerea* and common sandpipers *Actitis hypoleucos* in relation to the acidity of streams in south-west Scotland. *Ibis*, **133**, 178–85.

Walker, C.H., Newton, I., Hallam, S.D. and Ronis, M.J.J. (1987) Activities and toxicological significance of hepatic microsomal enzymes of the kestrel (*Falco tinnunculus*) and sparrowhawks (*Accipiter nisus*). *Comp. Biochem. Physiol. C.*, **86**, 379–82.

Walsh, P.M. (1990) The use of seabirds as monitors of heavy metals in the marine environment, in *Heavy Metals in the Marine Environment* (eds R.W. Furness and P.S. Rainbow), CRC Press, New York, pp. 183–204.

Weseloh, D.V., Teeple, S.M. and Gilbertson, M. (1983) Double-crested cormorants of the Great Lakes: egg-laying parameters, reproductive failure, and contaminant residues in eggs, Lake Huron 1972–1973. *Can. J. Zool.*, **61**, 427–36.

Wiens, J.A., Ford, R.G. and Heinemann, D. (1984) Information needs and priorities for assessing the sensitivity of marine birds to oil spills. *Biol. Conserv.*, **28**, 21–49.

Wise, S.A. and Zeisler, R. (eds) (1985) International review of environmental specimen banking, *National Bureau of Standards, Spec. Publ. 706.*, Washington, D.C.

Wong, C.S., Green, D.R. and Cretney, W.J. (1974) Quantitative tar and plastic waste distributions in the Pacific Ocean. *Nature*, **247**, 30–2.

Wren, C.D. and Stephenson, G.L. (1991) The effect of acidification on the accumulation and toxicity of metals to freshwater invertebrates. *Environ. Pollut.*, **71**, 205–41.

Zinkl, J.G., Henny, C.J. and Shea, P.J. (1979) Brain cholinesterase activities of passerine birds in forests sprayed with cholinesterase inhibiting insecticides, in *Animals as Monitors of Environmental Pollutants*, Natn. Acad. Sci., Washington D.C, pp. 356–65.

Zinkl, J.G., Mack, P.D., Mount, M.E. and Shea, P.J. (1984) Brain cholinesterase activity and brain and liver residues in wild birds of a forest sprayed with acephate. *Environ. Toxicol. Chem.*, **1**, 79–88.

4
Birds as monitors of radionuclide contamination

I. Lehr Brisbin, Jr

4.1 INTRODUCTION

Of the major classes of environmental contaminants, radionuclides tend to be generally less well known and less frequently studied than heavy metals or organic compounds. This may be because in the past elevated levels of environmental contamination with radionuclides have generally only been a problem in a relatively few and limited geographical localities. Most of these localities have either prohibited or greatly restricted public access because of safety and security considerations (e.g. nuclear weapons production or testing sites). As a result, the contamination of free-living flora and fauna with radionuclides has tended to be less apparent to both the scientific community and the public than contaminations with such substances as agricultural herbicides and pesticides. The latter, for example, may often produce the spectre of sick or dying birds in areas where they are commonly encountered by the public.

All of this changed with the disastrous explosion and fire at the Chernobyl nuclear power station in the Soviet Union in April 1986 (Hohenemser *et al.*, 1986). The global distribution of radioactive contaminants released by that accident quickly created a universal awareness of how the radioactive contamination of both natural and agricultural ecosystems can result in the rapid uptake, concentration and transport of these substances through both abiotic and biotic pathways – often leading directly to the human food chain

Birds as Monitors of Environmental Change. Edited by R.W. Furness and J.J.D. Greenwood. Published in 1993 by Chapman & Hall, London. ISBN 0 412 40230 0.

Introduction

(Anspaugh *et al.*, 1988). Subsequent efforts to study these contaminations relied on the extensive and yet relatively little-known body of literature in the area of radioecology. Many of the principles of this field were developed as early as the mid-1960s to document the global fate and effects of radioactive fallout from the atmospheric testing of nuclear weapons (Whicker and Schultz, 1982).

The concept of using natural flora and fauna to document and monitor radionuclide contamination grew out of the developing body of radioecological knowledge. This information has primarily focused on three areas: (1) studies of the pattern and rate of radionuclide uptake, concentration and turnover in the bodies of individual organisms; (2) studies of the pathways by which such organisms might either themselves move, or otherwise affect the movement of these contaminants in the environment; and (3) studies of the effects of radioactive contaminants upon these organisms and their body organs and tissues. Work on radiation effects has largely been conducted under highly controlled, often laboratory, conditions (e.g. Zach and Mayoh, 1982; 1984; 1986a; 1986b; Brisbin, 1991) and seldom have the resulting data been applicable to monitoring efforts under free-ranging or field conditions. Recently, however, advances in the use of techniques such as cell flow cytometry have begun to provide the opportunity to use the responses of free-ranging, so-called sentinel organisms to monitor the effects of environmental radioactive contaminants under field conditions (Dallas and Evans, 1990). George *et al.* (1991), for example, successfully used this technique to monitor changes in the cellular DNA contents of free-ranging mallards (*Anas platyrhynchos*) residing in the abandoned cooling reservoir of a nuclear reactor. With these exceptions, however, virtually all direct monitoring of environmental radioactive contamination through the use of birds has been based on measuring radioactive contaminants actually contained within the individuals' bodies. It is this approach that will form the basis for the remainder of this review.

The close relationship between the principles of the developing field of radioecology and basic ecology itself (Odum, 1965) is fundamental to any effort to use organisms or any environmental medium to monitor the fate and effects of radioactive contaminants in the environment. This approach emphasizes the importance of understanding basic natural history and acquiring descriptive ecological information on the organisms themselves, as well as developing a knowledge of the chemical nature, etc. of the contaminants involved, as discussed in a recent overview of the sub-discipline of avian radioecology (Brisbin, 1991). The basic principles of the field can now form a basis for evaluating the use of birds as monitors of environmental radionuclide contamination. As will be demonstrated by this review, however, any attempt to use the basic principles of avian radioecology to evaluate birds as monitors of radioactive contamination must be based on an understanding of the basic characteristics of the contaminants themselves as well as of the basic principles of how they cycle within the environment.

4.2 CHARACTERISTICS OF ENVIRONMENTALLY IMPORTANT RADIOACTIVE CONTAMINANTS

As a group, radionuclides are much more variable in their chemical characteristics, and hence behaviour in natural food webs, than other contaminants such as heavy metals or organics like polycyclic aromatic hydrocarbons. However, only a relatively few radioisotopes are important as environmental contaminants (Table 4.1). As discussed in detail by Whicker and Schultz (1982) and more briefly by Brisbin (1991), these particular isotopes are important for three reasons: (1) the frequency with which they are encountered as components of environmental releases from weapons testing or nuclear industrial accidents; (2) the degree to which they are taken up and incorporated within living organisms; and (3) the length of time that they persist in the environment or the organisms into which they have been incorporated. The last is, in turn, a function of both the physical half-life of the isotope (i.e. the rate at which it continues to undergo radioactive decay to its daughter products) and its physiological or biological half-life (i.e. the rate at which the isotope, like its stable elemental analogue, is turned over and eliminated from an organism's body). The ecological half-life on the other hand, describes the rate at which the established steady-state level of a contaminant in organisms, in equilibrium with the levels of that substance in the environment, gradually declines as the contaminant is eliminated or rendered biologically unavailable in that environment, thus resulting in the establishment of a new and lower equilibrium with the body-burden.

The chemical, biological and ecological characteristics of each of the environmentally important radioactive contaminants are discussed in detail by Whicker and Schultz (1982). They can be inferred from the corresponding characteristics of their stable chemical analogues (Table 4.1). Strontium-90 for example, as a chemical analogue of calcium, is particularly concentrated in bone, while caesium-137, as an analogue of potassium, is concentrated in skeletal muscle. This is important in attempts to use birds as monitors of environmental contamination with these radioisotopes because the specific distribution of the isotope within organs and tissues, together with an assessment of the type of radioactive emission made by the isotope as it decays (Table 4.1),

Table 4.1 Important characteristics of some of the radioactive isotopes most likely to be targeted by environmental monitoring programmes (modified from Brisbin, 1991, as adapted from Whicker and Schultz, 1982)

Isotope	Physical half-life	Biological retention time	Form of emission	Target organ	Chemical analogues
Tritium	12 years	Days	Beta	Total body	Hydrogen
Iodine-131	8 days	Weeks	Beta	Thyroid	Iodine
Caesium-137	30 years	Weeks	Gamma	Whole body, muscle	Potassium
Calcium-45	160 days	Months-yrs	Beta	Bone	Calcium
Strontium-90	28 years	Years	Beta	Bone	Calcium
Cobalt-60	5.2 years	Weeks	Gamma	Digestive tract, lungs	Cobalt
Zinc-65	245 days	Months-yrs	Gamma	Liver, lungs	Zinc
Plutonium-239	24 000 years	Years	Alpha	Bone, lungs	None

will determine the sampling protocol and type of counting equipment that will have to be used to detect or measure the amount of isotope in the body. Strontium-90, for example, chiefly emits beta radiation which is almost completely shielded from detection by even several mm of living tissue. In a living bird therefore, it would be difficult if not impossible to detect the presence of strontium-90 contamination because of its tendency to concentrate in bones, where its beta emissions would be largely shielded from external detection by surrounding tissues. The same situation is true for the actinide elements such as plutonium which emit alpha radiation. Alpha radiation is even weaker than beta emissions and would also be almost completely shielded from external detection when concentrated in the bird's bone or internal organs such as the lungs (Table 4.1). The use of birds to monitor such alpha- or beta-emitting isotopes would therefore require the birds to be killed and dissected, to isolate bone or organs or tissues. These samples would then have to be subjected to a complex (and invariably expensive) series of chemical preparations before the contaminating isotopes could be detected and quantified.

Gamma and X-radiations, on the other hand, are only slightly shielded by living tissue and can therefore be readily detected and reliably quantified in living birds, which can later be released unharmed. Detailed descriptions of some of the sampling protocols and types of counting equipment used in such studies of gamma radiation emissions from living birds have been provided by Levy *et al.* (1976) and Morton *et al.* (1980). This ability to detect gamma-emitting radioisotopes without harm to the living subjects is an important feature from a conservation point of view and also because it allows one to conduct studies that require one to measure changes in the isotopic body-burdens of individual birds over time (e.g. Brisbin *et al.*, 1990). Such studies based on the monitoring of the levels of one or more gamma-emitting isotopes, can, in many cases, be used to make inferences concerning the uptake and distribution of the more difficult to detect alpha- or beta-emitting isotopes or even other classes of environmental contaminants such as organics or heavy metals. Clay *et al.* (1980), for example, showed how data from the whole-body counting of the gamma-emitting isotope caesium-137 were able to provide information relevant to a study of mercury contamination in the same birds.

4.2.1 Radiocaesium

Of all the environmentally important radioisotopes, probably none is more appropriate for monitoring in living birds than is the gamma-emitter caesium-137, generally termed 'radiocaesium'. As explained above, this isotope can be readily quantified in living animals and studies describing equipment for this purpose, designed specifically for use with living birds, have been referred to above. Radiocaesium has also been one of the most common and widely distributed radioisotopes released from tests of nuclear weapons and from nuclear-industrial accidents (e.g. Whicker and Schultz, 1982; Anspaugh et al., 1988). The relatively long physical half-life of radiocaesium (30 years; Table 4.1) results in it persisting in the environment for long periods of time. A general rule of thumb is that a given amount of radioactivity will, on the basis of physical decay alone, remain detectable at significant levels above background for five half-lives. (That is 5 × 30 = 150 years for radiocaesium.) Finally, since it is specifically concentrated in the skeletal muscles of vertebrates (Narayanyan and Eapen, 1971; Brisbin and Smith, 1975; Potter et al., 1989) this contaminant has a high probability of entering directly into the food chain of humans – either through domestic livestock or through fish and game.

4.2.2 Plutonium and the actinide elements

The actinide elements present, in many ways, a pattern of characteristics converse to radiocaesium. Rather than being highly soluble and readily transported within environmental food webs, they tend to be decidedly insoluble in natural environments and often exist in aqueous solution only in complex chemical states (Whicker and Schultz, 1982). As a result, they tend to be taken up by animals to a much lesser degree than radiocaesium, and then largely through routes of external contamination and inhalation. When they are taken into the digestive tract it may be largely through the inadvertent ingestion of soil particles rather than through assimilation of food (Whicker and Schultz, 1982).

The actinide element plutonium has created particular concern as an environmental contaminant associated with both weapons-testing and nuclear-industrial activities. It is often touted as 'the most toxic substance known to man' (Whicker and Schultz, 1982) and it is true that, as an alpha-emitter, extremely minute particles of plutonium when inhaled and deposited in the lung, for example, can produce intense and localized damage in the alveolar tissue immediately surrounding it. However, despite high interest in the fate and effects of plutonium in the environment, the tendency of this and other actinide elements not to become incorporated readily into plants or animals and thus to enter natural food chains suggests that this particular group of radionuclides would be very difficult to monitor in birds. A few studies have described the behaviour of these radionuclides in birds under laboratory conditions (e.g. Mullen et al.,

Table 4.2 Concentrations of several radioisotopes in soil, vegetation and several tissue compartments of feral bantam chickens foraging in a contaminated floodplain forest while given free access to a hanging feed dispenser[a]

	Caesium-137[b]	Strontium-90[b]	Plutonium-238[b]
Soil[c]	179.3[d]	0.53[e]	5.72[e]
Vegetation[f]	468.6[g]	10.35[h]	0.067[h]
Egg shell[i]	0.52[i]	0.041[i]	N.D.[i]
Egg white[j]	0.28[j]	0.24[j]	N.D.[j]
Egg yolk[k]	0.34[k]	0.44[k]	N.D.[k]
Muscle[l]	3.89[m]	0.0010[m]	N.D.[m]
Liver[n]	1.09[o]	0.0032[o]	N.D.[o]
Feathers	0.24[p]	0.0023[q]	0.0012[q]
Bone[r]	0.99[p]	0.50[s]	0.00038[s,t]

[a] Birds were released on 5 August 1982, and allowed to forage in a 320 m² enclosure in contaminated habitat on the US DOE Savannah River Site, until being collected for tissue analysis on 22 December 1982. While in the enclosure, birds were given free access to uncontaminated drinking water and a commercial poultry feed which was provided in a hanging feeder inside a small roosting shelter.
[b] All values are given as pCi/g dry weight. N.D. = non-detectable levels.
[c] Sampled at 49 grid locations inside the enclosure before the birds were released, with two replicate samples being taken at each grid location, using a 0.001 m² × 15 cm long corer. Samples included organic litter but not standing vegetation.
[d] Mean of all 98 samples as described in footnote c above.
[e] Mean of a random subset of 25 samples, with each sample including both replicate cores from a grid location.
[f] Sampled at 11 random points within the enclosure before the birds were released. At each sampling point, all standing vegetation was harvested within a 1.0 m radius circle of each sampling point.
[g] Mean of all 11 samples as described in footnote f above.
[h] Mean of eight samples, including six of the 11 samples described in footnote g plus two pooled samples made by combining sampling material from two and three adjacent sampling points due to a lack of sufficient material for analysis at these locations.
[i] Value for a pooled sample of shells from 28 eggs collected during the last 23 days of the exposure period.
[j] Value for a pooled sample of whites from 28 eggs collected during the last 23 days of exposure.
[k] Mean of three pooled samples of the yolks of 10, 42 and 28 eggs each, all collected during the last 23 days of exposure.
[l] Pectoral and leg muscles from males sacrificed at the end of the 115 day exposure period.
[m] Mean of two pooled samples from three birds each and four pooled samples from two birds each.
[n] Whole livers from the same birds as in footnote l above.
[o] Mean of two pooled samples of the livers of five birds each and one pooled sample from four birds.
[p] Mean of samples from 33 birds.
[q] Mean of three samples, composed of pooled feathers from five, five and four birds respectively.
[r] Pooled samples including the humerus, radius, ulna, femur and bibiotarsus, from the same birds as in footnote l above.
[s] Mean of four pooled samples from two birds each and two pooled samples from three birds each.
[t] Four of the six samples were below the detection limits and were entered as 0 pCi/g in calculating the mean.

Table 4.3 Concentrations of several radioisotopes in several tissue compartments of feral bantam chickens foraging in a contaminated floodplain forest while provided with supplementary uncontaminated feed scattered on the ground[a]

	Strontium-90[b]	Plutonium-238[b]
Muscle[c]	N.D.[d]	N.D.[d]
Liver[c]	N.D.[d]	0.0017[d]
Feathers[c]	0.043[d]	0.0092[d]
Bone[c]	1.24[d]	0.00074[d]

[a] Birds were released on 25 August 1983, and allowed to forage in a 320 m² enclosure, as described in Table 4.2, until being collected for tissue/organ analysis on 1 February 1984. While in the enclosure, birds were supplied with free access to supplementary uncontaminated feed grains which were sprinkled on the soil of the enclosure.
[b] All values are given as pCi/g dry weight.
[c] Sampled as described in Table 4.2.
[d] Mean of four pooled samples, two which were composed of tissues from eight females each, one of tissues from nine females and one of tissues from eight males.

1976) and some limited data also exist on their occurrence in free-living wild birds (e.g. Markham et al., 1988). However, no studies have yet been carried out that would allow a direct comparison of the pattern and extent to which these radionuclides might be taken up by free-ranging birds. In particular, there is no published information for free-ranging birds, comparing the uptake of actinide elements with the uptake of more environmentally mobile radionuclides such as strontium-90 and caesium-137, in situations where both forms of radionuclides are present in the same environmental medium (e.g. soil or vegetation).

In view of the lack of published information, some original data are presented here (Tables 4.2 and 4.3). They summarize the results of a study of the simultaneous uptake and concentration of plutonium-238, strontium-90 and caesium-137 by free-ranging ground-foraging birds (feral bantam chickens) from both soils and vegetation of a contaminated floodplain forest. Of particular importance to this study is the fact that we first determined precisely the levels and patterns of distribution of each of these radionuclide contaminants in the soils and vegetation that were available to the foraging birds. The general characteristics of the habitat in which this study was conducted and procedures for sample preparation and analysis have been described by Pinder et al. (1984), and a more thorough description of the feral bantam chickens used and a history of their development as a strain which has been specifically selected for an ability to forage and survive under free-ranging conditions in contaminated habitats has been provided by Latimer (1976). The use of these birds as a surrogate ground-foraging sentinel species for environmental monitoring in terrestrial habitats will be described below, while efforts to determine whether these introduced exotic birds are capable of accurately describing the environmental behaviour of a radionuclide such as caesium-137 in native species are described by Brisbin (1991).

The results of this study (Tables 4.2 and 4.3) show clearly that strontium-90 and, to an even greater extent, caesium-137 were readily assimilated by these birds, the first accumulating in their bones, the second in their muscles. However plutonium-238 was not accumulated at any level detectable above background, even though it was present in both the soils and vegetation of the habitat (Pinder *et al.*, 1984; Tables 4.2 and 4.3). The data also show that measured plutonium accumulation primarily reflects external contamination of the feathers. Higher levels were shown in the birds when supplementary feed was scattered on the ground (Table 4.3) than when feed supplements were provided in a controlled-access feeder (Table 4.2), suggesting the importance of the inadvertent ingestion of soil (with the feed scattered on the ground) in the uptake of this largely insoluble radionuclide.

The statistical summaries of data in Tables 4.2 and 4.3 show that the levels of some radionuclides (e.g. caesium-137 and strontium-90) in the tissues and eggs of such free-ranging sentinel birds can be used to detect and monitor the presence of these contaminants in the habitat in which the birds are foraging. However, other isotopes (such as plutonium) are difficult if not impossible to monitor in this way because they scarcely enter food chains. Because the data in Tables 4.2 and 4.3 were collected from birds which supplemented their natural foraging with uncontaminated commercial poultry feed, the body-burdens indicated are almost certainly less than those that would have been attained if such supplementary feed were not available. However there are clear differences in the degrees to which the birds concentrated the various isotopes in different tissues and also in how these patterns differed for different isotopes, depending on their chemical characteristics. Each of these factors must be considered carefully in determining whether body-burden data from birds (or other organisms) could be used to monitor environmental contamination with such isotopes, as well as in determining tissue sampling protocols for such a monitoring effort.

4.3 EVALUATING BIRDS AS MONITORS OF RADIONUCLIDE CONTAMINATION

Although birds have long been proposed as useful monitors of environmental contamination, few studies have specifically evaluated the usefulness of avian body-burdens in predicting the levels of various contaminants in their environments. One study which provides data that can be used in such an evaluation and which deals specifically with radionuclides is that of Straney *et al.* (1975), which will be described here in some detail. In this study, birds were sampled on the United States Department of Energy's Savannah River Site (SRS; formerly known as the Savannah River Plant). The large size (75 km^2) and several other features of the SRS make it particularly useful for studies of this kind.

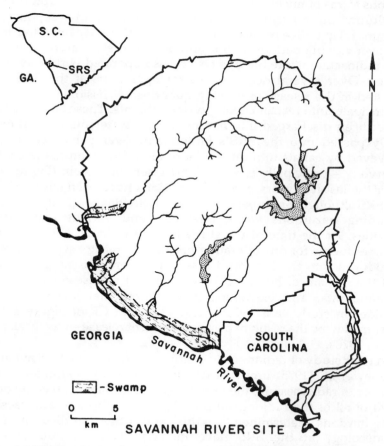

Figure 4.1 The location and major geographic features of the US Department of Energy's Savannah River Site in the south-eastern United States. (From Brisbin, 1991.)

4.3.1 The Savannah River Site

The SRS is a nuclear production and research facility which was closed to public access in 1952. The site occupies portions of Aiken, Allendale and Barnwell Counties in South Carolina in the southeastern United States (Figure 4.1). The site includes within its boundaries portions of both the Upper Coastal Plain and Sandhills Physiographic Provinces. The major vegetational communities of the SRS have been described by Workman and McLeod (1990) and the site's avifauna has been described by Norris (1963) and Mayer *et al.* (1986).

Since the time of its establishment, the SRS has been used for various forms of nuclear-industrial activities including the production of plutonium and tritium for national defence and weapons programmes. Up to five nuclear production reactors have operated at the site over varying periods of time and nuclear fuel separation facilities and radioactive waste burial sites are also operated within its boundaries. Over several decades, reactor operations at the SRS have resulted in the release of varying quantities of fissionable materials – principally into reactor cooling-water effluents. These effluents have either been discharged into natural stream watercourses and eventually passed into the SRS's bottomland swamp habitat or were conveyed by canals into one or more of several man-made cooling reservoirs – principally the Par Pond reservoir system (Figure 4.2). Over the last few years, reactor operations have been suspended at the SRS although a number of industrial activities continue at the site, including hazardous waste treatment and various construction activities. At the time of the study by Straney *et al.* (1975), cooling water effluents for one operating production-reactor were still being introduced into the Par Pond reservoir. There was no evidence at that time or since, however, of any current releases of radioactive contaminants, and elevations in environmental levels of radionuclides in study areas SC, CSRV, CSH and CR of Figure 4.2 are presumed to be the result of past contaminant release – mostly all prior to the 1970s.

In their study of radionuclide distribution within the SRS avifauna, Straney *et al.* (1975) limited their analyses to the gamma-emitting isotopes of radiocaesium, the most broadly distributed and easily quantified of all of the radionuclides at the site. Although radiocaesium body-burdens determined in this and other studies of avian radioecology at the SRS have included both caesium-137 and caesium-134, the shorter half-life of the latter (2.1 years) has resulted in the majority of the contribution to total body-burdens being that of caesium-137. Marter (1970) for example, found that even as early

Evaluating birds as monitors 155

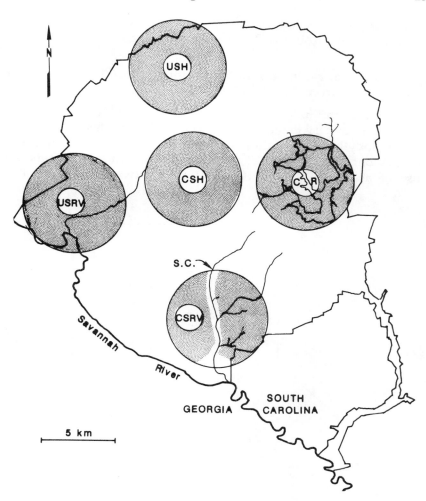

Figure 4.2 Areas used for the collection of birds in a study designed to determine the levels of radiocaesium in birds from both contaminated and uncontaminated habitats on the Savannah River Site (Figure 4.1). Sampling areas in contaminated habitats included: the floodplain and delta of Steel Creek, a stream which formerly received nuclear production reactor effluents (SC); Savannah River Valley lowland habitat surrounding Steel Creek (CSRV); upland Sandhills habitat surrounding a nuclear waste burial area (CSH), and habitat surrounding the Par Pond nuclear reactor cooling reservoir (CR). Similar but uncontaminated habitats were also sampled in the Savannah River Valley (USRV) and Sandhills (USH) physiographic provinces of the site. (From Brisbin (1991), as redrawn from Straney *et al.* (1975)).

as the late 1960s, the ratio of caesium-134 to caesium-137 was only 1:20 in waterfowl from the Par Pond reservoir (area CR; Figure 4.2).

4.3.2 Patterns of radiocaesium distribution within the avifauna of the Savannah River Site

Because adult and juvenile birds often differ in both their foraging patterns and dietary habits, Straney et al. (1975) analysed the radiocaesium body-burdens of these two groups separately. As expected, the body-burdens of juveniles were significantly more variable than those of adults ($F_{68,138} = 2.48$, $P<0.01$ for two-tailed test). The

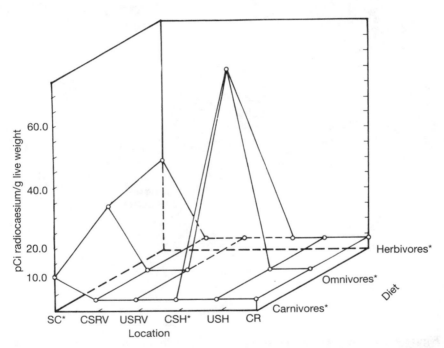

Figure 4.3 'Response surface' analysis of whole-body radiocaesium concentrations of 225 individual adult birds representing 53 species, collected during the summers of 1971 and 1972 from contaminated and uncontaminated habitats on the US DOE Savannah River Site, as indicated in Figure 4.2. Lines are drawn between collecting sites and trophic categories to simplify comparisons of contaminant body-burdens and do not imply functional relationships. Differences between trophic categories were not the same at all collecting sites (interaction $F_{10,138} = 5.09$, $P < 0.01$). Individual trophic categories and sampling sites within which significant differences were detected ($P \leq 0.05$) are marked with asterisks. (From Straney et al., 1975.)

limited mobility of juveniles compared to that of the more widely-ranging adults was undoubtedly a factor causing some juveniles from nests located close to particular sources of contamination (e.g. the burial grounds of area CSH; Figure 4.2) to show the highest body-burdens of any of the birds examined. Other juveniles from nests within this same collecting area but more distant from the burial grounds, failed to show any detectable levels of radiocaesium contamination (determined in this study to be less than 4.0 pCi/g live weight; Figure 4.3). The maximum level of radiocaesium determined for any bird collected in this study was 2992.7 pCi/g live weight for an eastern kingbird *Tyrannus tyrannus* from a boundary fence surrounding a nuclear waste burial ground in the centre of sampling area CSH (Figure 4.2). Adult birds, which individually tended to forage over broader areas of habitat than juveniles, were more likely to encounter both contaminated and uncontaminated habitats and thus showed less variance in their body-burdens and fewer extreme cases of contamination than juveniles. The higher within-site variance of the juveniles tended to obscure differences due to collection site or diet, so Straney *et al.* (1975) limited their analyses of these factors to adult birds.

Domby (1976) described an even more striking example of the variability that can be shown between the radiocaesium body-burdens of juvenile birds, even from nests located closely together within a single colony. In his study of little blue herons *Florida caerulea* nesting on a small island in the SRS's radiocaesium-contaminated Pond B reservoir (Figure 4.4), Domby (1976) documented differences between juveniles from different nests within the same colony, not only in the overall levels of radiocaesium, but also in the way that these levels changed during the nestling period. He associated these differences with temporal changes of the foraging sites of the adults tending these different nests (Domby and McFarlane, 1978). Thus, although all juveniles examined were taken from nests located within a reservoir of known elevated radiocaesium contamination (Whicker *et al.*, 1990), some of the herons were feeding their nestlings on prey obtained from uncontaminated sites away from the reservoir. Thus, reliance solely upon the body-burdens of these juveniles as biomonitors could have produced quite different, and in some cases erroneous, conclusions concerning the biological availability of this radionuclide in the food chains of the reservoir itself. It has been proposed that colonial wading birds such as these herons can be used as biomonitors of various forms of environmental contamination (Custer *et al.*, in press). However, the data presented by Domby (1976) and Domby and McFarlane (1978) suggest limitations to the conclusions that can be drawn from such monitoring, particularly if based on juveniles alone. Although no data are available for radiocaesium levels of adult herons nesting on the

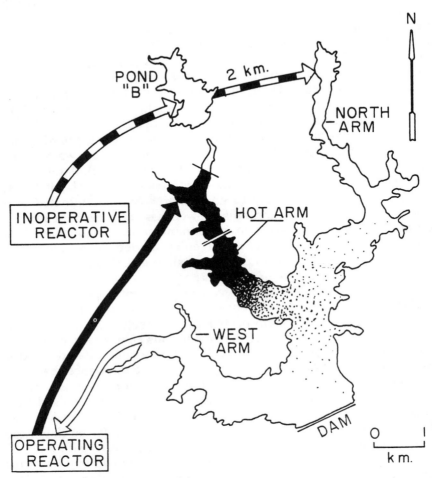

Figure 4.4 Map of the Par Pond reactor–cooling reservoir system of the Savannah River Site (area CR of Figure 4.2). Arrows showing flows of cooling water between the reactors and the reservoir are drawn diagrammatically. Heated effluents (as indicated by the intensity of stippling) discharged into the North Arm of Par Pond 1961–1964 and into the Hot Arm until 1988. The map thus depicts the situation that existed at the time when a number of studies of radiocaesium contamination levels in waterfowl and other aquatic birds were conducted (e.g. those of Brisbin *et al.*, 1973; Brisbin and Swinebroad, 1975; Brisbin and Vargo, 1982; Potter, 1987; Potter *et al.*, 1989; Brisbin *et al.*, 1990). (From Brisbin, 1991.)

Pond B reservoir, they would presumably have been less variable than those of the juveniles and, as suggested by Straney *et al.* (1975), might better have reflected the general level of radiocaesium availability within that study area (CR: Figure 4.2), considered as a whole.

In their analyses of radiocaesium body-burdens in adult birds of the SRS, Straney et al. (1975) found a significant interaction of diet (herbivores vs. omnivores vs. carnivores) with the locality of the birds' collection on the SRS site (Figure 4.3). The data indicated that only in area SC were there elevated levels of radiocaesium in birds of all three dietary types. In the only other area showing elevated levels of contamination (CSH), only the omnivores showed elevated body-burdens. Birds collected from the two uncontaminated habitats (USRV and USH) and from around the contaminated Par Pond reservoir (CR) had radiocaesium body-burdens that were uniformly low and indistinguishable from background (Figure 4.3).

Taken together, these data from adult birds from the SRS indicate that, unlike the case with other forms of environmental contaminants (e.g. Carson, 1962; Risebrough et al., 1967; Clay et al., 1980; Kendall and Lacher, in press), birds that occupy the top-carnivore position cannot always be assumed to have the most elevated levels of radiocaesium contamination. Anderson et al. (1973) also found in area SC of the SRS that higher trophic levels of arthropod food chains did not have higher concentrations of radiocaesium than lower levels. This same phenomenon has also been documented in the community of wintering migratory waterfowl of the Par Pond reservoir (area CR; Brisbin et al., 1973). Reichle et al. (1970) also reported this behaviour of radiocaesium as being a general characteristic of contaminated invertebrate food chains. On this basis then, insectivorous birds would generally not be expected to show the highest levels of radiocaesium concentration, in contrast to the expectations for the biomagnification of various organic contaminants in these same birds (Kendall and Lacher, in press). Nevertheless, under appropriate circumstances (e.g. in area SC), insectivorous and carnivorous birds also showed some elevation in their levels of radiocaesium contamination, though higher levels were shown by omnivorous or to an even greater extent, herbivorous species (Figure 4.3).

The greater radiocaesium contamination of omnivores in area CSH was mostly due to the particularly high body-burdens shown by crows (including both *Corvus brachyrhynchos* and *C. ossifragus*). Straney et al. (1975) related this to the larger home-range of these birds compared with smaller passerines, which would increase the likelihood that some of these crows had spent time feeding within the actual confines of the nuclear waste burial ground itself, in the centre of area CSH (Figure 4.2).

This raises the issue of the effect of habitat discontinuities, especially interfaces between wetlands and surrounding drier upland habitats, on the degree to which bird body-burdens reflect environmental contamination with radiocaesium. Even though elevated levels of contamination were shown by aquatic birds sampled from the Par Pond reservoir itself (Brisbin et al., 1973) and from the marsh and floodplain of the contaminated Steel Creek effluent stream (area SC; Figure 4.3), birds sampled from the surrounding upland areas (areas CR and CSRV, respectively; Figure 4.2) were not similarly contaminated (Figure 4.3) even though some of them were sampled from locations as close as 200 m to the boundary with the contaminated wetland (Straney et al., 1975). This was not the case with area CSH, however, where both the contaminated site and the surrounding area were upland habitats, with no abrupt habitat discontinuity. This, together with the particular abundance of more widely-ranging species such as the crows, apparently resulted in an increased tendency of the birds from this area to reflect the elevated levels of environmental radiocaesium contamination.

4.3.3 Monitoring of radionuclide transport by migratory waterfowl

Most sites with significant levels of environmental radionuclide contamination are found on lands that are closed to public access. Since most of them are surrounded by extensive buffer lands from which the public is also excluded, there may appear to be little opportunity for contaminants in these areas to enter the food chain of man. However, migratory waterfowl and game birds that may be consumed by hunters may spend time within such contaminated areas. If these birds move elsewhere, they may be harvested and eaten as food by hunters before eliminating much of the contaminant body-burden that they had acquired. Since most inadvertent releases of radionuclides occur through cooling water or other reactor effluents, aquatic systems are among the most likely habitats to show such contamination. Waterfowl are therefore of particular concern as possible carriers of released radioactive contaminants into the food chain of man. This could occur on either a local or regional scale or, in the case of some migratory species, over global distances. At the same time, if sampled after being collected by hunters, these birds could provide an opportunity to monitor both the biological availability and spatial spread of the contaminants released (e.g. Hanson and Case, 1963; Glover, 1967).

Waterfowl populations, including both resident and migratory forms, have been studied and monitored for a number of years at both the Savannah River Site and the U.S. Department of Energy's Idaho National Engineering Laboratory (INEL). At the INEL, various gamma- as well as beta- and alpha-emitting radionuclides were monitored in both wild waterfowl and in captive-reared mallard *Anas platyrhynchos* which were released on radioactive leaching ponds at the site for periods of 43–145 days (Halford et al., 1981, and Markham et al., 1988, respectively). Both of these studies showed the highest contaminant concentrations to occur in the gastrointestinal tract, followed by the feathers. A similar situation was reported by Cadwell et al. (1979) for American coots *Fulica americana* collected from caesium-137-contaminated waste ponds of the U.S. DOE's Hanford Reservation in Washington state. As discussed by these authors, radiocaesium concentration in the gut was over 20 times higher than

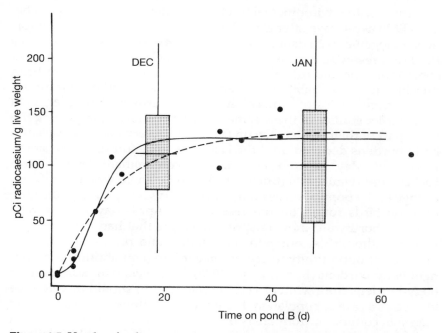

Figure 4.5 Uptake of radiocaesium by American coots following their capture from uncontaminated habitats and subsequent release on the contaminated Pond B reservoir of the Savannah River Site (Figure 4.4). Solid circles: body-burdens of individual birds; the entry at day = 0 is the average initial body-burden of all 13 coots that were subsequently recaptured. Corresponding whole-body-burdens of 10 wild free-ranging coots sampled in December 1987 and January 1988 are shown by horizontal lines (means), shaded rectangles (± two standard errors) and vertical lines (ranges) for each monthly sample. Curves represent least-squares fits to the data points for the transplanted coots, using either a reparameterized Richards sigmoid uptake model (solid line) or the generally-accepted 'negative exponential' model for radionuclide uptake, as explained by Brisbin et al. (1989). (From Brisbin (1991) as redrawn from Brisbin et al. (1990), and based on studies by Potter (1987).)

in muscle, which suggests that the birds being monitored had only recently arrived at the contaminated site. Conversely, birds showing higher levels of radiocaesium in muscle than in the gut contents would be more likely to have been in residence in the contaminated habitat for sufficient lengths of time for their total body-burdens to approach equilibrium with concentrations in their environment. Potter et al. (1989) report gut/muscle radiocaesium concentration ratios of 0.7–1.3 for coots collected from reactor cooling reservoirs of the SRS.

A study of the temporal dynamics of radiocaesium uptake by coots on the SRS (Brisbin et al., 1990) has now confirmed that an equilibrium is reached with environmental levels in less than 25 days and possibly in as little as 14 days after arrival at this site (Figure 4.5). This information, together with data on resightings of marked coots on the SRS's Par Pond reservoir system (Potter, 1987), have confirmed that most coots found on the SRS during December and January have been present in these habitats for enough time to attain maximum (equilibrium) levels of radiocaesium body-burdens. However, data from earlier studies of these same coots have shown that during the latter part of their winter stay on these SRS reservoirs, radiocaesium body-burdens declined until the last birds to be found on the site in April and May showed little if any contaminant levels above background (Figure 4.6) (Brisbin et al., 1973). There was a significant change in the population sex ratio during this same period, indicating that the birds found on the reservoir in April–May were almost certainly northward-moving spring migrants that had not spent the winter in the SRS's contaminated habitats and presumably had an insufficient opportunity to acquire maximum (equilibrium) contaminant body-burdens (Brisbin et al., 1973). The decline in radiocaesium body-burdens from winter to spring (Figure 4.6) should therefore be viewed as a case of population dilution with northward-moving spring migrants coming in from areas off of the SRS site, rather than as a case of a decline in the contaminant levels of individual birds (and therefore, by inference, in their environment). Whole-body radiocaesium determinations of an individual coot by Brisbin and Swinebroad (1975) did in fact confirm an increase rather than a decrease in its body-burden during the period of December through March – a time when the average body-burden was declining because of population turnover.

The above situation, explained in terms of migratory movements of population cohorts of birds that have wintered in different parts of the species' overall range, would be intuitively obvious to most ornithologists. It should be emphasized, however, that without determining the sex ratios of the monthly samples of coots contributing

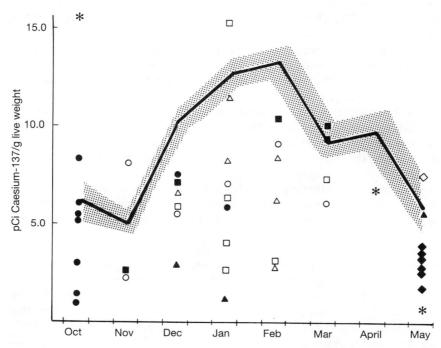

Figure 4.6 Seasonal changes in radiocaesium whole-body burdens of American coots wintering on the Par Pond reservoir at the Savannah River Site (Figure 4.4). The solid line connects means of monthly samples of 30 coots each, while the shaded area represents the 95% confidence intervals about these means. Body-burdens of individuals of other species are indicated by asterisks for common gallinules (moorhens) *Gallinula chloropus*, solid circles for pied-billed grebes *Podilymbus podiceps*, open circles for horned (Slavonian) grebes *Podiceps auritus*, solid triangles for ring-necked ducks *Aythya collaris*, open triangles for lesser scaup *Aythya affinis*, solid squares for bufflehead *Bucephala albeola*, open squares for ruddy ducks *Oxyura jamaicensis*, solid diamonds for common terns *Sterna hirundo*, and an open diamond for a black tern *Chlidonias niger*. (From Brisbin (1991), as taken from Brisbin *et al.* (1973).)

to the study shown in Figure 4.6 the observed pattern of seasonal changes in body-burdens could easily have been misinterpreted. This could have led, in turn, to the incorrect assumption that there was little or no net transport of radiocaesium to offsite habitats by the departing birds. Furthermore, the determination of sex ratios in population samples of coots is not a simple matter and requires either internal examination of the gonads or application of a complex series of analyses of external body measurements (Gullion, 1952; Eddleman, 1975). These procedures are unlikely to be used by persons not experienced in basic studies of birds, which argues strongly for the inclusion of ornithological expertise and the collection of all relevant natural history data in any programme designed to use birds as monitors of environmental contamination with radioactive materials. This basic principle, although well illustrated by this particular case, is not limited to the monitoring of radioactive contaminants alone.

Radiocaesium levels in coots at the SRS provide a striking example of how concentrations of this isotope do not increase progressing up the food chain (Brisbin et al., 1973). As for other species of SRS birds cited earlier (Straney et al., 1975), radiocaesium data for the Par Pond community of wintering waterfowl show the coot, which is almost completely herbivorous on its wintering grounds, consistently to have higher levels of this contaminant than other more carnivorous, omnivorous or piscivorous species (Figure 4.6). On the other hand, Clay et al. (1980) have shown that the Par Pond coots have relatively lower levels of mercury, which tends to biomagnify in the more carnivorous species of this same waterfowl community. On the basis of its relatively higher level of radiocaesium then, the coot would seem to be the best species in which to monitor this contaminant in this wintering waterfowl community. In addition, studies by Potter (1987) have shown that once on the reservoirs of the SRS, marked individual coots show a high degree of site fidelity, often remaining throughout the winter within several hundred metres of the original capture location on the reservoir shoreline. This is displayed over a number of years: Potter (1987), for example, reported marked individual coots returning to the same arm of the Par Pond reservoir (Figure 4.4) for up to three consecutive winters, after intervening trips to and from their northern breeding grounds. Because of this high site fidelity, the contaminant body-burdens of individual coots are more likely to reflect contaminant levels within the immediate area of the places in which they are collected than are those of other species of waterfowl that are less site-faithful (e.g. Fendley, 1978). Brisbin et al. (1973) demonstrated this by showing that, even within the Par Pond reservoir itself, there were significant differences in the radiocaesium body-burdens of coots, depending on the particular reservoir arm from which the birds were collected. Coots from the reservoir's North Arm, for example, showed significantly higher levels of radiocaesium than did those from either the Hot Arm or West Arm. It was in fact into the North Arm that radiocaesium-contaminated effluents were introduced both during and after the failure of a now inoperative reactor (Figure 4.4; Brisbin et al., 1973; Whicker et al., 1990).

The high wintering-site fidelity of coots in this reservoir has resulted in this spatial pattern of variation in radiocaesium body-burdens being maintained over several years. Brisbin and Vargo (1982), for example, found that the spatial mosaic of coot radiocaesium body-burdens in Par Pond, as described by Brisbin et al. (1973), was stable across a four-year sampling period, with no tendency for birds with the highest radiocaesium levels to be found further 'downstream' in the reservoir in succeeding years. This finding is consistent with the observations of Whicker et al. (1990), who found that most of the radiocaesium in this reservoir system tends to be held in bottom sediments. As long as no catastrophic event results in the sudden resuspension or downstream movement of these sediments, the spatial distribution of the contaminant would not be expected to change much over time and this is indeed consistent with the evidence from the coots' body-burdens (Brisbin and Vargo, 1982; Brisbin, 1991). Activities such as reservoir draw-downs or abnormally severe storms could of course significantly alter this situation.

Using the data from Brisbin and Vargo (1982) and from Potter (1987), the ecological half-life of radiocaesium in coots wintering on Par Pond was estimated to be over 20 years (Brisbin, 1991). This would suggest that it would be at least 100 years before the contamination levels in these birds would be indistinguishable from background. This situation contrasts sharply with the radiocaesium contamination of SRS's Steel Creek effluent stream and its delta swamp (area SC; Figure 4.2). Here, although the radiocaesium is again found mainly in the stream and floodplain sediments, considerable and continuing erosion of sediments has resulted in contaminated sediments both being shifted downstream and being covered by less contaminated sediments from upstream (Anderson et al., 1973; Brisbin et al., 1989).

The more rapid loss of radiocaesium from this area has been reflected in a more rapid decline in the body-burdens of this contaminant in wood ducks *Aix sponsa* collected and analysed by Fendley (1978). Brisbin (1991) estimated the ecological half-life of this isotope in these wood ducks in the Steel Creek delta swamp as being only 1.9 years – less than 10% that of the same contaminant in coots on the Par Pond reservoir. Moreover, the ecological half-life of radiocaesium in the Steel Creek wood ducks was half that of the same isotope in rat snakes *Elaphe obsoleta*, which were found to prey on the nests of these birds in the Steel Creek study area (Bagshaw and Brisbin, 1984). This contrasts with a 200-fold difference in the physiological half-lives of radiocaesium: 5.6 days in wood ducks (Fendley et al. 1977) vs. 900 days in rat snakes (Staton et al., 1974). In contrast, although the Steel Creek wood ducks and Par Pond coots differed by 10-fold in their ecological turnover rates of radiocaesium (above), they showed similar physiological half-lives: 5.6 days (Fendley et al., 1977) vs. 7.2 days (Potter, 1987), respectively. As indicated above, these findings suggest that it is the characteristics of the habitat (e.g., lentic reservoir impoundment vs. lotic effluent stream and swamp delta), rather than the physiological characteristics of the birds (or snakes), that are the important factors in determining the long-term rates at which the isotope will either be lost from or rendered biologically unavailable within the environment (and thus increasingly difficult to detect in birds and other organisms in the areas; Brisbin, 1991).

These studies suggest that aquatic birds such as wood ducks and particularly coots can be used effectively to monitor the rates at which radiocaesium levels are declining in the biota of such areas. Data of this kind become particularly important when an accidental release, such as the Chernobyl nuclear accident, causes radionuclides in waterfowl or game birds to exceed levels that are considered safe for human consumption (European Economic Community, 1986), creating a need to estimate the amount of time that must pass before such birds can again be safely harvested and used as food.

4.3.4 The monitoring of radioactive contaminants released by the Chernobyl nuclear accident

Studies of the cycling of radioactive contaminants in birds at sites such as the SRS, INEL and Hanford have generally indicated that the off-site transport of radionuclides from these locations by birds, as well as the consumption of waterfowl or other game birds from these areas by local hunters, do not pose significant risks to either human health or safety (Brisbin et al., 1973; Cadwell et al., 1979; Halford et al., 1981). Halford et al. (1981), for example, indicated that a hunter would have to eat 40 ducks or coots, including each bird's entire liver and muscle mass, 'to exceed the permissible whole-body dose'. Moreover, coots, the most highly contaminated species at each of these sites, are seldom consumed by the hunting public in North America and rarely in large numbers. The April 1986 nuclear accident at the Chernobyl power station in the Ukraine, however, released several orders of magnitude more radioactive fission products, particularly radiocaesium, than had been released by any of the sources contributing to the environmental contaminations of the above-mentioned North American sites (Hohenemser et al., 1986; Medvedev, 1986; Anspaugh et al., 1988). Furthermore, the contamination from the Chernobyl accident was transported through the atmosphere into global food chains in inhabited areas, rather than being confined to areas of restricted public access (Anspaugh et al., 1988). Finally, the extensive wetlands adjoining the site of the Chernobyl accident are important for migratory waterfowl (Isakov, 1966; Brisbin, 1991), raising the possibility of significant contamination occurring in birds that could be consumed by persons in many parts of the local region as well as along international migratory flyways. These flyways would mainly include countries in the south-west regions of Asia, southern Europe, North Africa and the Mediterranean region (Isakov, 1966). In a number of these areas, coots and other rallids are more likely to be consumed by humans than would be the case in North America (Ripley, 1976). All of these factors would argue for the importance of immediately developing a global strategy for the monitoring of the dispersal of radioactive contaminants from the site of such accidents by birds, particularly waterfowl.

While no such monitoring programme has yet been established on an international scale, some preliminary efforts have been made to determine the contamination levels of European birds resulting from the Chernobyl accident (e.g. Ruiz et al., 1987; Baeza et al., 1988; Baeza et al., 1991). Some of these studies (e.g. Baeza et al., 1988) have found that even within flyways which include the site of the Chernobyl accident, certain species, such as the mallard, may be represented by local non-migratory populations which, unlike their migratory counterparts, would not be exposed to radionuclide uptake in the contaminated wetlands adjacent to the Chernobyl site. In other species, such as the song thrush *Turdus philomelos*, regional differences in levels of radioactive contaminants could be shown dependent on the degree to which the birds may have encountered Chernobyl-contaminated habitats either on their breeding grounds or at migratory stopover sites (Baeza et al., 1991). Specific 'signatures' of the ratios of contamination levels of two or more radioisotopes were used in certain cases to establish the specific source of the contamination. A ratio of 0.5 in the concentration of caesium-134 to that of caesium-137, for example, indicated that the Chernobyl accident was the source of the birds' contamination (Ruiz et al., 1987).

Brisbin (1991) used uptake measurements from studies of waterfowl using habitats contaminated by radiocaesium on the SRS, as described above, to estimate the degree to which radiocaesium might be transported along migratory flyways by waterfowl using wetlands adjoining the Chernobyl site. These calculations demonstrated the importance of being able to estimate correctly the length of time that individual birds would stay in these contaminated wetlands during their migratory passage. The importance of being able to describe correctly the shape of the contaminant uptake curve for individual birds was also demonstrated by these calculations.

With regard to this, Brisbin *et al.* (1990) found a three-parameter Richards sigmoid model to provide a significantly better fit to radiocaesium uptake data for free-ranging coots on the SRS's Pond B reservoir than the usual negative exponential uptake model (Figure 4.5). Without the above information on the natural history, migratory patterns and behaviour of the birds themselves, and without information on the pattern and rate of contaminant uptake under field conditions, it would be difficult if not impossible to use contaminant body-burden data alone to infer the degree to which birds from such habitats might be a threat to the health of persons consuming them as food. Neither would it be possible to use such body-burden data as the basis for monitoring the amount, distribution and bioavailability of the contaminant in the environment from which the bird was collected. In the case of Figure 4.5, for example, the collection of coots showing body-burdens of 50 pCi/g live weight would indicate a level of environmental contamination too low to cause concern for the health of people eating waterfowl from this habitat, so long as one assumes that the coots collected had been in the contaminated habitat for 40 or more days and had therefore had sufficient time to acquire maximum body-burdens. If, however, these same coots showing 50 pCi/g live weight had been in the contaminated habitat for less than two days, such body-burdens would suggest a much higher level of environmental contamination (according to the uptake pattern of Figure 4.5) – one that would eventually result in asymptotic body-burdens that would pose a threat to human health if long-term resident waterfowl from that area were to be eaten as food.

4.3.5 The 'sentinel animal' approach to environmental monitoring of radionuclide uptake and concentration

Because of the inherent difficulty in making inferences concerning environmental contamination on the basis of body-burden data from wild birds whose migratory status and movements in the recent past are not known, an alternative 'sentinel animal' approach has been developed for the monitoring of environmental contaminants, using birds that are sufficiently tame or otherwise capable of being identified and repeatedly recaptured as they range and forage freely in contaminated habitats. Levy *et al.* (1976), for example, used the multiple-recapture of small passerines, which were routinely mist-netted as part of a bird-banding (ringing) programme, to monitor caesium-137 (determined by whole body gamma-counts) in the vicinity of a nuclear power station. Potter (1987) captured wild coots from uncontaminated habitats on the SRS and, after verifying that they had no detectable body-burdens of radiocaesium, clipped the primary flight feathers of these birds, marked them with plastic neck collars for future identification in the field, and then released them on the SRS's contaminated Pond B reservoir. Subsequent recaptures of these birds, each of which had been on the contaminated reservoir for a known number of days, produced the data for Figure 4.5.

An alternative approach to such sentinel animal studies would be the use of tamed 'surrogate species' which although not native to the habitat being monitored, would be analogous to native wild birds occupying a similar niche. The use of tamed and recapturable feral bantam chickens, described earlier, to monitor radiocaesium distribution and bioavailability in terrestrial habitats of the SRS's Steel Creek floodplain (Brisbin, 1991; Figure 4.2) and the use of tamed and imprinted game-farm mallards to similarly monitor radiocaesium in the Pond B reservoir at the same site (George et al., 1990) are two examples of this approach. Although the ease of recapture and handling of such tamed surrogate birds offers certain advantages over the use of wild birds which have been rendered flightless (and occasionally radio-tagged), the former method rests on extrapolating from such surrogates to their wild counterparts (e.g. from feral bantam chickens to wild galliform species such as the bobwhite quail *Colinus virginianus* or wild turkey *Meleagris gallapavo*). Brisbin (1991) discusses this problem and describes such methods as the comparisons of frequency distributions of contaminant body-burdens, to help ensure that the surrogate is providing realistic representations of the pattern of uptake and concentration of the contaminant by the wild species.

4.4 SELECTING A MONITOR TO DESCRIBE ENVIRONMENTAL RADIONUCLIDE CONTAMINATION

The distribution of a contaminant such as a radionuclide within even the most simple of environmental systems is generally much too complex to be measured directly. The concept of selecting a single ecosystem component, or group of components such as birds, to serve as biomonitors of that contaminant is based on the need for a rapid and effective means of providing an index that will assess the extent of the distribution and bioavailability of the contaminant(s) in question in all components of that environment. It is on the basis of these biomonitoring indices that management decisions related to human health and safety, as well as environmental quality, must often be made. Implicit in this process, but seldom directly addressed, is the assumption that the level of the contaminant in the monitored component(s) is related in some quantitative and predictable way to the level of that contaminant in all other components of the same ecosystem (soil, vegetation, insects etc.). Sometimes, however, the monitored component is of direct concern in its own right (e.g. an endangered species, human food or man himself).

Two attempts have been made to assess the degree of interrelatedness of radioactive contaminant levels within different component parts of natural ecosystems (Anderson et al., 1973; Brisbin et al., 1989). Unfortunately, neither of these studies involved birds. However, the approaches they have adopted and the procedures used to quantify the value of any one component part of the ecosystem as a predictive monitor of the distribution of a radioactive contaminant in other ecosystem components should be relevant to the selection of a given bird species or avian community rather than other ecosystem components for such monitoring purposes. Brisbin and Smith (1975) address the question of selecting the single most appropriate organ or tissue to monitor the levels of radiocaesium contamination in the bodies of white-tailed deer *Odocoileus virginianus* killed by hunters on the SRS. They reasoned that the monitoring value of a given organ or tissue would be in direct proportion to the degree to which its contaminant content was correlated with those of all other parts of the body – particularly those parts which, by virtue of their large size and high contaminant concentration per unit mass, represented proportionally large amounts of the total contaminant contained within the deer's body. Using a system based on the calculation of a quantifiable 'predictive index' from a correlation matrix between the contaminant concentrations in all candidate components under consideration as monitors, these authors concluded that skeletal muscle would be the most appropriate component to monitor radiocaesium in the bodies of these deer, since it not only comprises the largest proportion of the deer's total body mass but also contains the highest mass-specific concentration of the isotope. What was not intuitively obvious, however, was the fact that brain tissue proved to be the second most appropriate monitoring component. Brain tissue did not comprise a large proportion of the total body mass and neither did it show a particularly high concentration of radiocaesium, but radiocaesium levels in brain tissue proved to be the most closely correlated (of any of the components tested) with those of skeletal muscle. If such a second-ranked predictor also ranked high as a predictor of other contaminants, such as metals or organics, and if some of these were poorly monitored by muscle alone, it might be a better component for overall monitoring of contaminants than the muscle.

When this thinking is applied to the choice of an appropriate component to monitor the distribution of one or more contaminants within an ecosystem it quickly becomes obvious that birds, like other flora and fauna, will often rank quite low if evaluated solely on the basis of the amount of the ecosystem's total contaminant burden that they contain. Whicker et al. (1990) for example, found that all of the macrophyte plus animal components of the SRS's Pond B reservoir together contained only 0.66, 2.75 and 0.019% of that reservoir's total inventories of caesium-137, strontium-90 and transuranic isotopes respectively, with the sediments containing 98% or more of all radionuclide inventories except for strontium-90. Even within the biota, these authors found that the majority (80–99%) of the radionuclide inventories were associated with the macrophytes. On the basis of the radionuclide inventories and concentrations in this reservoir and other contaminated ecosystems that have been described in the literature, summarized by Whicker et al. (1990), birds would seem on these grounds alone to be a rather poor choice as environmental monitors of radionuclide contamination, compared with components of the physical environment, such as soils or sediments, or macrophytes and other vegetation. When considering the degree to which the radiocaesium concentrations of various organisms in the Steel Creek floodplain ecosystem were correlated with each other, Brisbin et al. (1989) found that two animals, the shrimp *Palaemonetes* and the fish *Etheostoma*, acually ranked higher than any form of vegetation as predictors of radiocaesium distribution within biotic components of the ecosystems overall, when quantified by applying the ranking formula of Brisbin and Smith (1975). The findings of Straney et al. (1975) for the patterns of radiocaesium distribution in birds in this same habitat (areas SC and CSRV of Figure 4.2), together with the findings of Domby (1976) and Domby and McFarlane (1978), would argue that, in comparison to other organisms, such as vegetation, fish or invertebrates, the high mobility of birds would make them rather poor indicators of the spatial distribution of radioactive contamination on scales of several metres to perhaps 1–2 km. Even within these distances, however, it is possible that for limited periods of time, smaller territorial passerines, while on their breeding grounds for example, might be capable of showing spatial or temporal patterns of radionuclide contamination in their environment.

Such a situation has been described by Willard (1960), who showed differences in the gross beta-radioactivity of summer birds between the inner and the outer zones of vegetation on and around a dried lake bed that had been contaminated by low-level radioactive wastes. Similarly, Levy et al. (1976) reported that passerine birds mist-netted at a nuclear power station site showed a transient increase in iodine-131 during a period when the reactor's primary containment vessel (normally sealed) was opened at the start of a repair and refuelling operation. In this case, again, the detection of differences in the levels of radionuclides in the birds was made over a relatively brief and specific time period (approximately three days), and within a community of birds that were probably frequenting well-defined and limited areas of habitat.

Other factors need to be considered in choosing an ecosystem component to measure radionuclide contamination across scales of many kilometres, such as those distances between sampling areas used by Levy et al. (1976) and by Straney et al. (1975) (Figure 4.2). In such cases, the greater mobility of birds as compared to smaller or less mobile biota should enable them to integrate across the smaller-scale spatial mosaics of contamination and, by thus reducing the variance of body-burdens within such sampling areas, provide a better opportunity to detect differences between sampling areas. Levy et al. (1976), for example, were able to demonstrate significant between-site variation in the radionuclide burdens of bobwhite quail and blue jays *Cyanocitta cristata* from sites separated by as little as 1 km. The findings of Straney et al. (1975) can be compared to those of Brisbin et al. (1974), who determined whole-body radiocaesium contents of snakes collected from approximately the same study areas of the SRS as those used by Straney et al. (1975) (Figure 4.2). Although no snakes were collected from area CSH (Figure 4.2), snakes from areas CR and SC (Figure 4.2) showed significantly higher radiocaesium body-burdens than did those from areas corresponding to USH and USRV (Figure 4.2). Thus, the body-burdens of both the birds and the snakes were shown by independent studies to be capable of revealing elevated bioavailability of radiocaesium in the vicinity of wetlands where accidental releases of this isotope had been made to both lotic and lentic habitats. Since all snakes are carnivorous, only the data for the birds provided an opportunity to demonstrate the effect of dietary habits on radiocaesium concentrations (Figure 4.3).

4.5 CONCLUSIONS – CAN BIRDS BE USED AS EFFECTIVE MONITORS OF ENVIRONMENTAL CONTAMINATION WITH RADIONUCLIDES?

The preceding discussion suggests that except under unusual circumstances, birds would probably not be useful for monitoring patterns of radionuclide contamination on relatively small (several metres to 1–2 km) spatial scales. For such purposes, some component of the physical environment (e.g. soils or sediments), vegetation or less mobile fauna (e.g. rodents, reptiles, amphibians or invertebrates) would almost certainly be more appropriate. However, over larger areas ranging from many kilometres to regional or even global scales, birds may not only be appropriate but might actually be among the best candidates for monitoring the presence and bioavailability of radionuclides. This conclusion would only be valid, of course, for those radionuclides, such as caesium, which readily enter and move through natural food chains. It would not apply to those radionuclides, such as plutonium, which are scarcely if at all taken up by organisms when released into the environment (Tables 4.2 and 4.3). The high metabolic rates of birds relative to other biota of comparable body size would suggest that they should come to equilibrium more quickly with those radionuclides which do move through natural food chains. Birds' high rates of ingestion would further suggest that they would be more likely to feed at contaminated sites they happened to visit, even if only for short periods of time.

The usefulness of birds as monitors of radionuclides or any other form of environmental contaminant is proportional to the degree of knowledge of their basic biology, ecology, natural history and, particularly, movement and behaviour in the area being studied. The great mobility of birds makes the need for this information even more acute than for less mobile animals which can often be assumed to have resided for a considerable time in the vicinity of the site where they were collected. This assumption cannot be made in the case of birds, and there are a number of instances (e.g. Figure 4.6) where failure to include the input from experienced ornithologists could have resulted in a complete misinterpretation of the data obtained from even a well-designed study of bird body-burdens.

Nowhere is this more true than when using migratory birds as monitors of the movement of radioactive contaminants on a regional or global scale. Hancock and Woollam (1987) for example reported either low or nondetectable body-burdens of caesium-137 in Bewick's swans *Cygnus columbianus bewickii* wintering in the British Isles following the Chernobyl accident. However, this species utilizes a flyway which is geographically isolated from the contaminated wetlands adjoining the site of that accident in the Ukraine (Isakov, 1966; Brisbin, 1991). As indicated by these latter studies, waterfowl using the flyway which includes the Chernobyl site and its adjoining wetlands, are most likely to be found wintering in north or northwest Africa or along the Mediterranean coast of western Europe rather than in the British Isles. The data of Hancock and Woollam (1987) thus provide a valuable negative case, verifying the likelihood of the differential flyway contamination patterns suggested by the work of Isakov (1966) and Brisbin (1991). It should be further noted, however, that in addition to the direct contamination of waterfowl and other birds visiting the actual site of the Chernobyl accident, a more indirect but nonetheless important route of contamination has resulted from the rapid physical transport of radioactive contaminants from the accident site by atmospheric circulation. This physical atmospheric transport, unlike the flyway biotic transport of contaminants to the south-west (and/or north-east), deposited contamination through precipitation, to the north and north-west of the accident site, over many of the Baltic countries, Scandinavia and the British Isles, in addition to certain north-central portions of the European continent (Hohenemser et al., 1986; Medvedev, 1986; Mascanzoni, 1987; Anspaugh et al., 1988).

Finally, a suite of other less tangible, although in some cases even more weighty, factors must be considered in any decision as to whether birds can or should play a role in any programme designed to monitor environmental contamination with radionuclides. Even though birds as a whole may contain only an infinitesimally small proportion of the total environmental inventory of a given contaminant (e.g. Whicker et al., 1990), the contaminants that they do contain and transport within their bodies often have a disproportionately high degree of 'political' importance – if for no other reason than because the public is interested in, and concerned about, birds (Temple and Wiens, 1989). This is particularly true for endangered species and those which are commonly eaten by humans. Perhaps, then, it is on these grounds more than any other that one can conclude that, as long as proper knowledge of their limitations as monitors is taken into consideration in proper sampling designs, and as long as the proper selection of species, appropriate isotopes and collection protocols are used, birds can indeed provide important information concerning radioactive contamination of the environment. This 'canary in the mines' should perhaps continue to alert us to environmental hazards as we move into and even beyond the nuclear age.

ACKNOWLEDGEMENTS

This manuscript is dedicated to the memory of the late William E. Odum, whose scholarly insight and enthusiasm for ecological research have positively influenced my own thinking over the years. Support for the research at the Savannah River Site has been provided by the United States Department of Energy and the University of Georgia. Much of this work was undertaken by a number of graduate students including Peyton R. Williams, Tim Fendley, Deborah Clay Harris, Catherine Potter and Linda George and I also acknowledge the hard work and important contributions of my technician colleagues including Paul Johns, Richard Geiger, Clarence Bagshaw, Susan McDowell, Howard Zippler, Robert Kennamer and Merlin Benner. Helpful guidance in the transuranic studies was provided by John Pinder and Robert Watters and support and encouragement of my work in avian radioecology has been provided by the staff of the Savannah River Ecology Laboratory, particularly Michael Smith, Domy Adriano, Carl Strojan, William McCort and Tony Towns. Helpful comments on this manuscript have been provided by Eric Peters and Tony Towns.

REFERENCES

Anderson, G.E., Gentry, J.B. and Smith, M.H. (1973) Relationships between levels of radiocaesium in dominant plants and arthropods in a contaminated streambed community. *Oikos*, **24**, 165–70.

Anspaugh, L.R., Catlin, R.J. and Goldman, M. (1988) The global impact of the Chernobyl reactor accident. *Science*, **242**, 1513–19.

Baeza, A., del Rio, M., Miró, C. *et al.* (1988) Radiocaesium concentration in migratory birds wintering in Spain after the Chernobyl accident. *Health Phys*, **55**, 863–7.

Baeza, A., del Rio, M., Miró, C. *et al.* (1991) Radiocaesium and radiostrontium levels in song-thrushes (*Turdus philomelos*) captured in two regions of Spain. *J. Environ. Radioactivity*, **13**, 13–23.

Bagshaw, C. and Brisbin, I.L., Jr. (1984) Long-term declines in radiocaesium of two sympatric snake populations. *J. Appl. Ecol.*, **21**, 407–13.

Brisbin, I.L., Jr. (1991) Avian Radioecology, in *Current Ornithology*, *8* (ed. D.M. Power), Plenum Press, New York, pp. 69–140.

Brisbin, I.L., Jr. and Smith, M.H. (1975) Radiocaesium concentrations in whole-body homogenates and several body compartments of naturally contaminated white-tailed deer, in *Mineral Cycling in Southeastern Ecosystems*, (eds F.G. Howell, J.B. Gentry and M.H. Smith), *ERDA Symp. Ser. (CONF-740513)*, pp. 542–56.

Brisbin, I.L., Jr. and Swinebroad, J. (1975) The role of banding studies in evaluating the accumulation and cycling of radionuclides and other environmental contaminants in free-living birds. *EBBA News*, **38**, 186–92.

Brisbin, I.L., Jr. and Vargo, M.J. (1982) Four-year declines in radiocaesium concentrations of American coots inhabiting a nuclear reactor cooling reservoir. *Health Phys*, **43**, 266–9.

Brisbin, I.L., Jr., Geiger, R.A. and Smith, M.H. (1973) Accumulation and redistribution of radiocaesium by migratory waterfowl inhabiting a reactor cooling reservoir, in *Proceedings of the International Symposium on the Environmental Behaviour of Radionuclides Released in the Nuclear Industry*, IAEA Symp., *IAEA-SM-17272*, pp. 373–84.

Brisbin, I.L., Jr., Staton, M.A., Pinder, J.E. III, and Geiger, R.A. (1974) Radiocaesium concentrations of snakes from contaminated and non-contaminated habitats of the AEC Savannah River Plant. *Copeia*, 1974(2), 501–6.

Brisbin, I.L., Jr., Breshears, D.D., Brown, K.L. *et al.* (1989) Relationships between levels of radiocaesium in components of terrestrial and aquatic food webs of a contaminated streambed and floodplain community. *J. Appl. Ecol.*, **26**, 173–82.

Brisbin, I.L., Jr., Newman, M.C., McDowell, S.G. and Peters, E.L. (1990) The prediction of contaminant accumulation by free-living organisms: applications of a sigmoidal model. *Environ. Toxicol. and Chem.*, **9**, 141–9.

Cadwell, L.L., Schreckhise, R.G. and Fitzner, R.E. (1979) Caesium-137 in coots (*Fulica americana*) on Hanford waste ponds: contribution to population dose and offsite transport estimates, in *Proceedings Health Physics Society Twelfth Midyear Topical Symposium on Low-Level Radiation Waste Management*, pp. 1–7.

Carson, R. (1962) *Silent Spring*, Houghton Mifflin Co., Boston.

Clay, D.L., Brisbin, I.L., Jr., Bush, P.B. and Provost, E.E. (1980) Patterns of mercury contamination in a wintering waterfowl community. *Proc. Ann. Conf. S.E. Assoc. Fish & Wildl. Agencies*, **32**, 309–17.

Custer, T., Rattner, B.A., Ohlendorf, H.M. and Melancon, M.J. (in press) Colonial waterbirds as indicators of estuarine contamination in the United States, in *Proc. Twentieth Internat. Ornithol. Congr.*, Christchurch, New Zealand.

Dallas, C.E. and Evans, D.L. (1990) Flow cytometry in toxicity analysis. *Nature*, **345**, 557–8.

Domby, A.H. (1976) Radiocaesium dynamics in herons inhabiting a contaminated reservoir system. M.S. Thesis, Univ. of Georgia, Athens.

Domby, A.H. and McFarlane, R.W. (1978) Feeding ecology of little blue herons at a radionuclide-contaminated reservoir, in *Wading Birds, Nat. Audubon Soc. Res. Rept.* 7, pp. 361–4.

Eddleman, W.R. (1975) A study of migratory American coots, *Fulica americana*, in Oklahoma, Ph.D. Dissertation, Oklahoma State Univ., Stillwater.

European Economic Community (1986) Derived Reference Levels as a Basis for the Control of Foodstuffs Following a Nuclear Accident: a Recommendation from the Group of Experts Set Up under Article 31 of the Euratom Treaty, *EEC Regulation 1707/86*, Commission of Economic European Community Printing Office, Brussels.

Fendley, T.T. (1978) The ecology of wood ducks (*Aix sponsa*) utilizing a nuclear production reactor effluent system, Ph.D. Thesis, Utah State University, Logan.

Fendley, T.,T., Manlove, M.N. and Brisbin, I.L., Jr. (1977) The accumulation and elimination of radiocaesium by naturally contaminated wood ducks. *Health Phys*, **32**, 415–22.

George, L.S., Dallas, C.E., Brisbin, I.L., Jr. and Evans, D.L. (1990) Radiocaesium (^{137}Cs) uptake in mallards at the Savannah River Site (SRS) and effects on DNA cell cycle in red bood cells. *Toxicologist*, **10**, 248.

George, L.S., Dallas, C.E., Brisbin, I.L., Jr. and Evans, D.L. (1991) Flow cytometric DNA analysis of ducks accumulating ^{137}Cs on a reactor reservoir. *Ecotox. and Env. Safety*, **21**, 337–47.

Glover, F.A. (1967) Distribution of mallards from the Columbia Basin region as indicated by the presence of zinc-65 in birds shot by hunters in the Pacific and Central Flyways, *Final Prog. Rept. to the U.S. Atomic Energy Comm., Contract AT(1101-1514)*, Colorado State Univ., Fort Collins.

Gullion, G.W. (1952) Sex and age determination in the American coot. *Jour. Wildl. Mgt.*, **16**, 191–7.

Halford, D.K., Millard, J.B. and Markham, O.D. (1981) Radionuclide concentrations in waterfowl using a liquid radioactive waste disposal area and the potential radiation dose to man. *Health Phys.*, **40**, 173–81.

Hancock, R. and Woollam, P.B. (1987) Radioactivity Measurements on Live Bewick's Swans, *Berkely Nuclear Laboratories, Report TPRD/B/0897/R87*, Central Electricity Generating Board, Gloucestershire.

Hanson, W.C. and Case, A.C. (1963), A method of measuring waterfowl dispersion utilizing phosphorus-32 and zinc-65, in *Radioecology, Proc. First Nat. Symp. on Radioecology*, (eds V. Schultz and A.W. Klement Jr.), Reinhold, New York, pp. 451–3.

Hohenemser, C., Deicher, M., Ernst, A. et al. (1986) Chernobyl: an early report. *Environ.*, **28**, 6–43.

Isakov, Y.A. (1966) MAR project and conservation of waterfowl breeding in the USSR, in *Proceedings of the Second European Meeting on Wildfowl Conservation*, Publ. Intern. Waterfowl Res. Bur., Slimbridge, England, pp. 125–38.

Kendall, R.J. and Lacher, T.E., Jr (in press) *The Population Ecology and Wildlife Toxicology of Agricultural Pesticide Use: A Modeling Initiative for Avian Species*, SETAC Special Publication. Lewis Publishers. Chelsea, MI.

Latimer, B.E. (1976) Growth and mortality responses of five breeds of chickens to gamma radiation stress, M.S. Thesis, University of Georgia, Athens.

Levy, C.K., Youngstrom, K.A. and Maletskos, C.J. (1976) Whole-body gamma-spectroscopic assessment of environmental radionuclides in recapturable wild birds, in *Radioecology and Energy Resources* (ed. C.E. Cushing), Dowden, Hutchinson and Ross, Stroudsburg, Pennsylvania, pp. 113–22.

Markham, O.D., Halford, D.K., Rope, S.K. and Kuzo, G.B. (1988) Plutonium, Am, Cm and Sr in ducks maintained on radioactive leaching ponds in southeastern Idaho. *Health Phys.*, **55**, 517–24.

Marter, W.L. (1970) *Radioactivity in the Environs of Steel Creek, Savannah River Lab., Publ. DPST-70-435*, Aiken, SC.

Mascanzoni, D. (1987) Chernobyl's challenge to the environment: a report from Sweden. *Sci. Total Env.*, **67**, 133–48.

Mayer, J.J., Kennamer, R.A. and Hoppe, R.T. (1986) Waterfowl of the Savannah River Plant, *Savannah River Ecology Lab., Publ. SREL-22UC-66e*, Aiken, SC.

Medvedev, Z.A. (1986) Ecological aspects of the Chernobyl nuclear plant disaster. *Trends Ecol. and Evol.*, **1**, 23–5.

Morton, J.S., Halford, D.K. and Parker, D. (1980) A confinement device for the determination of whole-body radionuclide concentrations in live ducks. *Health Phys.*, **38**, 234–6.

Mullen, A.A., Lloyd, S.R. and Mosley, R.E. (1976) Distribution of ingested transuranium nuclides in chickens and subsequent transport to eggs in *Transuranium Nuclides in the Environment – Proceedings of the Symposium on Transuranium Nuclides in the Environment, IAEA Symp., IAEA-SM-199/68*, pp. 423–33.

Narayanyan, N. and Eapen, J. (1971) Gross and subcellular distribution of caesium-137 in pigeon (*Columba livia*) tissues with special reference to muscles. *J. Rad. Res.*, **12**, 51–5.

Norris, R.A. (1963) *Birds of the AEC Savannah River Plant Area*, Publ. of the Charleston Museum, Charleston, SC.

Odum, E.P. (1965) Feedback between radiation ecology and general ecology. *Health Phys.*, **11**, 1257–62.

Pinder, J.E. III, McLeod, K.W., Alberts, J.J. and Adriano, D.C. (1984) Uptake of ^{244}Cm, ^{238}Pu and other radionuclides by trees inhabiting a contaminated floodplain. *Health Phys.*, **47**, 375–84.

Potter, C.M. (1987) Use of reactor cooling reservoirs and caesium-137 uptake in the American coot, M.S. Thesis, Colorado State Univ., Fort Collins.

Potter, C.M., Brisbin, I.L., Jr., McDowell, S.G., and Whicker, F.W., (1989) Distribution of ^{137}Cs in the American coot (*Fulica americana*). *J. Env. Radioactivity*, **9**, 105–15.

Reichle, D.E., Dunaway, P.B. and Nelson, D.J. (1970) Turnover and concentrations of radionuclides in food chains. *Nuclear Safety*, **11**, 43–55.

Ripley, S.D. (1976) Rails of the world. *Am. Scientist*, **64**, 628–35.

Risebrough, R.W., Menzel, D.B., Martin, D.J., Jr. and Olcott, H.S. (1967) DDT residues in Pacific sea birds: a persistent insectide in marine food chains. *Nature*, **216**, 589–91.

Ruiz, X., Jover, L., Llorente, G.A. *et al.* (1987) Song-thrushes (*Turdus philomelos*) wintering in Spain as biological indicators of the Chernobyl accident. *Ornis Scand.*, **19**, 63–7.

Staton, M.A., Brisbin, I.L., Jr. and Geiger, R.A. (1974) Some aspects of radiocaesium retention in naturally-contaminated captive snakes. *Herpetol*, **30**, 204–11.

Straney, D.O., Beaman, B., Brisbin, I.L., Jr. and Smith, M.H. (1975) Radiocaesium in birds of the Savannah River Plant. *Health Phys.*, **28**, 341–5.

Temple, S.A., and Wiens, J.A. (1989) Bird populations and environmental changes: can birds be bioindicators? *Am. Birds*, **43**, 260–70.

Whicker, F.W. and Schultz, V. (1982) *Radioecology: Nuclear Energy and the Environment* (2 vols.), CRC Press. Boca Raton, FL.

Whicker, F.W., Pinder, J.E. III, Bowling, J.W., Alberts, J.J. and Brisbin, I.L., Jr (1990) Distribution of long-lived radionuclides in an abandoned reactor cooling reservoir. *Ecol. Monogr.*, **60**, 471–96.

Willard, W.K. (1960) Avian uptake of fission products from an area contaminated by low-level atomic wastes. *Science*, **132**, 148–50.

Workman, S.W. and McLeod, K.W. (1990) Vegetation of the Savannah River Site: Major Community Types, *Savannah River Site National Environmental Research Park Programme*, Publ. SRO-NERP-19, Aiken, SC.

Zach, R. and Mayoh, K.R. (1982) Breeding biology of tree swallows and house wrens in a gradient of gamma radiation. *Ecology*, **63**, 1720–8.

Zach, R. and Mayoh, K.R. (1984) Gamma radiation effects on nestling tree swallows. *Ecology*, **65**, 1641–7.

Zach, R. and Mayoh, K.R. (1986a) Gamma radiation effects on nestling house wrens: A field study. *Rad. Res.*, **105**, 49–57.

Zach, R. and Mayoh, K.R. (1986b) Gamma irradiation of tree swallow embryos and subsequent growth and survival. *Condor*, **88**, 1–10.

5
Birds as indicators of changes in water quality

S.J. Ormerod and S.J. Tyler

5.1 INTRODUCTION

Throughout the world, birds are abundant, conspicuous and diverse components of freshwater ecosystems and of the wetlands and riparian areas around them. Although less than 4–5% of the Earth's land surface is covered by standing freshwater, as many as 11–23% of all bird species use inland waters or their margins at some time during their annual cycle (Table 5.1). In part, this diversity reflects the fundamental importance, productivity and wider ecological influences of water in all biomes. It reflects also the riparian influence of running waters: Britain alone has 80 000 km of streams and rivers, yet river drainage from the whole of Europe still represents only 8% of the world's total (Chorley, 1969).

Like all ecosystems, however, freshwaters and their catchments are subject to natural and man-induced change through physical, chemical and energetic processes (Figure 5.1). These modifications, in turn, have many biological repercussions and may alter conditions away from those that we value. The assessment of change is thus of central importance in understanding freshwater ecosystems and germane to the management of their associated resources.

Surveillance and monitoring in freshwaters provide some of the most firmly established of all the examples of environmental assessment (Kolkwitz and Marsson, 1909; Hynes, 1960; Hellawell, 1986). Of particular importance in the context of this chapter, biological themes

Birds as Monitors of Environmental Change. Edited by R.W. Furness and J.J.D. Greenwood. Published in 1993 by Chapman & Hall, London. ISBN 0 412 40230 0.

Table 5.1 The principal orders of birds represented in lakes or swamps (L/S) or rivers (R) in selected regions of the world, and the numbers of species of each. Birds from the estuarial and coastal sections of surface waters are excluded.

	Africa		W. Palaearctic		Asia		N. America		Australia	
	L/S[a]	R[b]	L/S	R	L/S	R	L/S	R	L/S	R
Gaviiformes			3		2		4			
Podicipediformes	3		5		3		7		4	
Pelecaniformes	6		5		8		5		6	
Ciconiiformes	26	1	13		36		12		18	2
Threskioniformes	5	2								
Phoenicopteriformes	2		2							
Anseriformes	34	5	34	6	41	3	47	1	22	1
Falconiformes/Accipitriformes	4	1	4		6	3	3		4	
Gruiformes	31	1	8		23		15		18	1
Charadriiformes	65	7	38	4	43	12	35	2	28	6
Strigiformes		3	1		2					
Caprimulgiformes	1									
Coraciiformes	5	6	2	1	4	6		3		2
Passeriformes	31	3	17	6	12	15	18	14	24	4
% of total Avifauna	11.7	1.5	18.3	2.3	13.4	3.0	21.4	2.9	17.1	2.2
% of all species excluding Passeriformes	22.0	2.9	27.1	3.4	26.0	5.9	36.3	4.9	30.6	3.9

[a] Some species use both lakes and rivers, but appear only in the lakes column.
[b] Species shown under rivers are more exclusive to this habitat. Species classed as from swamps include some which do not feed directly from the aquatic system, but are nevertheless dependent on emergent animals as food or plants for the provision of habitat.

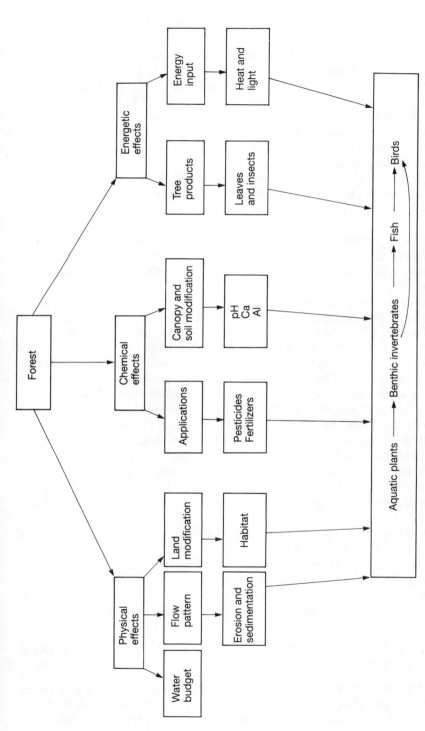

Figure 5.1 Schematic representation of the effects on river ecosystems of a change in catchment forest cover: an example of how freshwater ecosystems are often modified by human activity.

have figured prominently in assessments of water quality. It has long been recognized that, although they are chemical in nature, biologically damaging pollutants should be detected using biological, rather than solely chemical, criteria (Hynes, 1960). Biological indicators may have the added advantage of integrating exposure to total pollution over a span of time, whereas intermittent chemical sampling may miss short-lived and episodic bouts of pollution. Living organisms also have intrinsic importance, for example to conservation, economics or subsistence. Thus, not only does biological monitoring aid in the assessment of change, it also permits some understanding of the wider consequences of pollution for biological resources. Critics have argued, however, that biological measurements require lengthy, and hence expensive, processes. The indices used have often been unstandardized, the data difficult to interpret, and responses to pollutants variable within and between organisms (see Cullen, 1990; Spellerberg, 1991).

For the most part, biological assessments of freshwater quality have involved wholly aquatic organisms, such as algae, insects and fish (Hellawell, 1986; Flower and Battarbee, 1983). In all of these, measurement can involve lengthy laboratory procedures. By contrast, birds are conspicuous and relatively easily censussed on both running or standing waters. The realization that birds are integral components of freshwater ecosystems has also grown (Ormerod and Tyler, 1991a). Together, these features have prompted several authors to suggest a potential role for birds in assessing the status of freshwater ecosystems (Diamond and Filion, 1987). However, views vary about the general suitability of vertebrates for reliable indication of environmental change. Opponents point to the difficulties of unequivocally ascribing their responses to any single abiotic or biotic factor from the vast array which impinge on their ecology (Landres *et al.*, 1988).

In this chapter, we consider the potential role of aquatic birds as indicators or monitors of freshwater quality, placing our assessment in the context of these wider debates. In many cases, we illustrate our arguments and concepts by patterns revealed by our own research on two species of riverine passerines with contrasting ecology. In other cases, we use examples from elsewhere in the world.

5.2 POLLUTION AND OTHER SOURCES OF CHANGE IN LAKES AND RIVERS

Alteration in water qualty, in either time or space, is the dominant theme of this chapter. It is important to stress at the outset that this is not the only ecological factor affecting freshwater ecosystems and

Pollution in lakes and rivers

their birds. Taking the example of the effects on rivers of an alteration in catchment land use for forestry, it is possible to illustrate a wide array of chemical, energetic and physical impacts at the catchment and bankside level (Figure 5.1). Modification through any of these pathways might be sufficient to bring a detectable change in the ecology of a given water body, and hence also in the ecology of its avifauna. Assessments of change in the status of birds in relation to non-chemical modifications are clearly possible and populations are sometimes monitored following physical disturbances to river habitats (e.g. Raven, 1986). Although it has particularly profound ramifications, this forestry example demonstrates that ecological changes in freshwaters are often complex; the identification of causal links can be difficult. We return to this theme below.

We should be careful in our definition of 'water quality'. To the layman, a change in water quality is synonyous with a change in the degree of pollution, pollution being defined by Holdgate (1979) as:

> The introduction by man into the environment of substances or energy liable to cause hazards to human health, harm to living resources and ecological systems, damage to structure or amenity, or interference with legitimate uses of the environment.

Under this definition, a large array of pollutants in freshwaters can be recognized. They vary between those which have:

1. indirect effects on aquatic organisms through ecological change;
2. a mix of direct and indirect effects;
3. a direct biological effect on some organisms, from which other ecological effects can arise.

An example of pollutants in the first group would be organic matter that causes deoxygenation in surface waters as a result of microbial decay, the reduced oxygen concentration then acting directly on other organisms; increased loads of suspended solids might also occur, reducing water transparency while in suspension and occluding the river or lake bed upon settlement. Nutrients, such as nitrate and phosphate also fall into this group. An example from the second group would be oils, which can be toxic in their own right but also cause ecological changes through surface smothering, reduced water transparency and sedimentation. Waste heat is also a pollutant with direct and indirect effects. Pollutants in the third group include pathogens, radionuclides and an array of poisons (acids; alkalis; anions such as sulphide, sulphite and cyanide; metals such as cadmium, zinc and lead; pesticides; polychlorinated biphenyls (PCBs); and other organic chemicals such as phenolic compounds and formaldehyde). Most direct effects of such pollutants have further ecological consequences, since the organisms directly affected may be important to others in and around the ecosystem. In this group are various materials which show pronounced persistence and accumulate throughout the freshwater food web. Some of these contaminants may be scarcely soluble in water, but nevertheless are ingested or incorporated into organisms while adsorbed to organic matter.

In a temperate, industrial and densely populated country such as Britain, water pollution is widespread. At present, around 12% of Britain's 40 000 km of main river are classified as being of poor or bad quality on the basis of chemical criteria (for example, Biological Oxygen Demands (BOD) greater than 9 mg/l or oxygen saturation (DO) less than 40%; NRA, 1991). A further 23% are classified as only 'fair' (with BOD 5–9 mg/l or DO 40–60%). These data exclude small upland rivers,

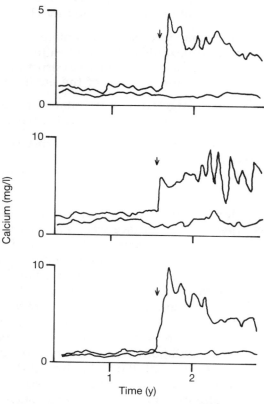

Figure 5.2 Changes in the calcium concentrations of three Welsh streams following the application of limestone to their catchments to mitigate acidification; trends in neighbouring reference streams are also shown. (After Ormerod *et al.*, 1990.)

some of which will be affected by diffuse pollution such as acidification, whose total length is considerable.

In an ecological context, the concept of water quality covers a wider array of factors than pollutants and their effects. A change in water quality is, essentially, a change in the chemical or energetic (heat) constitution of water. Not only does this include natural variations among substances that can also be pollutants (e.g. phosphates, nitrates, acidity) but also variations in substances seldom considered as such. For example, a change in the calcium concentration of lakes or streams can be of major ecological significance to their general biology and productivity (e.g. Krueger and Waters, 1983; Ormerod and Edwards, 1987). Calcium concentrations may vary from site to site through the natural influence of catchment soils and rocks (e.g. Edwards *et al.*, 1990) and through the agricultural activities of humans. Figure 5.2 shows how a pronounced change in water quality may be brought about by the application of agricultural limestone to stream catchments to mitigate the effects of acid deposition; modelling studies showed that calcium concentrations probably rose to levels unprecedented in the post-glacial history of the streams concerned (Ormerod *et al.*, 1990). Some organisms, such as fish, were affected positively, while others, such as stream macrophytes, were markedly reduced. In this example, there was sufficient modification in water quality to cause biological change, but only by making a value judgment can we say whether or not pollution is occurring (see Holdgate's definition, above). Birds can also show positive responses in density and biomass as a result of organic enrichment or eutrophication in lakes and estuaries, even though diversity sometimes declines (Campbell, 1978; Rutschke, 1987).

5.3 MONITORING AIMS

We must clearly set out our aims and describe the context when we question the role and scope of birds in assessing changing water quality (Chapter 1). There may be few instances, if any, in which birds figure in the derivation of water quality standards, and hence few instances in which they contribute to true compliance monitoring. Equally, few **long-term** mensurative or manipulative experiments (after Hurlbert, 1984) involve assessing the response of birds to trends in water quality. There are instances in which birds have been used to monitor controlled substances, however, and others where they have been used in surveillance. There will be many more where opportunities arise to monitor changes in bird populations before or after instances of pollution or its abatement.

A particular concept, important in the context of this chapter, is that of the biological indicator. Implicit here is the perception that a

Table 5.2 Potential response to generalized ecological perturbation at different scales, and examples of potential biological indicators through which change might be detected or monitored

Scale of perturbation	Potential indicators of change
Sub-physiological	Tissue concentrations of contaminants without biological effect
Physiological/morphological	Physiological dysfunctions/morbidity e.g. enzyme disturbance, reproductive impairment (genotoxicity, teratogenicity, malformation, embryotoxicity, altered shell structure, egg failure), tissue damage/organo-structural defects
Behavioural	Time–activity patterns, pair formation
Individual	Quality/condition, growth, energetics
Species	Presence/absence
Population	Density, dispersion, biomass, production, change, natality, recruitment, mortality, survival, distribution (dispersal, migration), demography, isolation, territoriality, life cycle
Community	Competition/coexistence, predation, parasitism, disease, diversity, equitability, biomass, connectedness, functional roles
Ecosystem	Production, respiration, succession
Landscape/biome	Cumulative responses across ecosystems
Global	Cumulative responses across biomes

biological measurement in some way **indicates** biological or physico-chemical conditions in the wider environment. Any one of a variety of biological parameters may be used over a variety of ecological scales (Table 5.2). In a paper specifically concerned with vertebrates, Landres *et al.* (1988) define an indicator species as

> an organism whose characteristics (e.g. presence or absence, population density, dispersion, reproductive success) are used as an index of attributes too difficult, inconvenient or expensive to measure for other species or environmental conditions.

In this respect, a biological indicator may contribute to any type of monitoring programme, and in a variety of ways. Landres *et al.* (1988) suggest three uses of indicator species, to assess:

Birds as indicators of contaminants in freshwaters

1. the presence and effects of environmental contaminants;
2. trends in the population of other species less easy to measure than the indicator, but nonetheless related to it ecologically;
3. trends in the habitat quality for other species or entire communities and ecosystems (the inference being that the status of the indicator somehow reflects these other features).

In this chapter, the habitat parameter in question is water quality, but this sphere also encompasses environmental contaminants. Thus, our discussions of birds as potential biological indicators in freshwaters expand on these two themes (1 and 3).

5.4 BIRDS AS INDICATORS OF CONTAMINANTS IN FRESHWATERS

5.4.1 Some background

The use of birds' eggs or tissues in assessing trace contaminants, notably some metals and the persistent organochlorines, is long established in a wide range of ecosystems (Chapter 3). Coincident with its monitoring purpose, such work has also important indicator value. The trophic level occupied by predatory birds means that their tissues can acquire, through bioaccumulation and biomagnification, significant concentrations of contaminants which are otherwise too dispersed for easy measurement. Moreover, the material used is often biological waste, either in the form of failed eggs or birds found dead. Both these sources permit measurement without the sacrifice of living organisms, although the strategy would carry an inherent bias if the samples contained systematically elevated or reduced contaminant burdens. These, and other, problems associated with the use of birds' eggs or tissues as monitors of pollutants are discussed in Chapter 3.

5.4.2 Birds and contaminants in freshwater systems

Stemming from the initial work in terrestrial systems, birds have figured in the monitoring of persistent contaminants in freshwaters. Here, toxicological effects have usually involved piscivorous birds, notably grebes, cormorants, pelicans, herons, gulls, terns, osprey *Pandion haliaetus* and the aquatic kingfishers. The scope and content of freshwater monitoring programmes has been variable, however, as have the criteria for selecting indicator species. Ease of organized sampling, evidence of contamination, and knowledge of feeding ecology, are all generally agreed requirements (e.g. Moore, 1966; Gibertson *et al.*, 1987) but views differ over the importance of ranging behaviour. Some authors suggest that wide-scale movements are important, so that the indicator species integrates general environmental patterns (e.g. Moore, 1966). By contrast, others suggest that movements should be more limited, so that local patterns of contamination are reflected faithfully (Phillips, 1978). The second view would hold if an indicator species was to be used in detecting specific pollution sources. Our examples reflect these contrasting views.

Around the North American Great Lakes, herring gulls *Larus argentatus* were chosen in 1973 for a monitoring programme in which eggs were routinely assessed for organochlorine contaminants. Several attributes of the species' ecology were highlighted as important in this choice, not least the fact that effects of DDT on breeding performance were being described (Hickey *et al.*, 1966). Herring gulls breed in colonies, so large samples of eggs can be easily and systematically collected each year, and breeding adults are year-round residents in the Great Lakes region, with comparatively little movement between water bodies. In addition, as piscivores, herring gulls could reflect contaminant levels in a food resource used by other animals, not least humans. Finally, the species has a holarctic distribution which would permit comparison with populations elsewhere, including coastal North America or Europe. Such comparisons have, indeed, been made by European workers, although not on a systematic basis (e.g. Lenmetynen *et al.*, 1984) and the predominantly coastal distribution of this species in Europe precludes its use as an indicator of pollution in freshwaters. In general, major contaminants declined markedly in Great Lakes herring gulls during the 1970s (Fox and Weseloh, 1987). The programme expanded during this time to cover other species, a range of other contaminants, and assessment of a wide range of physiological and morphological parameters (see Table 5.2 and Fox and Weseloh, 1987). These included assessments of the activities in embryos of enzymes responsible for detoxification, such as aryl hydrocarbon hydroxylase or benzyloxyresorufin-O-dealkylase. The expectation was that levels might be elevated under exposure to certain specific contaminants, for which the Great Lakes work provided some evidence (Fox and Weseloh, 1987). Enzyme measurements are credited with providing a cheap and effective early warning of contamination, which may be particularly important where non-persistent pollutants are involved in toxicity, but their use is still under evaluation, particularly with respect to the specificity of enzyme response (Custer *et al.*, 1991). The need to kill birds or embryos is a disadvantage. Other examples of contaminant assessment in Northern American water birds are given in Chapter 3.

Although there have been many opportunistic studies of contaminants in the eggs of freshwater birds in Europe (e.g. Cooke *et al.*, 1982; Burton *et al.*, 1986; Focardi *et al.*, 1988), long-term trends in the United Kingdom have for the most part been assessed using the livers of birds found dead (Newton, 1991). This British programme began in 1963, and the species involved are Eurasian kingfisher *Alcedo atthis*, grey heron *Ardea cinerea* and great crested grebe *Podiceps cristatus*. Despite the opportunism and potential bias in this approach, the derivation of the sample has been consistent throughout. Significant declines in mercury, HEOD, DDE and PCBs were apparent in all the species following legal and voluntary controls during the 1970s and 1980s (Figure 5.3), although the reduction in PCB levels in kingfishers was not statistically significant. Trends in the population and breeding performance of these freshwater species have not been as closely connected with contaminant burdens as in terrestrial raptors (e.g. Newton *et al.*, 1989). Perhaps other factors, such as the weather (e.g. Dobinson and Richards, 1964; Reynolds, 1979), affect their populations (heron, kingfisher), and in some cases their nests are not easily censussed (grebe, kingfisher).

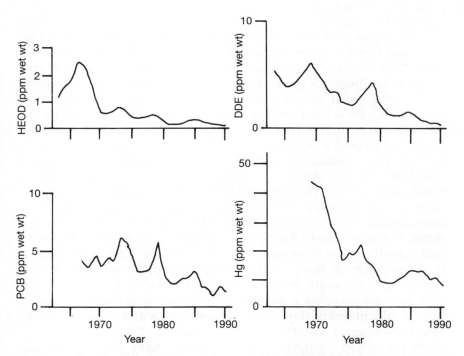

Figure 5.3 Trends in the mercury and organochlorine levels of herons found dead in Britain between 1963 and 1990 as shown by geometric moving means. (After Newton, 1991.)

5.4.3 Upland rivers

Monitoring involving kingfishers, herons and great crested grebes has several particular features. These species occur predominantly on the lower reaches of rivers, or on estuaries and lakes, so that upland rivers may have been poorly represented in the samples. These species also forage or migrate over large areas that sometimes include other parts of Europe, so that the sources of any contamination may not be precisely detected (Cramp and Simmons, 1977; Cramp, 1985). They are also relatively scarce, with only 5000–11 000 pairs of each breeding in Britain: trends for contaminants in great crested grebes and kingfishers are represented by an average of only 5–6 specimens per year (Newton, 1991). With these features in mind, we have recently assessed contaminants in the eggs of two common passerines, the dipper *Cinclus cinclus* and grey wagtail *Motacilla cinerea*, along upland rivers in the United Kingdom and Ireland.

There are important contrasts between dippers and grey wagtails that help us to understand the role of birds as ecological indicators generally. We return to this theme in the context of surface water acidification below, but the information is also important with respect to contaminants. Dippers occupy linear territories of 0.3–2 km along fast-flowing rivers (Ormerod and Tyler, 1987), seldom disperse more than 10–20 km from their birth places (Tyler *et al.*, 1990) and, except in northern Scandinavia, they are non-migratory (Cramp, 1988). Their diet of small, non-migratory stages of fish and aquatic invertebrates also reflects local conditions (Ormerod and Tyler, 1987), so any contaminant burden can only be acquired in a reasonably defined area. Their diet nevertheless places them in a trophic position that leads to detectable contamination by metals and organochlorines (Ormerod and Tyler, 1990b). Around 30 000 pairs nest in north and western Britain and Ireland, while representatives of the same species or genus have a global distribution that is wide enough to permit geographical comparisons. Grey wagtails also breed in territories along fast-flowing rivers, but in contrast to the dipper, feed opportunistically on terrestrial and aquatic insects (Ormerod and Tyler, 1991b). They differ further from the dipper in being partially migratory, with some British birds wintering as far south as France and Iberia (Tyler, 1979). In winter, grey wagtails occupy a wide range of habitats, which include ponds, river or lake margins, farmyards, sewage works and coastal marshes. There are around 25–50 000 pairs in Britain and Ireland.

Despite these contrasts, the two species have shown generally similar patterns in contaminant burdens over the duration of study. For example, total PCB residues in the eggs of both species declined by around 95% over the period 1988–1991 (Figure 5.4) patterns

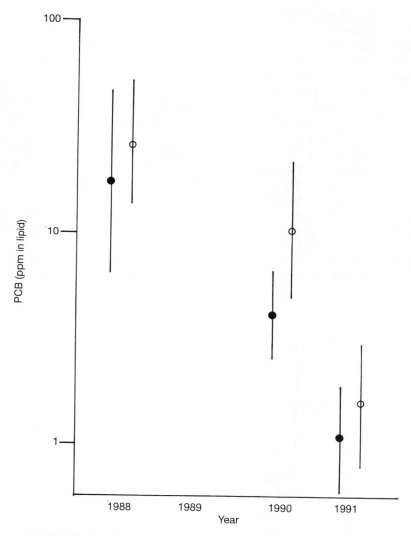

Figure 5.4 Trends in the total PCB residues (± S.D.) in the eggs of dippers (closed symbols) and grey wagtails (open symbols) in Wales, 1988–1991.

consistent with the reductions shown for the piscivores in Figure 5.3. However, a marked increase occurred in 1992. These annual changes in PCBs are thus great enough to mask the effects of differences between these species' ecology. Interspecific differences are apparent at more precise levels of analysis, however. For example, individual identification of some of the 209 congeners that make up the PCB group reveals some variability in the signature of the total PCBs present (Figure 5.5). In the dipper at least, the array of congeners is similar to those recorded for aquatic birds elsewhere in the world, while the dominance of PCB 101 in grey wagtail eggs is unusual (Ormerod and Tyler, 1992). Differences of a similar degree have been shown between other aquatic birds in Italy and may relate to their position in the food web (Focardi *et al.*, 1988). Whatever the reasons for these differences, they show how signals about ecosystem contamination can be affected by the choice of an indicator species: one lesson might be that species should not be taken in isolation as 'typical' indicators of contaminant trends in freshwaters, valuable information arising instead from inter-specific comparison.

Further data collected on dipper eggs show both the advantages and disadvantages of using eggs from territorial and site-faithful passerines as contaminant indicators. On the negative side, there is almost certainly a sampling problem in the extent to which any given passerine egg represents contamination in the clutch as a whole.

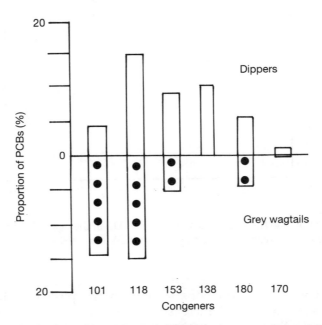

Figure 5.5 The contribution of different PCB congeners to the total PCB burden of the eggs of grey wagtails and dippers in Wales during 1990. Congeners are numbered after Ballschmitter and Zell (1980). (After Ormerod and Tyler, 1992.)

Variation in contaminant levels within dipper clutches is usually high (47–75%) relative to total variation, in marked contrast to raptors such as the sparrowhawk *Accipiter nisus* (5–15%; Newton and Bogan, 1978). One possibility is that, in passerines, the clutch mass represents a large proportion of the body mass (about 40% in the dipper as against about 15% in the sparrowhawk). Relatively more resources for egg formation will thus be drawn from recently ingested food so that subtle changes in diet, foraging location, or even episodes of pollution, might lead to differences in contaminant levels in different eggs from a clutch.

Table 5.3 Organochlorine concentration (ppm in lipid) of eggs of the dipper from different parts of Britain and Ireland in 1990; the values are geometric means (S.D., n) (from Ormerod and Tyler, 1992)

	E. Scotland	S.W. Ireland	Wales
DDE	3.54 (0.41,23)	0.63 (0.37,16)	1.91 (0.27,71)
HEOD	0.61 (0.33,18)	0.39 (0.29,6)	0.56 (0.34,22)
PCBs	10.47 (0.53,26)	3.99 (0.45,16)	4.18 (0.62,51)

On the positive side, the abundance and high productivity of passerines means that relatively large samples of eggs can be collected, even when they are only taken opportunistically after desertion or addling. Just one year's data from dipper eggs were sufficient to show gross differences in the contaminant burden from different regions of Britain and Ireland, with patterns clear enough to reappear in subsequent years (Table 5.3 and Ormerod and Tyler, 1992). Even more detailed geographical patterns were detected within regions. Figure 5.6(c) shows how the average concentrations of HEOD (from dieldrin) in dipper eggs from a range of Welsh rivers correlate with sheep densities on the river catchments. This suggests that sheep dip, of which dieldrin was a former active component, may have been responsible for at least some contamination. Figures 5.6(a) and 5.6(b) show how concentrations of one PCB congener decline on a SW–NE gradient through Ireland and Wales; this suggests that this contaminant has some westerly source and arrives by atmospheric deposition (Ormerod and Tyler, 1993a). Together, these features indicate that dipper eggs can aid in assessing local patterns of at least some contaminants. This is important, because toxic chemicals in rivers can arise from localized (sometimes illicit) sources that require detection at a local scale. The data also emphasize the integration and bioaccumulation measurements that birds' eggs can provide. In Wales, direct measurements of

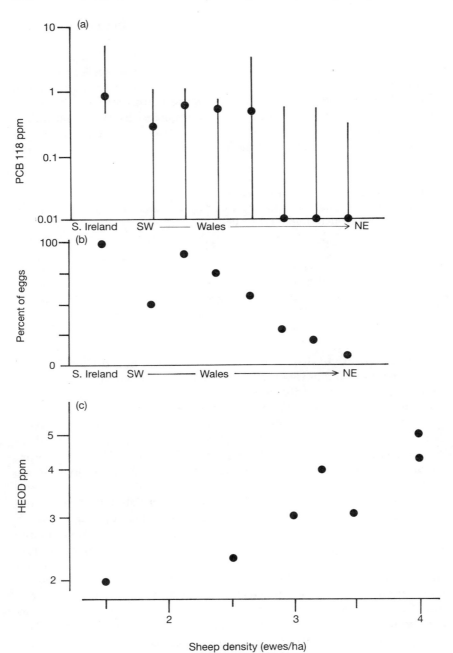

Birds as indicators of contaminants in freshwater

dieldrin, DDE or PCBs in raw surface waters are now limited to only 38 sites on main rivers which are only sampled monthly; few samples contain organochlorines at detectable levels. Yet, the dipper eggs show that significant contamination is still present in some catchments.

It thus seems that water birds can provide valuable information about spatial and temporal trends in persistent contaminants. But to what extent do they have a wider indicator value? Could such spatial patterns indicate risks to other aquatic and riparian organisms? For example, otters *Lutra lutra* in British freshwaters are sometimes affected by a variety of persistent chemicals, but are too rare to be sacrificed for monitoring (Mason et al., 1986). Interestingly, their populations were lowest in the period from 1950 to the 1970s, when organochlorines reached their highest levels in aquatic birds; since then organochlorine levels have decreased in birds, and otter populations have increased substantially (Mason, 1991). However, there have been few examinations of the degree to which differing groups of freshwater organisms reflect each other's contaminant burdens or population trends in time and space.

5.4.4 Other contaminants: metals and radionuclides

In many respects, the use of birds to assess metals in freshwaters has lagged behind work in marine and estuarine systems (e.g. Goede and de Bruin, 1984; Chapter 3). The exceptions are mercury, as described above, and lead, which has been involved in major effects on some freshwater birds as a result of its use in fishing and hunting (Mudge, 1983; Sears, 1989; Pain, 1991). Lead monitoring in the blood of mute swans *Cygnus olor* has shown the effectiveness of legislation to limit the use of lead weights by anglers (Sears, 1989), although even sublethal concentrations can predispose birds to some risks (O'Halloran et al., 1989).

Radionuclides have not been widely measured in freshwater birds, although substantial work was carried out on waterfowl and divers following the Chernobyl accident in 1986 (e.g. Petersen et al., 1986; Brisbin, 1991; Chapter 4). In Britain, geographical differences in the amount of fallout from this accident were detectable in otter spraints

Figure 5.6 Differences between Welsh catchments in residues of different chemicals in dipper eggs. (a) and (b) PCB 118 in lipid along a SW-NE gradient along the direction of prevailing wind, with values from Southern Ireland shown for comparison: (a) median and range; (b) percentages of eggs showing concentrations above the limit of detection. (c) Concentrations of HEOD in relation to sheep stocking densities. (After Ormerod and Tyler, 1993a.)

(Mason and Macdonald, 1988) but similar work on aquatic birds has not been reported. In the United States, work is currently in progress to use aquatic birds as 'sentinal indicators' of radiocaesium contamination (Brisbin, 1991; Chapter 4).

5.4.5 The future

Given the downward trend in the levels of organochlorines and mercury shown by freshwater birds in Britain and North America, questions about the future needs of contaminant monitoring are pertinent. As an example, our own work on organochlorines in dipper and grey wagtail eggs might be regarded as redundant because of the low concentrations recorded and the restricted evidence of any serious toxicological effects (Ormerod and Tyler, 1992). For several reasons, however, the need for monitoring continues, providing that modifications are made to suit the changing needs. These are as follows:

(a) Some controlled substances may still increase

Although the manufacture of PCBs has now ceased, only around 4% of their former production has been degraded or destroyed. Around 65% is either landstocked (in tips, dumps etc.) or in use in equipment and some authors have predicted future increase in contamination as these stored PCBs are released into the wider environment (e.g. Tanabe, 1988). Freshwater systems, particularly rivers, are liable to constitute sinks or transmission pathways for any such escaping material. PCB levels have continued to increase in the eggs of some British raptors, at the same time as levels in freshwater species declined (Newton, 1991). Continued vigilance is thus desirable, perhaps particularly in aquatic systems: there is evidence that piscivorous animals are vulnerable to PCB accumulation because they are deficient in certain enzyme types involved in PCB metabolism (Tanabe, 1988).

(b) The toxicological effects of some widespread contaminants are not yet wholly understood

Although evidence of pronounced toxic effects by PCBs on birds is still limited the belief is gaining acceptance that they are involved in subtle toxicity in mammals (Reijnders, 1986). Moreover, much of the past evidence on toxicity by PCBs has involved lumping all congeners as 'total PCBs'. Only recently have a small number of specific congeners been included in monitoring programmes, and basic research on their occurrence and effects in aquatic birds is still underway

(Ormerod and Tyler, 1992). It is increasingly recognized that the 20 PCBs that have chlorine atoms in coplanar positions (e.g. congeners 15, 37, 77, 81, 126, 169; numbered after Ballschmitter and Zell, 1980) have the greatest toxicity to vertebrates, since in structure and chemical behaviour they resemble dioxins (Jones, 1988; Kannan et al., 1989). Because of analytical difficulties they are not often measured specifically in biological tissues but they may be as widespread as other PCBs (e.g. Jones, 1988; Tanabe, 1988). They are also bioaccumulative in aquatic ecosystems (Kannan et al., 1989). Further data are required on the extent to which the more commonly measured PCB congeners indicate the presence in aquatic birds of these other, possibly more toxic, contaminants. More importantly, there is a need for basic data so that we can understand the trends and toxic effects of different PCBs, particularly the coplanar congeners, in the wider environment. In our work on dippers, PCB congener 118 occurred in significantly higher concentrations in eggs that failed to develop than in fertile eggs (Ormerod and Tyler, 1992). Some contribution to egg failure thus could not be ruled out, although the effects were minor and unlikely to contribute to population change because only a small percentage of eggs were affected (Newton, 1988). There remains, nevertheless, a need for similar data from a wider array of freshwater species.

(c) Illicit use or accidental spillage of controlled substances might still occur

Although controls for many uses of chemicals such as DDT and dieldrin began in the 1960s and 1970s, some uses of DDT were permitted until 1984 (Hardy et al., 1987). Dieldrin was not finally removed from agriculture until the late 1980s, the last legal use in Britain involving the protection of ornamental blooms. However, there have been several recorded instances of these substances being used illegally or spilled unintentionally after their withdrawal (e.g. Institute of Terrestrial Ecology, 1982; Royal Society for the Protection of Birds, 1987). The eggs or corpses of freshwater birds can help to reveal such instances on a local scale (e.g. Royal Society for the Protection of Birds, 1987; Ormerod and Tyler, 1992).

(d) Restrictions do not apply in all nations

Overwhelmingly, the data available on the occurrence of persistent chemicals in freshwater birds relates to the developed world. However, organochlorine pesticides are still used legally over large areas of Asia, Africa, and South America. India alone uses around 47 000 t of lindane and 19 750 t of DDT annually, and freshwater ecosystems show widespread contamination (e.g. Ramesh et al., 1989; 1990). Effects on aquatic birds have also been described (Douthwaite, 1992). Surveillance in these systems is likely to become important, and might benefit from the use of species which are closely related to those monitored elsewhere. Moreover, the persistence and propensity of pesticides and other organochlorines to circulate globally means that there is still a need for continued vigilance on our own continent. For example, the global atmospheric content of PCBs (500 t) is equivalent to around 25% of that in seawater and around 11% of the total global loading in biota, illustrating the reservoir of PCBs in atmospheric circulation (Tanabe, 1988). Some will fall out locally, although there are many examples in the literature of long-range transport of these and other persistent pollutants (Erikkson et al., 1989).

5.5 BIRDS AS ECOLOGICAL INDICATORS OF WATER QUALITY

Considerable progress has been made in the use of biological indicators of general water quality. Most assessments are made at the community level and usually involve aquatic invertebrates or lower plants (Hellawell, 1986; Mason, 1991). In many cases, 'biotic indices' have been derived from the ranges of water quality tolerated by different organisms. In such schemes, a score is calculated from the organisms present at a given site, and used to convey simple information about the water's biological and chemical status. Most schemes concentrate on one of the commonest pollutants, organic matter, although index scores can provide information about other influences on water quality (Mason, 1991). More recently, multivariate methods of classification have been used to identify indicator species that typify the response of animal or plant communities to altered water quality (Wright et al., 1984). We have used this approach in the British uplands to identify invertebrate and macro-floral indicators of water acidification (Ormerod et al., 1987; Rutt et al., 1990). For birds to be valuable indicators of water quality, therefore, they must at least complement the existing schemes. They can do so because they are easily surveyed and there are extensive historical records of their distribution and abundance. But how effective and reliable is the information that birds provide?

For a bird species, or group, to be a valuable indicator of water quality, it should have the following attributes:

1. Its status should reflect water quality and should be sensitive to change in quality through time and space.
2. Its status should be reflected by variables which are easily measured and informative.
3. Its response to changing water quality should be consistent in time and space.

Birds as ecological indicators of water quality

4. Its status should reflect water quality for other important biological resources. In this context, we might also wish the performance of our selected ornithological indicator to be better than, or at least complementary to, potential indicators from other taxa.
5. Change in its status caused by change in a given component of water quality should be readily separable from the effects of other components, and from changes in other habitat features.
6. Its ecology should be well understood, so that we can identify the causes of connections involved in criteria 1 to 5. This is important in understanding any instances where our indicator fails.
7. It is also advantageous if it is colourful, big, charismatic or unusual so that it attracts sufficient public interest for the monitoring programme to be sustained and the results heeded.

Here we use the example of surface water acidification to consider whether birds may meet these criteria.

5.5.1 Birds and surface water acidification

There is widespread evidence from Europe and North America of the acidification of surface waters in base-poor regions as a result of acid deposition (see Mason, 1991). The biological repercussions are pronounced, affecting organisms in most trophic levels (Ormerod and Wade, 1990). We also have an increasing understanding of how relationships between water acidity and ecology are modified by complicating factors such as changing land use (Figure 5.1; Ormerod et al., 1993). As a result, a wide variety of biological indicators of acidity have been proposed. Their value is particularly important in streams because those which suffer acidification are subject to marked episodic fluctuations in chemistry. Such episodes can be difficult to detect without appropriate (often expensive) chemical sampling but they are integrated in a meaningful way by stream organisms (Weatherley and Ormerod, 1991). In lake systems, some biological indicators are particularly valuable in tracing historical trends because their remains are preserved chronologically in lake sediments (Flower and Battarbee, 1983).

Because acidification affects birds that depend on freshwaters for fish and invertebrate food, both lacustrine and riverine birds have been among the organisms proposed for indicators (e.g. Eriksson, 1987; Ormerod and Tyler, 1987; McNicol et al., 1987; Vickery and Ormerod, 1991). So far, however, there has been no critical assessment of their potential value with respect to the criteria outlined above. Here we make such an assessment with emphasis on two upland river birds in Europe, the dipper and grey wagtail, whose general ecology we described earlier.

Some of the best supporting evidence available on the use of river birds as indicators of acidity arises from the case of the dipper in Wales and Scotland. A range of methods involving quantitative bank-side surveys (Ormerod and Tyler, 1987; Vickery, 1991), presence/absence over 1 km plots (Ormerod *et al.*, 1986) and catch per unit effort from ringing (Ormerod *et al.*, 1988a) all show pairs to be markedly less abundant where stream acidity and aluminium concentrations increase (Figures 5.7, 5.8); territories are also longer at low pH (Figure 5.7).

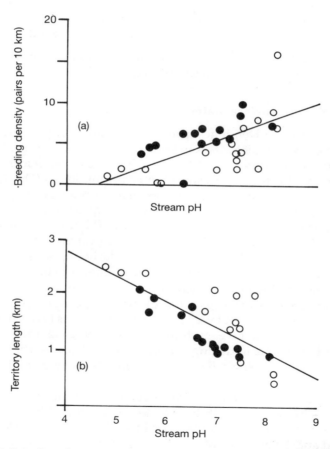

Figure 5.7 Breeding density (a) and territory length (b) of dippers in relation to mean stream pH in Wales (open symbols) and Scotland (closed symbols). (After Vickery and Ormerod, 1991.)

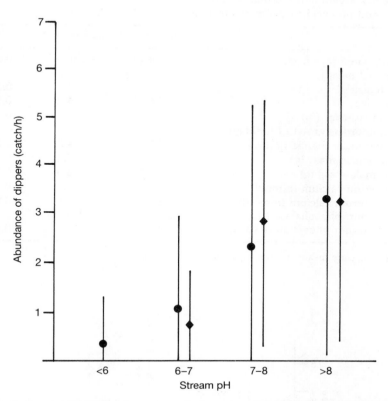

Figure 5.8 Abundance of dippers (mean, with S.D.) along streams of different pH during autumn in 1986 (circles) and 1987 (diamonds) as shown by the catch per unit effort of ringing activity. (After Ormerod et al., 1988a.)

Other easily measured variables reflecting deteriorating habitat quality at low pH include significantly delayed laying, reduced clutch size, reduced egg mass, reduced brood size, reduced nestling growth and inferior condition in adults, the latter shown by both live mass and condition indices (Table 5.4; Ormerod and Tyler, 1990b; Ormerod et al., 1991). The relationships between these variables and acidity have been generally consistent in time and space. For example, historical data from one Welsh river showed that the dipper population declined by 80–90% during a period of increasing acidity (Ormerod and Tyler, 1987). Comparisons between Scotland and Wales showed that the effects of acidity on abundance, territory length and breeding performance were consistent, despite some

Table 5.4 Breeding performance and characteristics of pre-breeding dippers in acidic and non-acidic catchments in Wales (after Ormerod et al., 1991)

	Acidic[a]	Non-acidic[a]	P
Laying date (days from annual mean)	+4 to +10	−2 to −3	0.001
Clutch size	4.0	4.8	0.001
Brood size	3.6	4.2	0.001
Second clutches (%)	0	21.9	0.001
Nestling tarsus growth (mm/day)	1.6	1.9	0.001
Nestling mass change (g/day)	3.7	4.3	0.01
Adult female mass (g)	55.9	57.4	0.01
Adult male mass (g)	67.2	68.9	0.01
Male serum calcium (mmol/l)	1.1	1.5	0.01
Female serum calcium (mmol/l)	1.4	1.5	0.05
Male serum phosphatase (IU/l)	251	159	NS
Female serum phosphatase (IU/l)	493	216	0.01

[a] The values are means, from sample sizes in excess of 22.

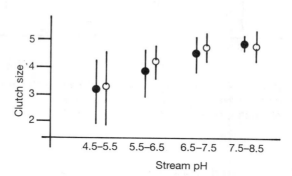

Figure 5.9 Clutch size (mean, with S.E.) of dippers in relation to stream pH in Wales (open symbols) and Scotland (closed symbols). (After Vickery and Ormerod, 1991.)

regional effects (Figures 5.7, 5.9; Vickery and Ormerod, 1991). There is also substantial evidence that these relationships are causal. Important and preferred food sources for dippers include mayfly nymphs and caddis larvae, these groups being scarce in acidic streams (Ormerod and Tyler, 1991a). Probably as a result, adults spend longer each day satisfying their energy requirement on acidic streams (O'Halloran *et al.*, 1990) and food delivery to dipper nestlings is impaired (Ormerod and Tyler, 1987; Vickery, 1988). Calcium-rich prey such as fish, molluscs and crustaceans are also scarce on acidic streams, and adult dippers of both sexes show a significant reduction in their concentration of plasma calcium (Ormerod *et al.*, 1991). Females show this difference to a lesser extent than males, possibly because they raise their serum alkaline phosphatase concentrations in order to mobilize structural bone (Table 5.4). Also reflecting calcium scarcity at low pH, shell-thickness is significantly reduced, even though organochlorine concentrations in eggs from acidic streams are low (Ormerod *et al.* 1988b; Ormerod and Tyler, 1990a). While it may be argued that all these differences between acidic and non-acidic streams reflect some adaptive response to the prevailing conditions, we can demonstrate that this is not so with reference to post-fledging survival. Figure 5.10 shows the probability of recapturing nestlings after fledging in relation to their time of hatching. These data show how the survival chances of birds from acidic streams would be maximized by earlier breeding, the reverse of the pattern that actually occurs.

In the absence of a large-scale experimental test, it seems reasonable to conclude that dippers are affected adversely by acidity; their

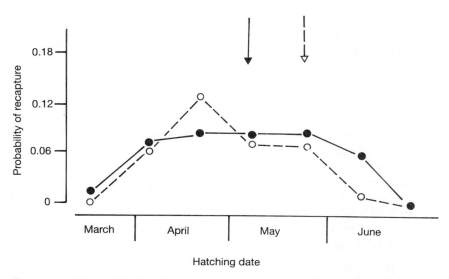

Figure 5.10 The probability of recapturing dippers after fledging from acidic (open symbols) and non-acidic (closed symbols) streams in upland Wales in relation to the time of hatching. The peak hatching periods on each stream type are shown by the arrows. (After Ormerod and Tyler, 1993b.)

status reflects acidity; the relationship with acidity is reasonably robust in time and space; and many of the variables that show this are not difficult to measure. Of our six major criteria outlined above, we can therefore accept numbers 1, 2 and 6, with qualified acceptance of criterion 3.

The results for the grey wagtail are in marked contrast. Work in neither Wales nor Scotland has revealed any relationship between acidity and grey wagtail abundance, which instead is largely determined by the availability of riparian shingle and broadleaved trees (Ormerod and Tyler, 1987; Vickery, 1991). Their food supply, often from terrestrial origins and predominantly consisting of adult Diptera, spiders and caterpillars, is unaffected by pH (Ormerod and Tyler, 1991b). Nor are there any pronounced effects of acidity on breeding performance (Tyler and Ormerod, 1991). As an indicator of acidity, therefore, this species fails on the most important of our criteria: its status is not related to water quality. The contrast with the dipper is nonetheless instructive: differences between these species in response to acidity appear to be a result of differences in their foraging ecology, reflecting the much greater role that the stream ecosystem plays for the dipper. But does this mean that dippers truly indicate biological conditions in the wider stream environment in ways that are more effective than other biological indicators?

To answer this question, we have used data from a large regional survey of water quality and biology undertaken to investigate stream acidity in upland Wales in 1984. The same survey initially provided data that showed the limited distribution of dippers at low pH (Ormerod et al., 1986), so our data are biased to produce a positive result for this variable. We have thus compared the effectiveness of the presence/absence of dippers in indicating acidity with information provided by two other indicators:

1. The stream macro-flora, where the species composition was classified using TWINSPAN (Hill, 1981) into four groups with nine indicator species (Ormerod et al., 1987).
2. The stream invertebrate fauna, where the species present at each site were used to place the sites into one of four groups derived by TWINSPAN from over 300 streams across the whole of upland Britain (Rutt et al., 1990). Fourteen invertebrate families figure as important indicators.

We have also assessed the effectiveness of each of the three indicators (dippers, macro-flora, macro-invertebrates) in showing differences between sites in the status of another important biological resource affected by acidity, the density of brown trout *Salmo trutta*. Finally, we examined the extent to which the presence and absence of dippers agreed with the classification of sites based on invertebrates.

As expected, Figure 5.11 shows that the presence or absence of dippers, but not grey wagtails, can indicate differences in pH ($P < 0.001$). However, a much stronger and more precise indication of

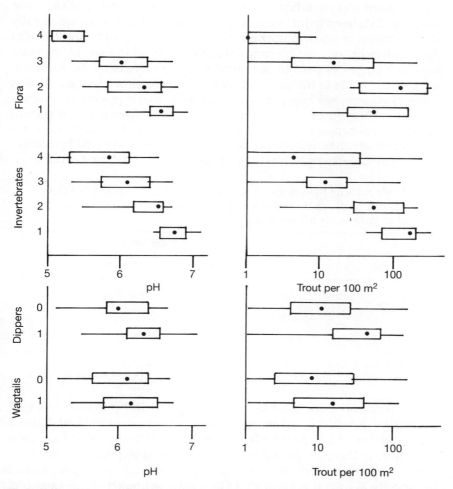

Figure 5.11 Differences in pH and trout density along 88 streams in upland Wales in relation to the presence or absence of grey wagtails and dippers, and the occurrence of various assemblages of aquatic invertebrates and macro-flora. The values are means, with 95 percentiles (boxes) and ranges (whiskers). The macro-floral and invertebrate assemblages are shown as four groups, defined by TWINSPAN classification.

pH was achieved by either macro-floral or invertebrate indicators, not least because the greater array of species allowed a four-way classification. Similar results were obtained with respect to trout densities. Nevertheless the presence or absence of dippers was highly concordant with a simple classification of sites based on invertebrates (Table 5.5), so dippers do provide information about ecosystem status. Despite these differences and concordances, major problems would arise if our requirement was to indicate conditions at a small number of sites: the absence of dippers in just one location does not indicate unequivocally that a stream is acidic and their presence does not indicate that it is not. However, our assessments would be strengthened by increasing the number of sites in our survey, or by increasing the information on the birds to include features such as territory length. At this stage, therefore, we might accept criterion 4, that dippers can indicate conditions for other ecosystem components. However, we should qualify our acceptance by suggesting that their performance does not match that of other indicators that are more directly affected by acidity, and whose diversity permits greater precision in indicating conditions.

Table 5.5 The presence and absence of dippers in upland Wales in relation to the characteristics of the invertebrate community (the χ^2 of association is 5.76, $P < 0.05$)

		Invertebrate community type:		
		1 or 2[a]	3 or 4[b]	
Dippers	Present	8	9	17
	Absent	8	38	46
	Total	16	47	63

[a] Groups 1 and 2 are diverse and typical of non-acidic streams.
[b] Groups 3 and 4 are typical of acidic streams (see Figure 5.11 and Rutt et al., 1990).

The remaining question is whether dippers indicate only changes caused by acidification, as opposed to other pollutants or habitat modifications. An important feature here, as with any other biological indicator, is that we should expect the species to perform well only under ecologically suitable conditions. We would not expect any of the species involved in Figure 5.11 to show the impact of, for example, acid mine drainage on a lowland stream because few of them would occur there anyway. This is important, and emphasizes the need to understand the biology of a species sufficiently well to define where it would normally be expected to flourish. The dipper in Britain chiefly inhabits the uplands of the north and west, where there are fast-flowing streams with gradients between 5 and 30 m/km; outside this range densities decline (Marchant and Hyde, 1980; Ormerod et al., 1985; Vickery, 1991). There is also an apparent effect on dippers by altitude, but this probably occurs simply because higher streams have gradients that are more suitable (Marchant and Hyde, 1980). There is a tendency also for pairs to prefer streams lined by broadleaved trees and to be in areas with adequate riffles for feeding (Ormerod et al., 1986). One other feature that causes year-to-year variation in the numbers of other river birds, adverse winter weather, has not so far been shown to affect dippers (Marchant et al., 1990). Other factors have not been identified as having major influences on dipper ecology. Providing that the major habitat factors identified are suitable or unchanged, it seems dippers respond mostly to changes in food abundance and hence water quality. But do they respond only to acidity?

We have a surprising dearth of quantitative data with which to answer this question: only one data-set is available. In Figure 5.12, we show how the density of dippers in streams in the Derbyshire Peak district generally decreases with organic pollution, as indicated by low values of Chandler's Biotic Score (Edwards, 1991). Low values of the score arise when mayflies, stoneflies and caddis are scarce in the invertebrate community, in other words when dipper prey are in low numbers. Grey wagtails show no similar correlation, as we might expect from their lesser dependence on aquatic insects. Thus spatial patterns in the abundance of dippers might result not only from patterns in acidity; they might reflect other water quality factors that

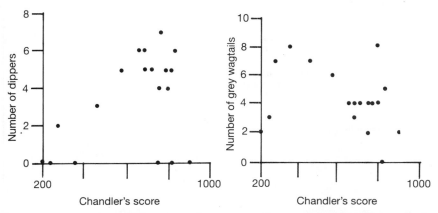

Figure 5.12 Numbers of dippers and grey wagtails recorded along streams in the English Peak District in relation to Chandler's Biotic Score. Low scores indicate the most polluted stretches. The correlation between bird numbers and score was significant for dippers ($P < 0.05$), but not grey wagtails. (After Edwards, 1991.)

208 Birds as indicators of changes in water quality

affect their food supply. A range of such factors could clearly operate. This is a general problem with biological indicators of water quality: absences or low scores can reflect a wide range of possible pollutants but only chemical analysis can establish which is important in a particular case. On the other hand, this is the 'early warning' function that we expect from biological indicators: biological and chemical measurements are complementary (Spellerberg, 1991). Moreover, in the case of acidification, the source of pollutants is diffuse and the streams at risk are in remote upland locations where other changes in water quality are unlikely. Furthermore, there is increasing understanding of the characteristics of geology, soils, rainfall chemistry and land use that render some areas susceptible to acidification (Ormerod et al., 1989; Edwards et al., 1990); all this information helps us to form a basis for judgement when we assess trends in dipper populations in either time or space. Nevertheless, acceptance of the final criterion (5) in our assessment of dippers as indicators of acidity must at least be qualified: a scarcity of dippers could only indicate acidity along streams that were otherwise suitable if other possible water quality trends could be excluded and if supporting chemical measurements had been made. This is represented formally in

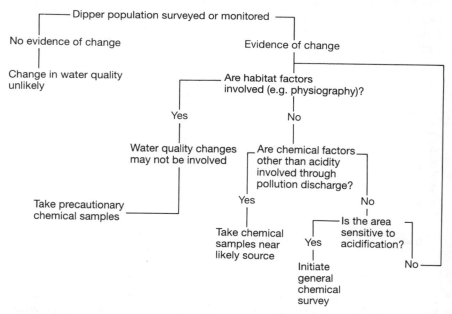

Figure 5.13 A conceptual model governing the use of dippers as indicators of acidity.

Figure 5.13, and in effect might be considered as the situation governing the use of any ecological indicator of a specific change in water quality.

5.5.2 General implications

On the basis of this example, we might conclude that birds can sometimes be useful biological indicators of water quality. However, as with other biological indicators, birds are valuable only within certain constraints and at best are only complementary to other methods of monitoring. A further lesson from the comparison between dippers and grey wagtails is that a biological indicator of water quality should be strongly connected ecologically to the aquatic ecosystem. There are limitations, too, when only one species is involved (see Figure 5.11). The use of whole bird communities as indicators might be more valuable than the use of single species, particularly where their ecological response to water quality differs between species (e.g. McNicol et al., 1987). For example, in Canadian or Scandinavian lakes, fish can act as competitors to goldeneye *Bucephala clangula* and hooded merganser *Mergus cucullatus*, so that these species increase if fish densities decline. Conversely, piscivores such as goosander (= common merganser) *Mergus merganser* or divers *Gavia* spp. decrease when fish decline (McNicol et al., 1987; Eriksson, 1987). Different effects of acidity on these species at least suggest the possibility of a community level response that might have indicator value, but such approaches are untested on the basis of the criteria used here. Moreover, habitat features such as lake size, shape and water transparency influence the distribution of all these species. Complex changes in diet during the growth of some divers (*Gavia arctica*), added to the risk of predation by fish on their young, also means that the relationships between distribution and food abundance are still unclear (Erikkson, 1987).

5.6 CURRENT PERSPECTIVES ON WATER QUALITY ASSESSMENT IN BRITAIN

The priorities implied in the foregoing discussion of birds as indicators of water quality may be misguided in some respects. One of the major aims of past work relating the biological status of surface waters to their chemical quality has been the derivation of water quality standards. Many such standards now govern water quality to protect resources such as salmonid fisheries or potable waters, although similar limits have yet to be applied for the conservation of aquatic birds. For example, our conclusions that dippers show a range of adverse biological reactions at pH values below pH 6 or aluminium concentrations over 0.1 mg/l might help in establishing these as limits which should not be exceeded if dippers are a resource we wish to protect. However, the data base from which to derive any such wider standards for aquatic birds is remarkably scant.

Even in Britain, with its histories of both widespread water pollution and ornithology, aquatic birds have never figured prominently in pollution research. Bird populations around waterways have been assessed only as part of general surveillance (e.g. Reynolds, 1979; Marchant and Hyde, 1980), with patterns infrequently related to water quality. In part, this reflects the dearth of polluted stretches in the array of sample plots (Marchant and Hyde, 1980), which in turn relates to the non-random and non-stratified procedures currently involved in plot selection.

The need for water quality standards for all components of freshwater ecosystems is important and pressing. In Britain, the National Rivers Authority (NRA) is responsible under the Water Act (1989) for the specification and achievement of water quality objectives for Britain's 40 000 km of main rivers and streams. For the first time, these objectives will soon have a statutory basis and will be the means by which a national plan for water quality improvement will be implemented (under the Water Resources Act, 1991). The main elements of this new scheme (National Rivers Authority, 1991) will be the identification of a set of 'use classes' (general ecosystem; special ecosystem; salmonid fishery and the like), each of which will have an accompanying set of chemical and biological standards specific to it. These standards will be linked, where appropriate, to relevant European Community Directives. However, only two of the use-related classes are of some importance to aquatic birds, these being the 'general ecosystem' and 'special ecosystem' classes; their aim is to safeguard the general integrity of aquatic ecosystems. In neither case, however, do birds figure explicitly in the criteria that are being derived: birds would be protected only through the assumption that standards set for other aquatic resources also have some wider value. This is unfortunate but reflects our restricted knowledge about the impact of pollution on the aquatic avifauna.

5.7 SYNTHESIS: ACHIEVEMENTS, PROBLEMS AND POSSIBILITIES

Of the two principal contributions by birds to the assessment of water quality reviewed here, the most progress has been made with respect to contaminants. Indeed, the low solubility and dispersion of some trace contaminants in freshwaters means that biological tissues are important monitoring media. Valuable information has been obtained from the eggs and corpses of freshwater birds, showing both spatial and temporal trends in contaminants, sometimes over spatial scales as small as individual headstream catchments. However, there is a need to adapt monitoring strategies to suit changing requirements: different geographical areas are now at risk from contaminants; and different chemicals are becoming important, with a need to understand their toxicological effects more fully. One additional pointer from our own work is that passerines, in view of their abundance and productivity, may be a valuable group in monitoring programmes. Moreover, suitably chosen pairs or groups of species with contrasting life styles can provide instructive information both on contaminants and on the ecological effects of changing water quality.

Partly because biological indicators of water quality are so well established among a wide range of plant and animal groups, a role for birds as an additional ecological indicator may be difficult to envisage. But we have shown that easily obtained information on the distribution and breeding performance of birds can reflect variations in water quality; demonstrable relationships between birds and acidity are robust in space and time, they reflect causal links, and they indicate the influence of acidity on other important organisms. On the negative side, birds may be less effective than invertebrates or plants as indicators of acidity, and their ecological responses to pollutants may not be specific. The latter is a general problem with ecological indicators: only chemical or physical analysis can reveal the real nature of chemical and physical change. Nevertheless, biological indicators can provide crude or early warnings of changing water quality and they place chemical and physical changes into an extremely important biological context. The effects of changing water quality on biological systems are often the reason for quality standards and for biological surveillance; these reasons are among the best for raising the profile of aquatic birds, as important conservation resources, in this sphere.

Judgment of the likely effectiveness of birds in water quality assessment also depends on the goals of monitoring. If we wish birds to be ecological indicators of chemical trends, the information will be at best equivocal, and will need to be supported by other work. If, on the other hand, we wish to monitor the status of birds in relation to pollution or to pollution abatement, many approaches are available, although little work of this type is in progress.

Perhaps most of all, the use of birds as wider indicators of water quality is hindered by a dearth of basic data. Even in Britain, where ornithology and water quality have been major environmental issues for many decades, an assessment of the relationships between birds and general water pollution is an important research need.

REFERENCES

Ballschmitter, K. and Zell, M. (1980) Analysis of polychlorinated biphenyls (PCB) by glass capillary gas chromatography. *Fres. Z. Anal. Chem.*, **302**, 20–31.

Brisbin, I.L. (1991) Birds as indicators of global contamination processes: the Chernobyl connection. *Acta XX Congressus Internationalis Ornithologici*, **IV**, 2503–8.

Burton, F.G., Marquiss, M. and Tullett, S.G. (1986) A note on eggshell porosity, nest humidity and the effects of DDE in the Grey Heron (*Ardea cinerea*). *Comp. Biochem. Physiol.*, **85c**, 25–31.

Campbell, L.H. (1978) Patterns of distribution and behaviour of flocks of seaducks wintering at Leith and Musselburgh, Scotland. *Biol. Conserv.*, **14**, 111–24.

Chorley, R.J. (1969) *Water, Earth and Man*. Methuen, London.

Cooke, A.S., Bell, A.A. and Haas, M.B. (1982) *Predatory Birds, Pesticides and Pollution*. Institute of Terrestrial Ecology, Cambridge.

Cramp, S. and Simmons, K.E.L. (eds) (1977) *The Birds of the Western Palaearctic*. Vol 1. Oxford University Press, Oxford.

Cramp, S. (ed.) (1985) *The Birds of the Western Palaearctic*. Vol. IV. Oxford University Press, Oxford.

Cramp, S. (ed.) (1988) *The Birds of the Western Palaearctic*. Vol. V. Oxford University Press, Oxford.

Cullen, P. (1990) Biomonitoring and environmental management. *Env. Mon. Assess.*, **14**, 107–14.

Custer, T.W., Barnett, R.A., Ohlendorf, H.M. and Melancon, M.J. (1991) Herons and egrets as proposed indicators of estuarine contamination in the United States. *Acta XX Congressus Internationalis Ornithologici*, **IV**, 2474–9.

Diamond, A.W. and Filion, F. (eds) (1987) *The Value of Birds*, ICBP Tech. Publ. 6, International Council for Bird Preservation, Cambridge.

Dobinson, H.M. and Richards A.J. (1964) The effects of the severe winter of 1962–63 on birds in Britain. *Brit. Birds.*, **57**, 373–434.

Douthwaite, R.J. (1992) Effects of DDT on the fish eagle *Haliaeetus vocifer* population of Lake Kariba in Zimbabwe. *Ibis*, **134**, 250–8.

Edwards, L.M. (1991) Water quality in rivers in the Peak District using macroinvertebrates, dippers and grey wagtails as indicator species. MSc Thesis, University of Manchester.

Edwards, R.W., Stoner, J.H. and Gee, A.S. (1990) *Acid Waters in Wales*, Kluwer, The Hague.

Erikkson, M.O. (1987) Some effects of freshwater acidification on birds in Sweden, in *The Value of Birds* (eds A.W. Diamond and F. Filion), ICBP Tech. Publ. 6, International Council for Bird Preservation, Cambridge, pp. 183–90.

Erikkson, G., Jensen, S. and Strachan, W. (1989) The pine needle as a monitor of atmospheric pollution. *Nature*, **341**, 42–4.

Flower, R.J. and Battarbee, R.W. (1983) Diatom evidence for recent acidification of two Scottish Lochs. *Nature*, **305**, 130–3.

Focardi, S., Leonzio, C. and Fossi, C. (1988) Variation in polychlorinated biphenyl congener composition in eggs of Mediterranean waterbirds in relation to their position in the food chain. *Environ. Pollut.*, **52**, 243–56.

Fox, G.A. and Weseloh, D.V. (1987) Colonial waterbirds as bio-indicators of environmental contamination in the Great Lakes, in *The Value of Birds* (eds A.W. Diamond and F. Filion), *ICBP Tech. Publ. 6*, International Council for Bird Preservation, pp. 209–16.

Gilbertson, M., Elliott, J.E. and Peakall, D.B. (1987) Seabirds as indicators of marine pollution, in *The Value of Birds* (eds A.W. Diamond and F. Filion) *ICBP Tech. Publ. 6*, International Council for Bird Preservation, Cambridge, pp. 231–48.

Goede, A.A. and de Bruin, M. (1984) The use of bird feather parts as a monitor of metal pollution. *Environ. Pollut. B*, **8**, 281–98.

Hardy, A.R., Stanley, P.I. and Grieg-Smith, P.W. (1987) Birds as indicators of the intensity of use of agricultural pesticides in the UK, in *The Value of Birds* (eds A.W. Diamond and F. Filion) *ICBP Tech. Publ. 6*, International Council for Bird Preservation, Cambridge, pp. 119–32.

Hellawell, J.M. (1986) *Biological Surveillance of Rivers*, Water Research Centre, Medmenham.

Hickey, J.J., Keith, J.A. and Coon, F.B. (1966) An exploration of pesticides in a Lake Michigan ecosystem. *J. Appl. Ecol.*, **3** (Suppl.), 141–54.

Hill, M.O. (1981) *TWINSPAN – a FORTRAN programme for arranging multivariate data in an ordered two-way table by classification of the individuals and attributes*. Department of Ecology and Systematics, Cornell University, New York.

Holdgate, M.W. (1979) *A Perspective of Environmental Pollution*, Cambridge University Press, Cambridge.

Hurlbert, S.H. (1984) Pseudoreplication and the design of ecological experiments. *Ecol. Monog.*, **54**, 187–211.

Hynes, H.B.N. (1960) *Biology of Polluted Waters*, Liverpool University Press, Liverpool.

Institute of Terrestrial Ecology (1982) *Report of the Institute of Terrestrial Ecology for 1981/82*, Institute of Terrestrial Ecology, Cambridge.

Jones, K.C. (1988) Determination of polychlorinated biphenyls in human foodstuffs and tissues: suggestions for a selective congener analytical approach. *Sci. Tot. Environment*, **68**, 141–59.

Kannan, N., Tanabe, S., Tatsukawa, R. and Phillips, J.H. (1989) Persistency of highly toxic coplanar PCBs in aquatic ecosystems: uptake and release kinetics of coplanar PCBs in Green Lipped Mussel *Perda viridis*. *Environ. Pollut.*, **56**, 65–76.

Kolkwitz, R. and Marsson, M. (1909) Okologie der tierischen Saprobien. *Int. Rev. ges. Hydrobiol.*, **2**, 125–52.

Krueger, C.C. and Waters, T.F. (1983) Annual production of macroinvertebrates in three streams of different water quality. *Ecology*, **64**, 840–50.

Landres, P.B., Verner, J. and Thomas, J.W. (1988) Ecological uses of vertebrate indicator species: a critique. *Cons. Biol.*, **2**, 316–28.

Lenmetynen, R., Rantamaki, P. and Karlin, A. (1984) Levels of DDT and PCBs in different stages of the life cycles of the Herring Gull. *Chemosphere*, **11**, 1059–68.

Marchant, J.H. and Hyde, P.A. (1980) Aspects of the distribution of riparian birds on waterways in Britain and Ireland. *Bird Study*, **27**, 183–202.

Marchant, J.H., Hudson, R., Carter, S.P. and Whittington, P. (1990) *Population Trends in British Breeding Birds*, British Trust for Ornithology, Tring.

Mason, C.F. (1991) *Biology of Freshwater Pollution*, 2nd edn, Longman, Harlow.
Mason, C.F. and Macdonald, S.M. (1988) Radioactivity in Otter scats in Britain following the Chernobyl reactor accident. *Water, Air and Soil Pollut.*, **37**, 131–7.
Mason, C.F., Ford, T.C. and Last, N.I. (1986) Organochlorine residues in British Otters. *Bull. Env. Contam. Toxicol.*, **36**, 656–61.
McNicol, D.K., Blancher, P.J. and Bendell, B.E. (1987) Waterfowl as indicators of wetland acidification in Ontario, in *The Value of Birds* (eds A.W. Diamond and F. Filion), *ICBP Tech. Publ. 6*, International Council for Bird Preservation, Cambridge, pp. 149–66.
Moore, N.W. (1966) A pesticide monitoring system with special reference to the selection of indicator species. *J. Appl. Ecol.*, **3** (Suppl.), 261–9.
Mudge, G.H. (1983) The incidence and significance of ingested lead pellet poisoning in British wildfowl. *Biol. Conserv.*, **27**, 333–72.
National Rivers Authority (1991), *Proposals for Statutory Water Quality Objectives*. National Rivers Authority, London.
Newton, I. (1988) Determination of critical pollutant levels in wild populations, with examples from organochlorines in birds of prey. *Environ. Pollut.*, **55**, 29–40.
Newton, I. (1991) Long-term monitoring of organochlorine and mercury residues in some predatory birds in Britain. *Acta XX Congressus Internationalis Ornithologici*, **IV**, 2487–93.
Newton, I. and Bogan, J.A. (1978) The role of different organochlorine compounds in the breeding of British Sparrowhawks. *J. Appl. Ecol.*, **15**, 105–16.
Newton, I., Bogan, J.A. and Haas, M.B. (1989) Organochlorines and mercury in the eggs of British Peregrines *Falco peregrinus*. *Ibis*, **131**, 355–76.
O'Halloran, J., Meyers, A.A. and Duggan, P.F. (1989) Some sublethal effects of lead on mute swans *Cygnus olor*. *J. Zool., Lond.*, **218**, 627–32.
O'Halloran, J., Gribbin, S.D., Tyler, S.J. and Ormerod, S.J. (1990) The ecology of dippers *Cinclus cinclus* in relation to stream acidity in upland Wales: time activity patterns and energy use. *Oecologia*, **85**, 271–80.
Ormerod, S.J. and Edwards, R.W. (1987) The ordination and classification of macroinvertebrate assemblages in the catchment of the River Wye in relation to environmental factors. *Freshwat. Biol.*, **17**, 533–46.
Ormerod, S.J. and Tyler, S.J. (1987) Dippers *Cinclus cinclus* and grey wagtails *Motacilla cinerea* as indicators of stream acidity in upland Wales, in *The Value of Birds* (eds A.W. Diamond and F. Filion), *ICBP Tech. Publ. 6*, International Council for Bird Preservation, Cambridge, pp. 191–208.
Ormerod, S.J. and Tyler, S.J. (1990a) Environmental pollutants in the eggs of Welsh dippers *Cinclus cinclus*: a potential monitor of organochlorine and mercury contamination in upland rivers. *Bird Study*, **37**, 171–6.
Ormerod, S.J. and Tyler, S.J. (1990b), Assessments of body condition in dippers *Cinclus cinclus*: potential pitfalls in the derivation and use of condition indices based on body proportions. *Ring. and Migr.*, **11**, 31–41.
Ormerod, S.J. and Tyler, S.J. (1991a) Exploitation of prey by a river bird, the dipper *Cinclus cinclus* (L.), along acidic and circumneutral streams in upland Wales. *Freshwat. Biol.*, **25**, 105–16.
Ormerod, S.J. and Tyler, S.J. (1991b) The influence of stream acidification and riparian land use on the feeding ecology of grey wagtails *Motacilla cinerea* in Wales. *Ibis*, **133**, 53–61.

Ormerod, S.J. and Tyler, S.J. (1992) Patterns of contamination by organochlorines and mercury in the eggs of two river passerines in Britain and Ireland with reference to individual PCB congeners. *Environ. Pollut.*, **76**, 233–43.

Ormerod, S.J. and Tyler, S.J. (1993a), Further studies on local and regional patterns in the organochlorine content of eggs of a river bird, the dipper *Cinclus cinclus*. *Bird Study* (in press).

Ormerod, S.J. and Tyler, S.J. (1993b) The adaptive significance of brood size and time of breeding in the dipper *Cinclus cinclus* as seen from post-fledging survival. *J. Zool., London.* (in press).

Ormerod, S.J. and Wade, K.R. (1990) The role of acidity in the ecology of Welsh lakes and streams, in *Acid Waters in Wales* (Eds J.H. Stoner, A.S. Gee, and R.W. Edwards), Kluwer, The Hague, pp 93–119.

Ormerod, S.J., Boilstone, M.A. and Tyler, S.J. (1985) Factors influencing the abundance of breeding dippers *Cinclus cinclus* in the catchment of the River Wye, mid-Wales. *Ibis*, **127**, 332–40.

Ormerod, S.J., Allinson, N., Hudson, D. and Tyler, S.J. (1986) The distribution of breeding dippers (*Cinclus cinclus* (L.), Aves) in relation to stream acidity in upland Wales. *Freshwat. Biol.*, **16**, 501–7.

Ormerod, S.J., Wade, K.R. and Gee, A.S. (1987) Macro-floral assemblages in upland Welsh streams in relation to acidity, and their importance to invertebrates. *Freshwat. Biol.*, **18**, 545–57.

Ormerod, S.J., Tyler, S.J. Pester, S.J., and Cross, A.V. (1988a) Censusing distribution and population of birds along upland rivers using measured ringing effort: a preliminary study. *Ring. and Migr.*, **9**, 71–82.

Ormerod, S.J., Bull, K.R., Cummins, C.P. *et al.* (1988b) Egg mass and shell thickness in dippers *Cinclus cinclus* in relation to stream acidity in Wales and Scotland. *Environ. Pollut.*, **55**, 107–21.

Ormerod, S.J., Donald, A.P., and Brown, S.J. (1989) The influence of plantation forestry on the pH and aluminium concentration of upland Welsh streams: a re-examination. *Environ. Pollut.*, **62**, 47–62.

Ormerod, S.J., Weatherley, N.S., Merrett, W.J. *et al.* (1990) Restoring acidified streams in upland Wales: a modelling comparison of the chemical and biological effects of liming and reduced sulphate deposition. *Environ. Pollut.*, **64**, 1–91.

Ormerod, S.J., O'Halloran, J., Gribbin, S.D. and Tyler, S.J. (1991) The ecology of dippers *Cinclus cinclus* in relation to stream acidity in upland Wales: breeding performance, calcium physiology and nestling growth. *J. Appl. Ecol.*, **28**, 419–33.

Ormerod, S.J., Rundle, S.D., Lloyd, E.C. and Douglas, A.A. (in press) The influence of riparian management on the habitat structure and macroinvertebrate communities of upland streams draining plantation forests. *J. Appl. Ecol.*

Pain, D.J. (1991) Lead poisoning in birds: an International Perspective. *Acta XX Congressus Internationalis Ornithologici*, **IV**, 2343–8.

Petersen, R.C., Landner, L. and Blanck, H. (1986) Assessment of the impact of the Chernobyl reactor accident on the biota of Swedish streams and lakes. *Ambio*, **15**, 327–31.

Phillips, O.J.H. (1978) The use of biological indicator organisms to quantitate organochlorine pollutants in aquatic environments: a review. *Environ. Pollut.*, **16**, 167–229.

Ramesh, A., Tanabe, S., Iwata, H. et al. (1989) Seasonal variations of organochlorine insecticide residues in air from Porto Novo, south India. *Environ. Pollut.*, **62**, 213–22.

Ramesh, A., Tanabe, S., Iwata, H. et al. (1990) Seasonal variation of persistent organochlorine insecticide residues in Vellar River waters in Tamil Nadu, south India. *Environ. Pollut.*, **67**, 289–304.

Raven, P.J. (1986) Changes in the breeding bird population of a small clay river following flood alleviation works. *Bird Study*, **33**, 24–35.

Reijnders, P.J.H. (1986) Reproductive failure in common seals feeding on fish from polluted waters. *Nature*, **324**, 456–7.

Reynolds, C.M. (1979) The heronries census: 1972–77 population changes and a review. *Bird Study*, **26**, 7–12.

Royal Society for the Protection of Birds (1987) *Evidence to the Royal Commission on Environmental Pollution: Freshwater Quality*. Royal Society for the Protection of Birds, Sandy.

Rutschke, E. (1987), Waterfowl as bioindicators, in *The Value of Birds* (eds A.W. Diamond and F. Filion), ICBP Tech. Publ. 6, International Council for Bird Preservation, Cambridge, pp. 167–72.

Rutt, G.P., Weatherley, N.S. and Ormerod, S.J. (1990) Relationships between the physicochemistry and macroinvertebrates of British upland streams: the development of modelling and indicator systems for predicting fauna and stream acidity. *Freshwat. Biol.*, **24**, 463–80.

Sears, J. (1989) A review of lead poisoning among the River Thames Mute Swan *Cygnus olor* population. *Wildfowl*, **40**, 151–2.

Spellerberg, I.F. (1991) *Monitoring Ecological Change*. Cambridge University Press, Cambridge.

Tanabe, S. (1988) PCB problems in the future: foresight from current knowledge. *Environ. Pollut.*, **50**, 5–28.

Tyler, S.J. (1979) Mortality and movements of grey wagtails. *Ring and Migr.*, **2**, 122–31.

Tyler, S.J. and Ormerod, S.J. (1991) The influence of acidification and riparian land use on the breeding performance of grey wagtails *Motacilla cinerea* in Wales. *Ibis*, **133**, 286–92.

Tyler, S.J., Ormerod, S.J. and Lewis, J.M.S. (1990) Breeding and natal dispersal amongst Welsh dippers *Cinclus cinclus*. *Bird Study*, **37**, 18–23.

Vickery, J.A. (1988) The effects of surface water acidification on riparian birds – with particular reference to the dipper. DPhil thesis, Oxford University.

Vickery, J.A. (1991) Breeding densities of dippers *Cinclus cinclus*, grey wagtails *Motacilla cinerea* and common sandpipers *Actitis hypoleucus* in relation to the acidity of streams in S.W. Scotland. *Ibis*, **133**, 178–85.

Vickery, J.A. and Ormerod, S.J. (1991) Dippers as indicators of stream acidity. *Acta XX Congressus Internationalis Ornithologici*, **IV**, 2494–502.

Weatherley, N.S. and Ormerod, S.J. (1991) The importance of acid episodes in determining faunal distributions in Welsh streams. *Freshwat. Biol.*, **25**, 71–84.

Wright, J.F., Moss, D., Armitage, P.D., and Furse, M.T. (1984) A preliminary classification of running water sites in Great Britain based on macroinvertebrates species and the prediction of community type using environmental data. *Freshwat. Biol.*, **14**, 221–56.

6
Birds as indicators of change in marine prey stocks

W.A. Montevecchi

6.1 INTRODUCTION

For millenia humans have followed birds at sea to locate fish and mammals. Seabirds are highly visible wide-ranging upper trophic level consumers that can indicate marine productivity and biotic interaction. Compared with fish, marine mammals, and other animals that live primarily or exclusively underwater, seabirds are easy to survey, census and study. This chapter reviews the types, utilities and limitations of avian indicators of the condition of fish stocks and recommends research needed to improve understanding of avian trophic relationships.

Morrison (1986) and Temple and Wiens (1989) have argued that avian population fluctuations are not useful monitors of environmental change. The turnover rates of vertebrate populations are usually too low to closely track environmental change. This is especially true for long-lived species that exhibit delayed maturity and low reproductive rates, such as seabirds (Lack, 1967). Behaviour and life-history responses buffer vertebrates from effects of environmental changes and often introduce time lags in population responses to such change (Wiens, 1989). Yet while population fluctuations provide little indication of environmental change, behavioural and reproductive buffers are highly responsive to environmental contingencies and can often generate useful environmental information including indication of the condition of fish stocks. Thus monitoring of seabird breeding numbers

Birds as Monitors of Environmental Change. Edited by R.W. Furness and J.J.D. Greenwood. Published in 1993 by Chapman & Hall, London. ISBN 0 412 40230 0.

at colonies may not provide a good picture of changes in prey abundance, but measures such as the territory attendance of off-duty breeding adults might show a stronger and more immediate response to changes in prey abundance.

Following the productive leads of early terrestrial zoogeographers (e.g. Merriam, 1898), marine biologists used planktonic species as indicators of current systems and water masses (e.g. Russell, 1935; Fraser, 1952; Hida, 1957; Aron, 1962; Cushing, 1982; Richard, 1987). Behavioural and ecological analyses of marine birds allow researchers to go beyond rudimentary associations of species and hydrography. Fluctuations in prey distribution and abundance evoke different behavioural responses by birds that can produce changes in tissue composition, physiology, diet and reproductive success. During extreme conditions, effects at the population level may be evident.

Oceanographic factors, such as water temperature, salinity, density, currents, mixing and large-scale climatic and hydrographic processes that affect these variables, generate variation in the production, distribution and abundance of the organisms on which birds feed (e.g. Leggett et al., 1984; Rothschild and Osborn, 1990) and hence of birds at sea (e.g. Brown, 1980; Blomqvist and Peterz, 1984; Kampp, 1988; Haney, 1991; Hunt, 1991; Piatt, 1987; Schneider, 1991). Seabirds prey primarily on small pelagic fish, crustaceans and squid and on the youngest age-classes of demersal fish. These animals are patchily distributed, highly mobile and difficult to survey (Clarke, 1977). They also exhibit large year to year fluctuations in recruitment (e.g. Shelton et al., 1985), and fisheries based on them are often prone to collapse (MacCall, 1979; McEvoy, 1986). Recruitment indices, which are useful for forecasting fishery conditions, can be enhanced with data derived from seabirds. Birds also prey on fish and invertebrates that are not commercially exploited and hence receive little research attention. These latter sorts of data are highly relevant to management considerations, because commercial fisheries are continually targeting new prey and thus changing their impact on marine food webs.

Fishery assessments rely on catch and catch per unit effort (CPUE) data obtained from commercial vessels. These data have been confounded with technical improvements in the finding and catching of fish, with market conditions, catch quotas, overharvesting and misreporting (e.g. Shannon et al., 1984). Research catches and acoustic surveys supplement this information, but pelagic fish and young demersal fish often occur in shallow inshore regions that are not surveyed (e.g. Hewitt and Brewer, 1983; Barrett et al., 1990) and in the upper water column that is hydroacoustically invisible (Hampton et al., 1979; Donmasnes and Rottigen, 1985). Measurements of prey availability derived from seabirds provide natural indices that

Figure 6.1 Different feeding methods and foraging sites used by marine birds. Note that almost all of the avian predators feed in groups and that all of the pelagic prey occur in schools. Drawn by D. Nelson.

complement fisheries data, assay prey in regions inaccessible to traditional surveys, and can be incorporated in stock-assessment models.

6.2 AVIAN ROLES IN MARINE FOOD WEBS

Densely-schooling small pelagic fish, moderately sized pelagic crustaceans and squid in the upper- and mid-water column are the main food of seabirds (e.g. Furness, 1978; Anderson and Gress, 1984; Croxall *et al.*, 1984a; Piatt and Nettleship, 1985; Figure 6.1). At high northern latitudes, Arctic cod *Boreogadus saida*, sandeels *Ammodytes* spp., capelin *Mallotus villosus* and herring *Clupea harengus* dominate the pelagic fish exploited by seabirds, whereas sardines and anchovies are primary seabird prey in temperate boundary currents (Rice, 1992). These fish are often over-fished and experience recruitment failures driven by oceanographic events (see also Sherman *et al.*, 1981; Rice, 1992). Many seabird species in coastal areas are benthic foragers and often feed on young demersal fish and on shellfish (e.g. Birt *et al.*, 1987; Goudie and Ankney, 1988; Barrett *et al.*, 1990), many of which (e.g. cod *Gadus morhua*; crabs) produce planktonic eggs or pelagic larvae that are also eaten by birds (Straty and Haight, 1979; Hunt and Butler, 1980; Briggs *et al.*, 1984).

6.2.1 Prey bases

Seabirds consume substantial quantities of prey (Table 6.1; see Swatzman and Haar (1983) on marine mammals). There may be considerable overlap between human and avian harvests, this being more common for small prey, e.g. anchovies *Engraulis mordax* preyed on by brown pelicans *Pelecanus occidentalis* (Anderson and Gress, 1984) and krill *Euphausia superba* preyed on by penguins (Croxall *et al.*, 1984), though there is often overlap for larger prey as well, e.g. mackerel *Scomber scombrus* preyed on by northern gannets *Sula bassana* (Montevecchi *et al.*, 1987a). Many seabirds depend on forage species, such as capelin or sandeels, which are major foods of other fish and are often harvested by humans (Monaghan *et al.*, 1989; Furness, 1990). These species are centrally involved in complex interactions in marine food webs (Harwood, 1983; Harwood and Croxall, 1988; Croxall, 1989). As a prey stock declines, specialist avian and mammalian predators may consume proportionally more of it, while generalists may consume equivalent or reduced amounts by switching to alternative prey (Bailey *et al.*, 1991). Most consumption during breeding seasons occurs at environmental hot spots around colonies and may affect local fisheries (Furness, 1990; Bailey *et al.*, 1991).

Table 6.1 Major energy models of fish harvests by seabird communities

Location	Major consumers	Estimated % pelagic production consumed	Source
Oregon Coast	Shearwaters Storm petrel Cormorant Guillemot	22	Wiens and Scott, 1975
Foula	Fulmar Guillemot Shag Puffin	11–64	Furness, 1978
North Sea	Fulmar Gulls Terns Guillemot Puffin	5–8	Bailey, 1986; Bailey et al., 1991
North Sea	Fulmar Shearwater Gannet Shag Gulls Kittiwake Terns Razorbill Guillemot Puffin	5–10	Tasker et al., 1989
Benguela region	Penguin Gannet Cormorant	29	Furness and Cooper, 1982

Until recently, bioenergetics models of consumption by predators relied on extrapolations from laboratory measurements of basal metabolic rate (BMR) or of artificial activity-specific metabolic rates (e.g. Tucker, 1974) of a limited range of species which were extrapolated to other species and free ranging animals for which activity time budgets were guessed. Usually, activities with different energy demands, such as flight, swimming, etc. were assigned multiples of BMR to estimate energy expenditure (requirement). Such crude approaches inevitably produced inaccuracies. Doubly labelled water techniques (DLW; Nagy, 1980, 1983) have generated direct estimates of energy turnovers of free ranging animals (e.g. Weathers et al., 1984).

These estimates provided the basis on which new allometric relationships of field metabolic rates (FMR) and average daily energy expenditures (ADEE) of species with different foraging modes and from different oceanographic regions have been extrapolated (Birt-Friesen et al., 1989). Estimates of energy utilization by nonbreeding birds are rare (Gaston, 1985) and measurements outside breeding seasons are, because of logistic difficulties, virtually nonexistent (Bailey, 1986; Cairns et al., 1990a; Harwood and Croxall, 1988; section 6.4.1).

With comparative data bases inadequate, it is usually assumed that seabird diets and FMRs and ADEEs are constant within species. However, diets (Ainley et al., 1984; Schneider and Hunt, 1984; Hatch and Sanger, 1992), foraging ranges (e.g. Gaston and Nettleship, 1981) and FMRs (Gales and Green, 1990; Montevecchi et al., 1992) vary between colonies, seasons and years, and such variation requires incorporation in large-scale ecological energetics models.

The biological significance of prey consumption by marine birds depends on their foraging ranges and on the movements, production and biomass of prey (Rice, 1992). Marine production figures are given on the basis of area (e.g. Steele, 1974), and estimates of the percentage of production consumed by birds are extrapolated from aerial coverages. The sea surface area over which a colony or community of seabirds forages is a function of the square of maximum foraging radius, so small errors in estimates of foraging range produce much larger errors in estimates of foraging area and thus of the percentage of fish production consumed. Telemetric measurements of foraging ranges (Jouventin and Weimerskirch, 1990; Wanless et al., 1991a) need to be integrated into community energetics models. Even within foraging ranges, seabirds may use only a very small proportion (e.g. about 10%; Wanless et al., 1991a) of their potential foraging area around a colony. These 'hot spots' (Cairns and Schneider, 1990) may be more relevant for benthic foragers and pursuit divers that exploit prey concentrations induced by bathymetric flow gradients. By contrast many surface foragers and pursuit divers in the Antarctic exploit less predictable prey aggregations in deeper water (Schneider et al., 1986; see also Brown et al., 1979; Schneider et al., 1990; section 6.3.1).

Models of consumption by marine homeotherms have by necessity used uniform and static extrapolations of standing prey stocks. Prey movements (e.g. Crawford, 1981; Montevecchi et al., 1987a) and physical influences, such as advection and diffusion, can, however, result in prey passing through foraging areas at rates that exceed their consumption by birds (Cairns and Schneider, 1990; Schneider et al., 1992; see also Springer et al., 1987). Therefore food supply around colonies in highly dynamic oceanographic regimes may depend more on prey movement than on production (Cairns and Schneider, 1990; see also Hunt, 1991). Dynamic prey distributions modify avian foraging ranges (Cairns *et al.*, 1987, 1990b; Piatt, 1987), and thus the energy requirements of breeding seabird populations during a reproductive season. During studies of seabird feeding ecology, prey movements necessitate concurrent measurements of food supply within colony foraging ranges (e.g. Burger and Piatt, 1990). Usually, prey abundance is extrapolated from stock assessments made over a larger spatial scale than the foraging ranges of birds under study and made before or after the ornithological investigations were conducted. Careful studies are needed to investigate the validity of such extrapolations and of the movements of nonbreeding and migrant seabirds in and out of prescribed foraging areas (Bailey, 1986; Cairns *et al.*, 1990a) in order to appreciate the spatial and temporal patterns of energy flux through seabird communities.

6.2.2 Assessment of fish stock abundance

Most models of fish stock abundance use retrospective Sequential Population Analysis (SPA) that requires three parameter estimates:

1. natural mortality
2. numbers caught at age in commercial fisheries
3. an integrated abundance index.

Natural mortality rates are basically unknown and usually assumed to be about 20% per annum (Cairns, 1992a). Abundance indices, used to assess rates of population change, are derived from three methods, namely commercial catch per unit effort (CPUE); research CPUE; and acoustic estimates. Research vessels control gear type, trawl depth and fishing effort and randomize locations within the ranges of fish stocks, though these coverages are much more limited than those of commercial vessels. Research and commercial CPUEs may be used in a combined index. Additionally, acoustic estimates of stock biomass are sometimes incorporated in SPA. Indices from different sources are combined with multiple regression techniques to produce integrated abundance indices. Age-specific population calculations are then interactively regressed against abundance indices to 'calibrate' stock estimates. Natural-mortality estimates (usually subjective) are factored into the model until the strength is maximized for the regression between the indices of abundance and population at age; maximum regression strength is used to estimate stock size (Cairns, 1992a; see also Harris, 1990). This estimate is used with the previous long-term mean to project the following year's stock size, on which total allowable catches are based.

Problems associated with data inputs and model structure limit the accuracy of fish population estimates. Assessments developed for single species of demersal fish generalize poorly to pelagic species (Rice, 1992). The models are also quite sensitive to errors in natural-mortality estimates (Hilden, 1988; see also Furness, 1990). The catch rates of densely-schooling pelagic fish are often high until stocks are almost depleted and hence are not well correlated with stock abundance (Rice, 1992). Traditional under-reporting of fish catches is also problematic (Anderson et al., 1980; Harris, 1990), and all these circumstances combine to produce large quantities of low-quality data (Cairns, 1992a) which are fed into population models. Fishery closures are currently being used to counter overfishing pressures and are likely to increase in future; in such circumstances information on prey availability derived from seabirds will be more useful. On the positive side, uncertainty associated with the output of SPA parameters can be quantified to permit more realistic presentations for resource managers (Pope and Gray, 1983; Hoenig et al., 1991). More interestingly for marine ecologists, multi-species models (Andersen and Ursin, 1977; Harwood, 1983; Sparre, 1991) can be used to probe effects of population fluctuations on food webs (Rice, 1992).

Ornithological inputs to SPA models are independent of changing technologies that bias conventional assessment measurements (Cairns, 1992a; section 6.7.1). Seabirds exploit many pelagic and demersal fish that are difficult to survey, that often occupy coastal habitats which research and commercial vessels do not cover, and that are excluded from commercial catches (e.g. Bergstad et al., 1987; Barrett et al., 1990; Montevecchi and Berruti, 1991; section 6.7.1.a). Birds are particularly useful samplers of pelagic fish and other prey that avoid ships and that often occur in the upper 5–10 m of the water column (Montevecchi and Berruti, 1991). Estimates of natural mortality which is in many cases substantial (Beddington et al., 1985; Furness, 1990) and in some cases limiting (Barrett et al., 1990; Hatch and Sanger 1992; section 6.7.3) can be enchanced with dietary and energetics data derived from marine birds and mammals (Cairns, 1992a). Present limitations to avian assays of fish larvae and eggs are the rapid digestion of these and the absence or small size of identifiable 'hard parts' (Straty and Haight, 1979).

(a) Single- and multi-species perspectives

Single- and multi-species analyses of predator–prey interactions are based on different assumptions and yield different results. Single species, 'surplus production' models (Ricker, 1975) assume that unharvested fish populations increase to reach a stable carrying capacity. Within such scenarios, predator removal is considered to 'free' some portion of prey consumption for fisheries exploitation, an argument that leads to culls of predators thought to compete with human harvesters. But many factors invalidate this simple view, from which other aspects of intra- and inter-specific interactions are excluded (Harwood, 1983), and 'surplus production' is rarely realized in multi-species associations (Steele, 1979; Kerr and Ryder, 1989). Comprehensive analyses of marine food webs benefit from inclusion of many avian species with different trophic relationships and foraging niches (section 6.7.5).

6.2.3 Prey abundance and distribution (availability)

Prey abundance (population numbers or biomass) and availability, as determined by distribution and density, are very different factors for foraging predators. While availability is a function of abundance, the two are independent to the extent that hydrography and behaviour affect the horizontal and vertical distributions and concentrations of prey relative to the foraging capabilities of predators. Variability in distribution and density or patchiness of prey influences foraging behaviour, effort and success. Huge swings in abundance are common among pelagic fish, squid and crustaceans, and as associations between abundance and availability are unlikely to be linear, predator indices may continue to be most useful when prey fluctuations are extreme (Croxall, 1989). Effects of changing prey availability on seabird predatory success complicate efforts to use avian data to assess changes in fish stock abundance (Bailey *et al.*, 1991), yet availability is the most fundamental determinant of the success of both human and avian fisheries (Montevecchi and Berruti, 1991).

6.3 MONITORING SEABIRDS

6.3.1 Scales of interaction and generalization

Predator–prey interactions in marine ecosystems are atuned to, and constrained by, spatial and temporal oceanographic discontinuities (Wiens, 1976, 1989; Hunt and Schneider, 1987; Rose and Leggett, 1990). Different behavioural, physiological and ecological patterns of seabirds integrate influences of food conditions over different spatial and temporal scales. Prey landings and foraging trip durations usually reflect food conditions over fine- to macro-scale areas during periods of hours or days and could provide real time assessments of prey availability. Egg and clutch sizes integrate feeding conditions over days, and growth and breeding success develop over weeks and months.

Population trends integrate activites over large areas that include migratory ranges, and over annual and decadal time scales (see also Cairns, 1987; Croxall et al., 1988a; Croxall, 1989).

Species with different foraging modes respond differently to oceanographic variation. Surface feeders and small shallow-diving feeders interact with horizontal and near-surface gradients, whereas large-bodied pursuit divers like the schooling pelagic fish that they pursue are less constrained by surface variation (Haney, 1991). Most northern-hemisphere seabirds forage over continental shelves where water movements often concentrate prey in predictable patches (Hunt, 1991), with pursuit divers dominating the energy flux to seabirds in shallower regions, and surface feeders dominating in offshore waters (e.g. Schneider et al., 1986; Haney, 1991). In contrast, Antarctic and tropical seabirds usually forage in deeper water where prey patchiness may be primarily controlled by behaviour rather than by physical processes and may be unpredictable (Hunt, 1991).

Predator–prey interactions during seabird breeding seasons are concentrated in colonies and include two-dimensional foraging areas for surface feeders and three-dimensional foraging volumes for pursuit divers (Gaston and Nettleship, 1981; Table 6.2). These categorizations are not as rigid as tabulation might imply, because foraging ranges can vary among individuals and regions and with season, prey movements and oceanographic events (e.g. Anker-Nilssen and Lorentsen, 1990) and because birds often exploit environmental hotspots (section 6.2.1).

Failure to match measurement scales used for predators with those used for prey can lead to erroneous conclusions (Rose and Leggett, 1990; Taggart and Frank, 1990). Synoptic studies of prey availability within colony foraging ranges using hydroacoustics (Safina and Burger, 1988; Burger and Piatt, 1990) and SCUBA transects of benthic prey (Birt et al., 1987) are useful, as are comparisons of fisheries and avian harvests within the same meso-scale regions (Montevecchi et al., 1987a; Montevecchi and Berruti, 1991). Montevecchi and Myers (1992) have recently shown that avian harvests of large pelagic fish and of squid correlate with fisheries landings from the vicinity of the colonies, from meso-scale regions adjacent to the colonies, and from the macro-scale regions in which the colonies are located.

6.3.2 Measurement variables: dependent and independent

Which species and which aspects of their behaviour and biology can be used as indicators usually becomes apparent during basic research. Aspects of seabird biology that correlate with prey ecology and oceanography can be used as 'independent variables' to generate

Table 6.2 General foraging domains and trophic levels of some colonial seabird species

Water-column depth (m)	Foraging Range (km)				
	Coastal (<10)	Inshore (10–25)	Offshore (25–50)	Pelagic (50–500)	Oceanic (>500)
Surface	Common tern (P)[a] Larus gulls (M)	Arctic tern (P) Larus gulls (P)	Gannets (P) Black-legged kittiwake (P)	Red-legged kittiwake (M) Storm petrel (M) Fulmar (M)	White-chinned petrel (M) Albatrosses (M)
5–50	Black guillemot (P) Shag (P) Crown cormorant (P) Gentoo penguin (P)	Cormorant (P) Rockhopper penguin (Pl)	Atlantic puffin (P) Razorbill (P) Macaroni penguin (Pl)	Dovekie (Pl)	
50–100	Cape cormorant (P) Bank cormorant (P) Blue-eyed shag (P)		Tufted puffin (P) Horned puffin (P) Guillemots (P)		
>100				Emperor penguin (P) King penguin (P)	

[a] P – piscivorous; Pl – planktivorous; M – mixed.
Sources: Adams and Brown, 1989; Croxall et al., 1988a, 1991; Pearson, 1968; Furness, 1978; Furness and Barrett, 1985; Piatt and Nettleship, 1985; Barrett and Furness, 1990; Prince and Harris, 1988; Wanless et al., 1991a, b; Byrd et al., 1992

testable predictions about those phenomena. The derivation of causal inferences and functional relationships requires the validation of avian indices with spatially and temporally concordant but independent measurements of prey availability taken across the full range of abundance (Hunt *et al.*, 1991; Cairns, 1992a).

(a) Functional responses, diet breadth and prey switching

Fluctuations in availability and abundance determine prey profitability (energy gain/unit effort in the case of predators; and, taking account of market conditions, dollars/unit effort in the case of the fishing industry) and hence the spectrum of avian and human prey selection. Seabird species can be generally and seasonally classified as dietary specialists or generalists. The functional responses of predators to changes in prey abundance (availability), and indices of prey abundance based on foraging effort (CPUEs), may be most easily obtained and most realistic for specialists (Cairns, 1992a; Monaghan *et al.*, 1989). Specialists are highly influenced by fluctuations in prey availability and are very vulnerable to stock depletion caused by over-fishing (e.g. Barrett *et al.*, 1987; Vader *et al.*, 1990a; Bailey *et al.*, 1991; Monaghan *et al.*, 1991). The diets of generalist and opportunistic seabirds reflect the seasonal availabilities of pelagic prey (e..g Crawford, 1981; Adams and Klages, 1989) and may be particularly useful in monitoring prey stocks when they are at low levels (Montevecchi and Berruti, 1991). These predators, however, may provide less useful indication of the range of prey abundance (Martin, 1989, Furness, 1990), because it is difficult to determine the conditions under which they switch among different prey (Montevecchi and Berruti, 1991). Comparisons of closely related sympatric species that show differences in dietary breadth (Elliot *et al.*, 1990; Hatch and Sanger, 1992) will aid in understanding the effects of changes in prey abundance on avian harvests and in some instances on reproductive success (Uttley *et al.*, 1989).

(b) Measurements at colonies

Parental attendance patterns, food provisioning rates, diet, dietary changes, breeding success, population fluctuations in study plots, and other reproductive parameters monitored at colonies can be compared with variation in feeding and oceanographic conditions (e.g. Hamer *et al.*, 1991; Klomp and Furness, 1992).

(c) Measurements at sea

We know much more about the behaviour of seabirds on land than at sea, where they live most of their lives. Studies of birds at sea have generally focused on the correspondences of bird distributions with oceanographic events and more recently with prey abundance (estimated acoustically). Fine scale associations of birds and their prey tend to be weak (Hunt, 1991), though stronger relationships are observed when predators are actually feeding (e.g. Rose and Leggett, 1990), and large prey aggregations are disproportionally important to predators (Heinemann *et al.*, 1990; Hunt *et al.*, 1990; Hunt, 1991). Stronger associations between predators and prey are found at larger scales of measurement (Schneider and Duffy, 1985; Schenider and Piatt, 1986; Heinemann *et al.*, 1990; Hunt *et al.*, 1990), as measuring over a larger scale reduces variation associated with prey patchiness, so such trends might be expected. Common guillemots *Uria aalge* show threshold-like aggregative responses to prey density (Piatt, 1990), indicating that guillemots act to maximize ingestion rates (Hassell and May, 1974). Alternatively, predators that pursue mobile marine prey might act to minimize variation in ingestion rates. The shapes of functional response curves are thus central to considerations of seabird indices of prey abundance or availability (section 6.7.1).

Telemeters and activity recorders enable studies to advance beyond correlative associations of predators, prey and oceanographic conditions to investigations of the behaviour of individual animals at sea (e.g. Wanless *et al.*, 1991a, b). Foraging behaviour, time and success (i.e. CPUE) can be quantified. Activity recorder studies can be combined with DLW techniques to derive activity-specific metabolic rates (Birt-Friesen *et al.*, 1989) that can be used to convert activity budgets to energy estimates.

Winter activity budgets are unknown but could be measured with activity recorders attached to birds departing from colonies in the fall, activity data being remotely sensed or read upon return to land the following spring. Studies are easier to propose than to execute, but could help determine the activity (and hence energy) budgets of migrants during non-breeding periods (Cairns *et al.*, 1990a). To date these data have not been quantifiable.

6.4 POPULATION RESPONSE TO FOOD SUPPLY

6.4.1 Food supply and population regulation

The regulation of seabird populations by density-dependent and density-independent factors has been debated extensively. Wynne-Edwards (1962) postulated that individual reproductive restraint created self-limiting populations and prevented depletion of food supplies. The group selection concept inherent in this hypothesis has been logically refuted (e.g. Lack, 1966). Ashmole (1963), focusing on tropical seabirds, contended that the greatest energy demands on adults occurred during breeding and that colony size in the absence of nesting and predatory constraints would be limited by food. In contrast, Lack (1966) focused on temperate seabirds and proposed that greatest energy demands occurred during thermal stress in winter, when, he thought, seabird mortality would be greatest. Of course, the seasons are not independent, as winter food supply may influence spring reproductive condition, and body condition after the breeding season may affect winter survival (Diamond, 1978; Birkhead and Furness, 1985; Furness and Monaghan, 1987).

Some evidence supports the extension of Ashmole's (1963) prey depletion hypothesis to temperate seabirds. Independent studies with diverse species have shown that breeding success, fledging mass and peak chick mass are all inversely related to colony size (Gaston et al., 1983; Birkhead and Furness, 1985; Hunt et al., 1986), which could result from prey depletion or from interference competition. Furness and Birkhead (1984) found that adjacent colonies of some species show a tendency to be inversely related in size, suggestive of density-dependent colony distributions (see also Cairns, 1989, 1992). Direct evidence for prey depletion comes from a demonstration that benthic fish abundances were lower near colonies than they were outside the foraging ranges of double-crested cormorants *Phalacrocorax auritus* (Birt et al., 1987). The prey depletion hypothesis predicts decreasing prey abundance within colony foraging ranges during the breeding season, a prediction so far untested. Density-independent factors associated with oceanographic and meteorological events often have major, widespread impacts on seabird populations (Wooler et al., 1992; section 6.4.2) and interact with density-dependent factors in the regulation of many seabird populations (Birkhead and Furness, 1985; Cairns, 1992b).

6.4.2 Effects of oceanographic perturbations

Oceanic conditions constrain the physiology and behaviour of marine animals. At low temperatures planktonic eggs take longer to hatch and the pelagic larvae of shellfish and demersal fish grow more slowly, prolonging their vulnerability to predators (Straty and Haight, 1979; Rice, 1992). Cold water also slows the burst swimming speeds of fish, which might make them more vulnerable to predation by pursuit-diving homeotherms. The concentrations of diving seabirds and marine mammals at high latitudes may reflect such large-scale oceanographic influences (Straty and Haight, 1979; Cairns et al., 1990a; section 6.7.5).

Year to year fluctuations in seabird reproductive success have been correlated with sea temperatures at both low (Boersma, 1978) and high latitudes (Murphy et al., 1986). Tropical blue-footed boobies *Sula nebouxii* are sensitive to water temperature regimes and, like black-legged kittiwakes *Rissa tridactyla* at high northern latitudes, exhibit large variation in reproductive success (Anderson, 1989; Hatch et al., in press). The most spectacular seabird population crashes are linked to El Nino–Southern Oscillation (ENSO) warm water events which make prey unavailable to surface-foraging avian predators (e.g. Schreiber and Schreiber, 1984, 1989; Boekelheide and Ainley, 1989). Widespread effects of ENSOs on seabirds can be found throughout the Pacific (e.g. Schreiber and Schreiber, 1984; Ainley et al., 1988; cf. Hatch, 1987; Murphy et al., 1991) and possibly in the Atlantic (LaCock, 1986; Schneider and Duffy, 1988).

Seabird reproductive failures at high latitudes have often been associated with over-fishing (section 6.4.3), though large-scale failures often occur outside the range of commerical fisheries (Harris and Wanless, 1990; Trivelpiece et al., 1990; Murphy et al., 1991) and in association with cold-water events and the late break up of ice in spring (e.g. Ainley and LeResche, 1973; Barrett and Schei, 1977; Barrett and Runde, 1980; Gaston and Nettleship, 1981; Heubeck,1989; Baird, 1990; Harris and Wanless, 1990; Hatch et al., in press). During cold-water events small pelagic fish often remain beyond the ranges of foraging birds, especially surface and near-surface feeders (Springer et al., 1984, 1987; Harris and Wanless, 1990). Comparisons of reproductive performance by surface and pursuit-diving species that feed on the same prey in the same region can reflect the horizontal and vertical distributions and availability of prey (e.g. Duffy et al., 1984). Oceanographic factors that influence surface feeders have less effect on pursuit divers, especially large ones (Harris and Wanless, 1990; Hatch et al., in press). Seabird failures at high latitudes appear more frequent than the spectacular ENSO crashes that have been better and longer documented. Reproductive failures among surface and near-surface feeders at high latitudes would be more comparable to ENSO events, if pursuit divers (which are rare in the tropics) were left out of consideration (section 6.7.5). Fluctuations in seabird production at high latitudes could prove useful in tracking climate and environmental change (e.g. Schneider and Duffy, 1988).

6.4.3 Effects of overfishing on seabirds

Commercial fishing can drastically affect avian food webs (e.g. Burger and Cooper, 1984; Furness and Ainley, 1984; Ryan and Moloney, 1988; Rice, 1992) and seabird populations (e.g. Schaefer, 1970). Fishing

Table 6.3 Correspondences between overfishing and the failure of seabird breeding or population decline

Fish	Bird	Location	Years	Sources
Herring	Atlantic puffin	Norway	1964–89	Barrett et al., 1987; Vader et al., 1990a.
Capelin	Guillemot	Barents Sea	1985–87	Vader et al., 1990a,b.
Sandeel Herring	Shag Great skua Black-legged kittiwake Arctic tern Common tern Guillemot	Shetland (North Sea)	1986–90	Furness, 1990; Uttley et al., 1989; Hamer et al., 1991; Bailey et al., 1991; Klomp and Furness, 1992; Monaghan et al., 1989.
Capelin	Atlantic puffin	N.W. Atlantic	1981	Brown and Nettleship, 1984.
Anchovy	Brown pelican	S. California Bight	1969–80	Anderson et al., 1982; Anderson and Gress, 1984.
Anchoveta	Peruvian brown pelican Guanay cormorant Peruvian booby	Humbolt Current	1950s–70s	Duffy, 1983.
Pilchard	Jackass penguin Cape gannet	Benguela	1956–80	Crawford et al., 1985; Burger and Cooper, 1984.

Population response to food supply

increases rates of change in prey stocks, which may influence predator populations, but the directions and magnitude of fisheries induced changes vary considerably. Fishing mortality can amplify natural fluctuations and drive prey stocks to very low levels (e.g. Rice, 1992).

Except with extreme cases, the effects of over-fishing on populations are difficult to document (Morrison, 1986). From 1956 to 1980, the estimated breeding population of cape gannets *Sula capensis* in South Africa decreased by about one half, a decline that has been attributed to the collapse of the regional stock of pilchard *Sardinops ocellata* and reduced availability of anchovies *Engraulis capensis* (Crawford *et al.*, 1980, 1983). Other examples of the effects of over-fishing on seabird populations include: anchoveta and Peruvian guano birds (Schaefer, 1970), anchovies and pelicans in the Southern California Bight (Anderson and Gress, 1984), sandeels and seabirds in the Shetlands (Heubeck, 1989; Monaghan *et al.*, 1989), and capelin and alcids in Norway (Vader *et al.*, 1990a, b; Table 6.3). The extended breeding failures of Atlantic puffins *Fratercula arctica* on Røst, Norway, have been linked to the over-fishing of North Sea herring (Barrett

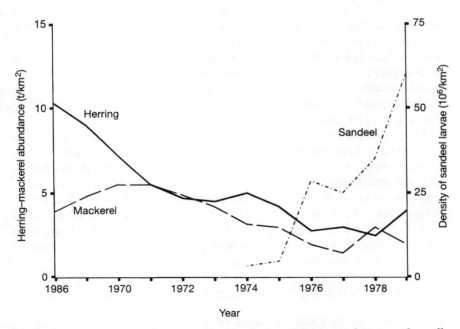

Figure 6.2 Relationships between the biomasses of large predatory and small pelagic fish. (Based on Sherman *et al.*, 1981.)

et al., 1987). Guano yields indicate trends in avian populations that are related to changes in fish abundance (Crawford and Shelton, 1978, 1981), and often provide pre-fishery indices of substantial fluctuations in avian and fish populations (e.g. Soutar and Isaacs, 1974; Crawford and Shelton, 1981; see also Montevecchi and Hufthammer, 1990).

In view of the many well known examples of fisheries adversely affecting seabirds, it may seem ironic that most marine fisheries generally enhance avian and mammalian food supplies (e.g. Hatch and Sanger, 1992). Catches of larger species and fish in older age classes eliminate competing predators and cannibals and thereby increase the abundance of small fish (e.g. Sherman et al., 1981; Swatzman and Haar, 1983; Alverson 1991; Figure 6.2). Furthermore, the commercial trawling of demersal fish has produced massive new artificial food sources for scavenging seabirds in offshore regions (e.g. Burger and Cooper, 1984; Hudson and Furness, 1988, 1989; Furness et al., 1992).

6.4.4 Population parameters as indicators of prey fluctuations

Populations are usually resistant to environmental change and so are of little use in monitoring prey conditions (Morrison, 1986; Temple and Wiens, 1989). Furthermore, seabird population counts often carry a 10–20% error with conspicuous large open-netting species and are even less accurate with inconspicuous species and crevice and burrow-nesting ones (e.g. Montevecchi et al., 1987b; Furness, 1990). As a result only relatively large population changes can be detected. Even when fluctuations in colony size are documented, it is often impossible to identify whether they are caused by changes in recruitment, in adult mortality, in dispersal (emigration, immigration) or by non-breeding (Klomp and Furness, 1992; see also Ricklefs, 1973; Temple and Wiens, 1989).

Despite these problems, the numbers of birds at colonies are in some circumstances associated with prey and oceanographic conditions (e.g. Murphy et al., 1991). These relationships are most evident for species that lay multi-egg clutches and whose populations are quite variable, e.g. Arctic terns *Sterna paradisaea* (Monaghan and Zonfrillo, 1986), elegant terns *S. elegans* (Schaffner, 1986) and pelicans (Anderson and Gress, 1984). The counting of birds at colonies has the important advantage of being unobtrusive and has provided early warning of the decline or collapse of fish stocks.

6.5 PHYSIOLOGICAL RESPONSE TO NUTRITION

6.5.1 Adult body condition and mass

Reproductive costs are often evident in mass fluctuations, and female mass has been correlated with laying date and clutch size (Drent and Daan, 1980). Adults often lose mass after egg-laying and hatching (Coulson *et al.*, 1983; Monaghan *et al.*, 1989; Montevecchi *et al.*, 1992). Such loss has been considered adaptive in facilitating greater flight efficiency (e.g. Monaghan and Metcalfe, 1986), especially after hatching when parental foraging trips may increase by two- or three-fold (Croll *et al.*, 1991). However, adults lose more mass in poor food years (Croll *et al.*, 1991), so mass loss also reflects food conditions. In very poor years, adults do not return to colonies (e.g. Murphy *et al.*, 1991), presumably due to poor body conditions, and, in extreme circumstances, adults die (Crawford *et al.*, 1980; Cushing, 1982; Schreiber and Schreiber, 1984; Vader *et al.*, 1990a, b).

6.5.2 Egg and clutch size

Eggs are, in general, energetically inexpensive (Montevecchi and Porter, 1980), though energy acquisition, egg production and transport could produce energetic bottlenecks and increase parental risks. Variation in food supply may (Drent and Daan, 1980; Pehrsson, 1991) or may not (Monaghan *et al.*, 1989) influence the egg size, which may be genetically constrained (e.g. Rhymer, 1988) and vary with female age (e.g. Coulson and White, 1958), courtship feeding (Nisbet, 1973) and laying date (e.g. Parsons, 1975; Montevecchi, 1978, Birkhead and Nettleship, 1982). Marked changes in food supply have only slight effects on egg size (Hamer *et al.*, 1991) but greater effects on clutch size (e.g. Pehrsson, 1991).

Clutch variation is very limited among seabirds, with many species laying obligate one-egg clutches (Table I in Montevecchi and Porter, 1980). Among species laying multi-egg clutches, egg number often varies in response to food supply (Drent and Daan, 1980) and to such factors as age (e.g. Pugesek, 1981), laying date (e.g. Coulson and White, 1961) and the number of clutches at a colony (Hatch and Hatch, 1990a; Murphy *et al.*, 1991). For species that lay one egg, variation in egg size is analogous to variation in clutch size (Birkhead and Nettleship, 1982), and flexibility in time budgets may compensate for their inability to adjust clutch and brood sizes (Burger and Piatt, 1990).

Inadequate intake of protein or of an essential amino acid (methionine) reduces egg production by domestic fowl (Leveille *et al.*, 1961). Links between nutritive intake and egg production are interesting, and though it seems unlikely that variation in egg and clutch size will prove useful as indictors of marine food supplies, more studies of egg size and food availability are warranted, especially with species that lay one-egg clutches.

Table 6.4 Associations among measurements of seabird biology, fish abundance and environmental indices

Prey[a]	Age[b]	Location	Avian[c]	Index Fish	Environment	Years	Sources[d]
H C	PR A	Norway	ATPU production population	acoustic index		1964–90	(1)
H	PR	Shetlands (North Sea)	SHAG GRSK BLKI ARTE COGU numbers production chick growth attendance	trawl surveys		1974–82	(2)
C	A	E. Newfoundland (N.W. Atlantic)	COGU time away from colony	fine scale acoustic index		1982–85	(3)
S,A		New York (N.W. Atlantic)	COTE ROTE laying date feeding rate growth production	fine scale acoustic index		1984–85	(4)
P	PR	Bering Sea	BLKI RLKI production	commercial catch research survey		1975–84	(5)

S	Bering Sea	BLKI production laying date	air temperature	1975–89	(6)
C, S	Gulf Alaska	BLKI GWGU production	water temperature dilution	1977–78	(7)
A	S. California Bight	WEGU numbers breed	aerial survey	1972–77	(8)
PR	S. California Bight	XAMU laying date	aerial survey	1976–78	(8)
A	S. California Bight	BRPE production	commercial catch trawls	1969–80	(9)

[a] A = anchovy; C = capelin; H = herring; S = sandeel
[b] A = adult; PR = pre-recruit
[c] ARTE = Arctic tern; ATPU = Atlantic puffin; BLKI = black-legged kittiwake; BRPE = brown pelican; COGU = common guillemot; COTE = common tern; GRSK = great skua; GWGU = glaucous-winged gull; RLKI = red-legged kittiwake; ROTE = roseate tern; SHAG = shag; WEGU = western gull; XAMU = Xantus' murrelet
[d] 1) Vader et al., 1990a,b; 2) Heubeck, 1989; Furness, 1990; Furness and Barrett, 1985; Barrett and Furness, 1990; Hamer et al., 1991; 3) Burger and Piatt, 1990; 4) Safina et al., 1988; 5) Springer et al., 1986; 6) Murphy et al., 1991; 7) Baird, 1990; 8) Hunt and Butler, 1980; 9) Anderson et al., 1982; Anderson and Gress, 1984

6.5.3 Timing of egg laying

Larger clutches and greater production tend to be associated with 'early years' (Coulson and White, 1961; Murphy et al., 1991), and the timing of laying may be sensitive to food supply and environmental conditions during the prelaying period (Perrins, 1970; Aebischer, 1986; Aebischer et al., 1990; Hatch and Hatch, 1990b). The earlier in the breeding season that reproductive failure occurs, the more severe it tends to be (Harris and Wanless, 1990).

6.5.4 Breeding success, chick growth and fledging mass

Studies of reproductive success can be useful in assessing fluctuations in prey availablility or abundance, when conditions are moderate to poor (Cairns, 1987). These relationships are most evident among species that lay multi-egg clutches (e.g. Anderson and Gress, 1984; Monaghan et al., 1989); species with one-egg clutches show less variation in productivity over a range of food conditions (Hatch and Hatch, 1990a). Guillemots, for example, showed no difference in egg and chick success during a study in which fish abundance varied by an order of magnitude (Burger and Piatt, 1990). Breeding success can also be affected by weather (Becker and Specht, 1991) and biased by age and intra- and inter-colony location (Coulson, 1968; Pugesek, 1981; Croxall et al., 1988a). Studies of reproductive success as an indicator of prey conditions seem most profitably directed at small surface-feeding species that produce multi-egg clutches and that have less flexible time and activity budgets than larger species (Pearson, 1968; Furness and Ainley, 1984; Table 6.4).

Numerous comparisons have demonstrated faster and higher levels of growth in 'good' versus 'poor' years, yet very few have related these to independent indices of prey availablility or abundance (e.g. Safina et al., 1988). Williams and Croxall (1990) argued that because only 'high quality' chicks survive in 'poor' years and because a higher proportion of lighter chicks fledge in 'good' years, there could be an inverse relationship between fledging mass and food supply and greater variation in fledging masses in good years. Fledging mass is also influenced by colony size, suggesting that breeding populations may deplete food supplies around colonies (Gaston et al., 1983; Hunt et al., 1986; section 6.6.1). Chick growth has also been associated with meteorological (e.g. Becker and Specht, 1991) and oceanographic variation. For instance, Ricklefs et al. (1984) reported that blue-footed booby chicks at colonies in warm water grew more slowly than chicks in cool water colonies and attributed this to differences in the productivity of warm and cool water. A similar effect might have occurred if parents from warm-water regions travelled extra distance to forage in cool water, though the boobies' near-shore foraging distributions (Anderson and Ricklefs, 1987) weigh against this possibility.

The use of chick growth and other aspects of breeding success as indicators of food conditions assume that parents work at maximum capacity (e.g. Ashmole, 1963; Diamond, 1978; Drent and Daan, 1980; Burger and Piatt, 1990; Cairns, 1992a). Yet parents often have considerable time (Montevecchi and Porter, 1980; Furness and Barrett, 1985; Cairns et al., 1987, 1990b), that can be used to increase foraging efforts when food is scarce (Croxall et al., 1988a, b; Burger and Piatt, 1990; Hamer et al., 1991; section 6.6.1). Flexibility in parental foraging behaviour buffers chicks over a range of food conditions and may preclude the use of growth as an indicator of food conditions except among species with inflexible time budgets or when food supplies are poor (Burger and Piatt, 1990; section 6.6.1). Chicks of pelagic seabirds often accumulate considerable lipid, which dampens effects of variability in food supply (Ricklefs, 1990). Overall, breeding success, growth of chicks and fledging mass can provide useful indications of food supplies during moderate to poor conditions (Cairns, 1987), and these indicators are most informative when multi-species comparisons are used.

6.5.5 Sources of reproductive failure

Breeding failure can occur at any stage of the reproductive cycle and can result from many causes. In general, laying failures and inadequate parental care of eggs have the greatest effects on production (Harris and Wanless, 1990; Hatch et al., in press). It is informative to compare different aspects of breeding failure and to consider interactive causes of such failure: e.g. food shortage can lead to increased foraging effort and hence to decreased parental attentiveness which results in increased predation on chicks and eggs.

6.6 BEHAVIOURAL RESPONSE TO PREY AVAILABILITY

Common guillemots exhibit considerable plasticity in the time that they devote to foraging activities, which are related to prey abundance (Burger and Piatt, 1990). Common guillemots breeding in southern Labrador made many more foraging trips during a 'poor' food year than during a 'good' one, though chick success and fledging masses were similar (Birkhead and Nettleship, 1987), suggesting that parental efforts buffered chicks during 'poor' food conditions.

Increased parental absence from eggs and chicks by black-legged kittiwakes, presumably indicative of increased foraging effort, is associated with reduced prey availability (Harris and Wanless, 1990; Hatch and Hatch, 1990a). However, smaller surface-feeding species, like kittiwakes and terns, being more constrained by energy demands associated with rearing offspring (e.g. Pearson, 1968) and by hydrographic variation (e.g. Hatch et al., in press) probably have little flexibility in their foraging budgets (Furness and Ainley, 1984), and their production is quite variable (e.g. Monaghan and Zonfrillo, 1986; Bailey et al., 1991; Monaghan et al., 1991). Thus, while flexibility in parental foraging behaviour can buffer chicks over a range of food conditions, tolerances vary among species. Parental foraging effort could also be driven to maximal levels by prey depletion around colonies during breeding seasons or under poor food conditions. For example, Hamer et al. (1991) found that adult great skuas *Catharacta skua* increased foraging effort during years of food shortage in order to maintain chick growth, but adult mortality was also associated with increased foraging effort, particularly among older birds that made the greatest increase in foraging effort (Hamer and Furness, 1991). Reductions in reproductive success became evident only after the buffering capacity of progressive changes in diet and increased foraging effort broke down. Such plasticity in parental foraging is at variance with hypotheses that parents maximize foraging time and effort, which are constrained by metabolic (Drent and Daan, 1980) and digestive physiology (Diamond et al., 1986), and by age (Pugesek, 1981). This may indicate that increased foraging effort generally incurs a cost in terms of adult survival, as seen in great skuas. Data on variation in foraging behaviour and success can be integrated to produce avian catch per unit effort indices (section 6.7.1).

6.6.1 Central place foraging (breeding birds)

Breeding birds are tied to colonies to defend territories, tend eggs and provision young. These circumstances greatly constrain foraging options and imply that colonies are located where prey tends to be abundant, particularly during chick rearing. Data from colonial seabirds reflect local food conditions around colonies (Montevecchi et al., 1987a), as well as those in larger adjacent regions (Montevecchi and Myers, 1992; see also Croxall, 1989; Cairns, 1992a, b).

6.6.2 Self-centred foraging (non-breeding birds)

When not breeding, birds often range widely at sea and migratory species may show little overlap of summer and winter areas (e.g.

Diamond, 1978: Elliott *et al.*, 1990). In contrast, some inshore species do not depart far from land, and others maintain similar foraging areas during breeding and non-breeding periods, e.g. cape gannets roost at colonies in winter (Berruti, 1987).

6.6.3 Parental food loads and chick provisioning rates

Variation in parental food loads and in chick provisioning rates are easily and unobtrusively studied in species that carry food in their bills. Among multi-prey loaders, inverse relationships between number of items carried per trip and total masses of food loads (Barret *et al.*, 1987; Montevecchi and Barrett, 1987) suggest that the number of items carried may yield a crude index of prey conditions, i.e. fewer, larger items when food supplies are good (cf. Orians and Pearson, 1979). The chick-provisioning rates of species with flexible foraging budgets remain relatively constant under a range of good food conditions but can be ranked according to good, intermediate and poor prey conditions (compare Burger and Piatt, 1990; Uttley *et al.*, submitted). The functional responses of birds to different levels of prey availability (Piatt, 1987, 1990) are essential determinants of avian indices of prey conditions (section 6.7.1).

6.6.4 Prey harvests

Harvests of prey by seabirds and humans can be compared directly (Table 6.5). The age distribution of small fish taken by birds and

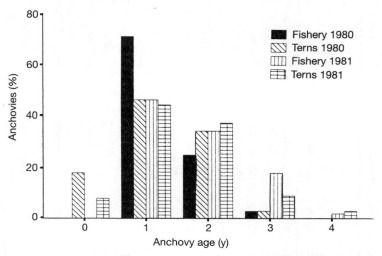

Figure 6.3 Age distribution of anchovies harvested by elegant terns and fisheries in southern California. (Based on Schaffner, 1986.)

Table 6.5 Associations between prey harvests by birds and humans and abundance indices

Prey	Age[a]	Bird	Location	Associations		Dates	Source
				Human fishery	Abundance index		
Sandeel	A	Great skua	Shetlands (North Sea)	+	+	1973–89	Hamer et al., 1991
Sprat Sandeel	PR	Atlantic puffin	Isle of May (North Sea)	na	+	1971–76	Hislop and Harris, 1985
Herring	PR	Atlantic puffin	Isle of May (North Sea)	na	+	1973–82	Hislop and Harris, 1985
Squid Mackerel	A	Northern gannet	Funk Is. (N.W. Atlantic)	+	+	1977–91	Montevecchi and Myers, 1992
Pilchard Horse mackerel	A	Cape gannet Jackass penguin Cape cormorant	Benguela	+	na	1958–83	Burger and Cooper, 1984; Berruti and Colclough, 1987
Anchovy	A,PR	Western gull	S. California Bight	na	+	1972–77	Hunt and Butler, 1980
Anchovy	A,PR	Elegant tern	S. California	+	na	1979–83	Schaffner, 1986
Pollack	PR	Tufted puffin	Semidi Is. (Gulf Alaska)	+	+	1985–87	Hatch and Sanger, 1992

[a] A = adult; PR = pre-recruit

fisheries often correspond closely (Schaffner 1986; Figure 6.3). The percentage of pilchard in the diets of cape gannets and commercial landings of these fish are significantly correlated over annual intervals (Berruti and Colclough, 1987; Montevecchi and Berruti, 1991), and similar relationships hold for harvests of short-finned squid *Illex illecebrosus* and mackerel by northern gannets and humans in the north-west Atlantic (Montevecchi et al., 1987a; Montevecchi and Myers, 1992). Relationships between prey harvests by seabirds and independent estimates of prey abundance have also been found (e.g. Hatch and Sanger, 1992; Montevecchi and Myers, 1992; Figure 6.4). Much of the strength of these correlations comes from poor years in which both avian and human fisheries for common prey fail. In some instances, when prey stocks are reduced, specialist predators may, owing to increased foraging effort, exhibit unchanged or even increased prey harvests (Barrett and Furness, 1990; Bailey et al., 1991). Such situations uncouple avian harvests from prey abundance, especially where alternative prey are not available.

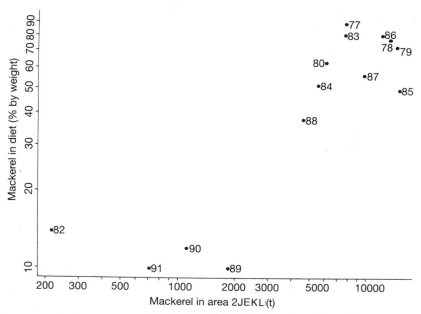

Figure 6.4 Relationship between the harvest of mackerel by northern gannets and estimated abundance of mackerel in the north-west Atlantic NAFO Fisheries Region 2JEKL. Each point represents a separate year; $r = 0.87$. (From Montevecchi and Myers, 1992.)

6.7 SEABIRDS AS FISHERIES INDICATORS

6.7.1 Avian indices of catch per unit effort

The development of avian catch per unit effort ratios (CPUEs) using remotely-sensed activity recorders can be derived from the foraging behaviour of large-bodied pursuit-diving dietary specialists (e.g. Kooyman *et al.*, 1982). CPUEs of breeding birds provide useful indices when prey conditions are relatively good, i.e. good enough to forage to feed offspring. Comparisons of common guillemots in the northeast and north-west Atlantic suggest that chick provisioning rates and provisioning rates as functions of parental foraging efforts reflect prey abundance over a range of poor, intermediate and good food conditions (Burger and Piatt, 1990; Cairns *et al.*, 1987, 1990a, b; Uttley *et al.*, submitted). These comparisons along with others, such as colony attendance, that are sensitive to variations in prey availability when food conditions are poor, (Murphy *et al.*, 1991), could be used in fisheries-stock assessment models (6.2.2; Cairns, 1988, 1992a).

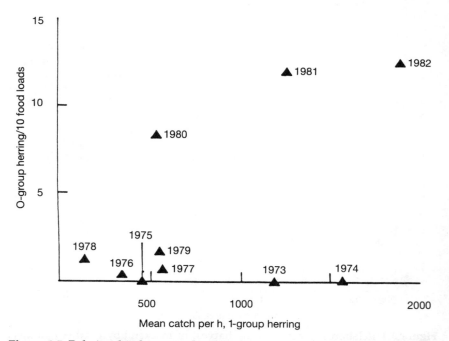

Figure 6.5 Relationship between the abundance of herring in the diets of puffin chicks on the Isle of May and an abundance index of larval herring in the North Sea. (Based on Hislop and Harris, 1985.)

(a) Recruitment indices

Due to extreme variability during the early life stages, the recruitment rates of fish are not well correlated with spawning stock biomass. It is therefore necessary to survey young fish in order to generate recruitment indices. Seabirds are useful here. For instance, the percentage of O-group saithe *Pollachius virens* in the diets of breeding shags *Phalacrocorax aristotelis* and subsequent fisheries indices in Norway were well associated (Barrett, 1991). The proportions of young of the year walleye pollock *Theragra chalcogramma* in the diets of tufted puffins *Fratercula cirrhata* on the Semidi Islands correlated positively with fisheries catches, trawl surveys and with estimates from stock-assessment models (Hatch and Sanger, 1992). The breeding success of Arctic terns in the Shetland Islands has been positively associated with estimates of O-group sandeel abundance (Monaghan *et al.*, 1989),

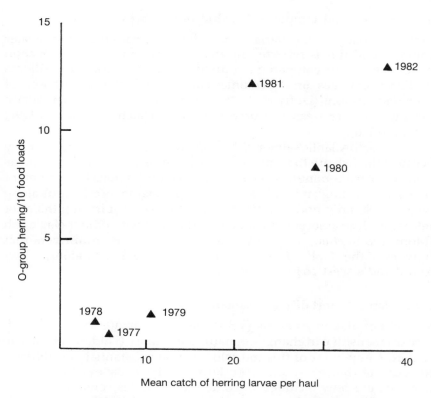

Figure 6.6 Relationship between the abundance of herring in the diets of puffin chicks on the Isle of May and trawl catch rates of 1-year and older herring during the following February. (Based on Hislop and Harris, 1985.)

and numbers of breeding terns were correlated with abundance indices of pre-recruit herring in the Firth of Clyde (Monaghan and Zonfrillo, 1986). The occurrence of sprat *Sprattus sprattus* and herring in the diets of Atlantic puffin chicks on the Isle of May correlated with independent estimates of fish abundance in the North Sea over a 10-year period (Hislop and Harris, 1985; see also Martin, 1989; Figure 6.5). The frequency of herring delivered to chicks and the trawl catch rates of one-year and older herring during the following February were also correlated (Figure 6.6). Information on the strengths, growth rates, distributions and movements of different age classes of fish can be gained in analyses of their size-at-age at different colonies and times (Barrett *et al.*, 1990; Barrett, 1991; Hatch and Sanger, 1992; section 6.7.2). Hislop and Harris (1985) suggested that seabirds could provide useful indication of trends in fish recruitment, and subsequent research has supported their suggestion.

6.7.2 Spatial and temporal distributions of prey

Prey availability is more variable near the sea surface than in the water column, and this is reflected in greater variation in the production of surface feeders compared to pursuit divers. Geographic gradients in dietary changes among colonies can be used in considerations of prey distributions (Duffy *et al.*, 1984; Berruti, 1987) and of larval and juvenile 'nursery areas' for spawning stocks (Hatch and Sanger, 1992; section 6.7.6).

Post-smolt Atlantic salmon *Salmo salar* have low rates of tag recovery because the fish are too small for capture in fishing nets. Tag returns from a seabird community have shown that post-smolts from different rivers in New England and eastern Canada migrate northwards along Newfoundland's north-east coast about three to four months after release in close association with each other and with other pelagic fish (Montevecchi *et al.*, 1988). In contrast, post-smolts from rivers that flow into the Gulf of St. Lawrence migrate north along Newfoundland's west coast.

6.7.3 Natural-mortality assessments

Virtual Population Analysis (VPA) used to estimate the size of fish stocks is sensitive to changes in natural mortality (Hilden, 1988). The seabird component of this mortality can be substantial and difficult to estimate (Barrett *et al.*, 1990; Furness, 1990; Bailey *et al.*, 1991). Seabirds are now included in some multi-species considerations of fish assessments, and natural mortality due to avian predation could be incorporated in VPA by running community bioenergetics models as subunits of larger VPA models (Cairns, 1992a).

6.7.4 Sampling non-commercial marine prey

Ashmole and Ashmole (1968) pointed out that seabirds can be especially useful as sampling agents for otherwise unstudied marine fauna. Food sampling from gannets showed that Atlantic saury *Scomber scomberesox* occurred regularly in inshore waters of eastern Canada and the Benguela region (Montevecchi and Berruti, 1991). Saury were abundant in the late 1980s in the north-west Atlantic, where they have apparently been commonly available, judging from the local name 'billfish' that fishermen apply to the species. Saury have been harvested for bait for tuna in the south-east Atlantic (Berruti, 1988) and for fish meal by Russians and Norwegians in the north-west Atlantic in the 1960s, and a small fishery was pursued in the 1940s in Nova Scotia (Leim and Scott, 1966). The related Pacific saury *Cololabis saira* is an Asian food delicacy, and as Asian markets purchase large quantities of pelagic prey that were formerly used locally as 'bait' (e.g. squid, capelin), fisheries for Atlantic saury might be developed to diversify commercial fishing activity. Seabirds also sample rare and unusually distributed pelagic and mesopelagic invertebrate fauna (Vermeer and DeVito, 1988; Steele and Montevecchi, 1993).

6.7.5 Multi-species comparisons

Multi-species comparisons of seabirds with different trophic relationships and foraging modes enhance avian indication of fish conditions. Surface and sub-surface foragers that feed on the same prey can provide information on the vertical distributions of prey. Many recent demonstrations of discordant reproductive success of surface feeders and pursuit divers were associated with cold sea surface temperatures (SSTs) and with unavailability of pelagic prey in surface waters (Table 6.6). Small surface-feeding planktivores and specialists are probably the species most sensitive to oceanographic discontinuities (Briggs *et al.*, 1984; Furness and Ainley, 1984; Brown and Gaskin, 1988). Pursuit divers tend to forage in stratified water (Haney, 1991), so cold SSTs are less problematic for them. Deep-foraging pursuit divers (e.g. common guillemots) usually prey on larger fish than do the smaller, shallower divers (e.g. least auklets *Aethia pusilla*) and very deep divers (e.g. penguins) often feed on crustaceans. The simultaneous reproductive failure of both surface and sub-surface feeders at the same site may indicate pervasive large-scale oceanographic or biological changes that may have been caused by human activities. To date, one situation has been documented (in Norway), where surface feeders (black-legged kittiwakes) exhibited good reproductive success while pursuit divers (Atlantic puffins) did not (R.T. Barrett, in prep.; Table 6.6).

Table 6.6 Comparative reproductive success of surface and sub-surface feeders in different regions and possible fisheries or oceanographic associations

Location	Species[a]	Foraging site[b]	Breeding success	Fish/ocean conditions	Years	Source
Alaska	BLKI	S	−	Cold air temp.,	1976, 77	Hatch et al., in press
	RLKI	S	−	Cold SST,[c]	84, 85, 89	Murphy et al., 1991
	COGU	SS	−	Prey at depth		
	BRGU	SS	+			
	PECO	SS	+			
	RECO	SS	+			
Gulf Alaska	BLKI	S	−	Cold SST	1977–78	Baird, 1990
	GWGU	S	−	Surface dilution		
	TUPU	SS	+	Capelin at depth		
N.W. Atlantic	BLKI	S	−	Ice; cold SST	1991	J. Neuman, unpubl.
	HEGU	S	−	Capelin at depth		Montevecchi et al., unpubl. data
	NOFU	S	+			
	LESP	S	+			
	COGU	SS	+			
	ATPU	SS	+			
Shetlands (N. Sea)	GRSK	S	−	Sandeels unavailable	1988, 90	Harris and Wanless, 1990; Monaghan et al., 1991; Furness, unpubl. data
	BLKI	S	−			
	ARTE	S	−			
	SHAG	SS	−			
	COGU	SS	+			
N. Norway	BLKI	S	+	Capelin overfished, poor production	1980s	R. Barrett, in prep.
	ATPU	SS	−			

[a] ATPU = Atlantic puffin; BLKI = black-legged kittiwake; BRGU= Brunnich's guillemot; COGU = common guillemot; GWGU = glaucous-winged gull; HEGU = herring gull; PECO = pelagic cormorant; RECO = red-faced cormorant; TUPU = tufted puffin
[b] S = surface feeder; SS = sub-surface feeder
[c] SST = sea surface temperature

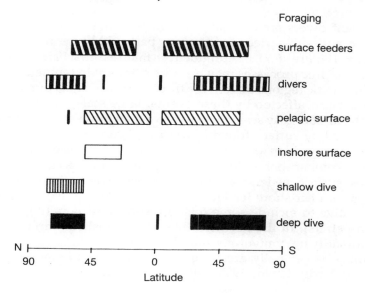

Figure 6.7 Latitudinal distributions of seabirds with different foraging modes. Note surface feeders dominate tropical and temperate regions with divers in colder regions. Population estimates of high-latitude surface feeders are probably low and some are missing but the different latitudinal distributions of sub-surface and surface feeders are apparent. (Based on data derived from Croxall *et al.* (1984b) and Croxall (1991).)

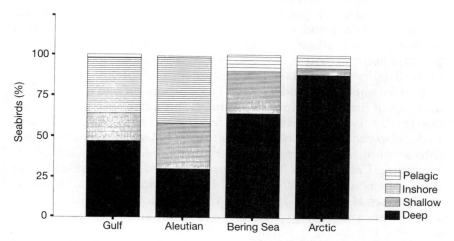

Figure 6.8 Feeding guild compositions (inshore and pelagic surface feeders and deep and shallow divers) of Alaskan seabird communities in different regions ordered from South to North: Gulf of Alaska, Aleutian Islands, Bering Sea, Arctic. (Data derived from Lensink, 1984.)

Pursuit divers are common at high latitudes, so comparisons of surface and sub-surface feeders are particularly germane to these regions (Figure 6.7). Reproductive failures that are driven by oceanographic processes may be more common in temperate, boreal and low arctic regions than in the tropics. However, as pursuit diving birds are less affected by these events, large scale breeding failures at high latitudes may seem less extensive than those in tropical regions, dominated by surface foraging species (Figure 6.7). Considerations of the composition of seabird communities are also important in regional comparisons of diving and surface feeding species (Figure 6.8).

Most surface feeders used in multi-species comparisons to date are inshore or near-shore feeders (terns, kittiwakes, gulls), so it will be informative to include some offshore surface feeders (e.g. fulmars, storm-petrels). In the northern hemisphere inshore surface feeders are primarily nektonivorous and planktivorous. Variation in production may be generally greater among inshore than among offshore foraging birds (Lloyd, 1985; Ollason and Dunnet, 1988).

6.8 CONCLUSIONS

Systematic monitoring of seabird behavioural ecology can usefully enhance fish-stock assessments by providing estimates of predator-induced mortality and technology-independent indices of fish abundance and availability. Avian indices can be particularly useful in sampling prey and sites that are not easily or usually surveyed by conventional means, i.e. pre-recruits, non-commercial prey, upper water column, inshore areas, etc. Avian data often yield early indication of large scale fluctuations in stocks of pelagic fish and oceanographic perturbations; of distributions of species, stocks and age classes; and of availability or abundance of unstudied, unexploited prey.

Seabird energetics models could be incorporated into multi-species analyses of fish populations to estimate mortality due to avian predation (Bailey, 1986, Furness, 1990; Bailey *et al.*, 1991; Cairns, 1992a) and to develop conservative harvest allotments that take natural predators into account. In a community energetics model of the Gulf of St. Lawrence, Cairns *et al.* (1990a) mapped seabird prey harvests by the NAFO areas used by Fisheries and Oceans Canada to map fisheries catches (see also Bailey, 1986). The incorporation of fisheries catches and the seasonal energy flux through avian and mammalian communities into such a scheme would permit comparisons of the magnitudes and spatial and temporal overlaps of the harvests of human and nonhuman predators. Geographical information systems (GIS; Shaw and Atkinson, 1990) could facilitate such analyses and help to pinpoint times and sites where inter-specific interactions would be expected to be most intense and where potential competition might occur.

Associations between avian and fisheries data have produced nominal (presence/absence) and ordinal (low, intermediate, high) relationships. These data can be highly useful for management adjustments (Duffy et al., 1984), in ecological studies of marine assemblages and food webs (e.g. Gomes, 1991), and in fisheries assessment models which usually contain no recruitment predictors other than the long-term mean (Cairns, 1992a). Marine ornithology is carried out by some national fisheries organizations (e.g. Portugal, South Africa), though the mandates of most relevant governmental agencies are so limited to stock assessment as to preclude ecological synthesis. The integration of seabird research into interdisciplinary oceanographic programmes will aid in long-term studies of biophysical interactions in marine food webs.

6.9 RESEARCH DIRECTIONS AND RECOMMENDATIONS

Some seabirds stand out as potentially useful broad-scale indicator species. For instance, in the northern hemisphere black-legged kittiwakes have a holarctic distribution, are relatively well monitored, and their reproductive parameters exhibit wide yearly and oceanographic variation (Harris and Wanless, 1990; Murphy et al., 1991; Hatch et al., in press). Kittiwakes have experienced large-scale breeding failure in the northern Pacific (Hatch et al., in press) and in the north-east Atlantic (Barrett and Runde, 1980; Harris and Wanless, 1990) and north-west Atlantic (J. Neuman and H. Reghr, unpublished data). The most extensive database of kittiwake breeding biology in the world is from a small colony in the north-east Atlantic that has exhibited little yearly variation in breeding success (Coulson and Porter, 1985; Coulson and Thomas, 1985). Comparative dietary studies from kittiwake populations in different regions are needed to fill large gaps in our knowledge of their trophic relationships.

For species, such as kittiwakes, with wide geographic ranges, efforts to standardize comparative studies will aid in understanding effects of fisheries, physical processes and ocean and climate change on avian behavioural ecology. When possible studies should be conducted at colonies within and outside spheres of commercial fishing activity (Croxall et al., 1988a,b; Murphy et al., 1991; Hatch et al., in press).

Many seabird species can be monitored at breeding colonies with minimal disturbance, and it is important to capitalize on unobtrusive studies whenever possible. Standardized counts of cliff-nesting kittiwakes and guillemots can be done non-disruptively from a distance, and birds that carry parental food in the bill, such as puffins, guillemots, razorbills *Alca torda*, black guillemots *Cepphus grylle* and terns can be visually monitored for dietary information with minimal disturbance. Species like puffins and gannets often nest in dense accessible colonies, where it is possible to quickly collect many samples of prey returned to chicks (Montevecchi *et al.*, 1987a; Barrett *et al.*, 1990; Hatch and Sanger, 1992; Montevecchi and Berruti, 1991). Studies like these and others that utilize stomach-pumping techniques (e.g. Gales, 1988) have largely eliminated the need to kill animals to gain dietary information, though collections of birds that do not normally regurgitate food and of birds at sea may still prove useful. Stable-isotopic analyses of blood, feathers or other tissue that can be removed from live animals offer a novel technology with which to study the trophic relationships of extant and extinct animals (Hobson, 1991; Hobson and Montevecchi, 1991).

Telemetric studies of the behaviour and success of foraging individuals (Anderson and Ricklefs, 1987; Burger, 1991; Wanless *et al.*, 1991a,b) need to be integrated with fine-scale acoustic and trawl surveys within foraging ranges around colonies to determine prey distributions and availability to surface and sub-surface feeders (Burger and Piatt, 1990). Such integration will enable study of functional responses by avian predators to a range of prey conditions and hence provide a basis with which to develop comprehensive seabird indices of fish abundance and availability.

Non-breeding and wintering birds are major energy consumers in marine food webs, and their poorly understood feeding ecology requires intensive study (Bailey, 1986; Erikstad, 1990; Furness, 1990) so that comprehensive annual energy budgets of avian communities can be calculated. Non-breeders account for almost all the energy flux through seabirds during most of the year and often account for a large portion or even the bulk of the energy flux through seabird communities during breeding seasons (Cairns *et al.*, 1990a). Stable isotopes can be used to analyse the seasonal and oceanographic variability of trophic relationships.

Differential or selective predation by seabirds on different age classes and genders of particular prey species can be usefully incorporated into monitoring schemes. To assess the effects of avian predation on different age classes and on fish populations, analyses of yield per recruit over the age classes of a fish cohort could be used to supplement biomass estimates of harvest tonnages (Bailey *et al.*, 1991; Cairns, 1992a). Seabirds feed primarily on younger, smaller fish, so these analyses will indicate much greater impacts of avian predation than estimates of the tonnages of prey harvested (Barrett *et al.*, 1990; Hatch and Sanger, 1992). When such exercises are carried out, they need to be presented in the context of multi-species food webs and of density dependent population mechanisms.

Marine ornithology is an integral aspect of biological oceanography. Conceptual advances will emerge as seabird studies are more fully integrated into large multi-disciplinary biophysical research programmes focused on trophic relationships in a dynamic ocean.

ACKNOWLEDGEMENTS

I am grateful to Bob Furness for providing the opportunity, patience and encouragement for me to prepare this chapter, to David Cairns, Bob Furness and Jeremy Greenwood for comments on an earlier draft, to Pat Monaghan for access to unpublished manuscripts and to John Horn for discussion. Research was supported by the Natural Sciences and Engineering Research Council of Canada (NSERC).

REFERENCES

Adams, N.J. and Brown, C.R. (1989) Dietary differentiation and trophic relationships in the sub-Antarctic penguin community at Marion Island. *Mar. Ecol. Prog. Ser.*, **57**, 249–58.

Adams, N.J. and Klages, N.T. (1989) Temporal variation in the diet of the Gentoo Penguin *Pycoscelis papua* at sub-Antarctic Marion Island. *Colonial Waterbirds*, **12**, 30–36.

Aebischer, N.J. (1986) Retrospective investigation of an ecological disaster in the shag, *Phalacrocorax aristotelis*: a general method of long term marking. *J. Anim. Ecol.*, **55**, 613–29.

Aebischer, N.J., Coulson, J.C. and Colebrook, J.M. (1990) Parallel long term trends across four marine trophic levels and weather. *Nature*, **347**, 753–755.

Ainley, D.G., Carter, H.R., Anderson, D.W. *et al.*, (1988) Effects of the 1982–83 El Nino–Southern Oscillation on Pacific Ocean seabird populations. *Acta XIX Congressus Internationalis Ornithologici*, 1747–58.

Ainley, D.G. and LeResche, R.E. (1973) The effects of weather and ice conditions on breeding in Adelie penguins. *Condor*, **75**, 235–55.

Ainley, D.G., O'Connor, F.F.O. and Boekelheide, R.J. (1984) The marine ecology of birds in the Ross Sea, Antarctic. *Amer. Ornithol. Union Monogr.*, **32**.

Alverson, D.L. (1991) Commercial fisheries and the Steller sea lion (*Eumetopias jubatus*): the conflict arena. *Fish. Res. Inst. Rep. FRI-UW-9106*, Fish Research Institute, Seattle.

Andersen, K P. and Ursin, E. (1977) A multispecies extension to the Beverton and Holt theory of fishing, with accounts of phosphorus circulation and primary production. *Medd. Dan. Fisk. Havunders*, **7**, 319–435.

Andersen, D.J. (1989) Differential responses for boobies and seabirds in the Galapagos to the 1986–87 El Nino–Southern Oscillation event. *Mar. Ecol. Prog. Ser.*, **52**, 209–16.

Anderson, D.J. and Ricklefs, R.E. (1987) Radio tracking masked and blue footed boobies (*Sula* spp.) in the Galapagos Islands. *Nat. Geog. Res.*, **3**, 152–63.

Anderson, D.W. and Gress, F. (1984) Brown pelicans and the anchovy fishery off southern California, in: *Marine Birds: Their Feeding Ecology and Commercial Fisheries Relationships*. (eds D.N. Nettleship, G.A. Sanger and P.F. Springer), Canadian Wildlife Service, Ottawa, p. 128-35.

Anderson, D.W., Gress, F. and Mais, F.K. (1982) Brown pelicans: influence of food supply on reproduction. *Oikos*, **39**, 23-31.

Anderson, D.W., Gress, F., Mais, K.F. and Kelly, P.R. (1980) Brown pelicans as anchovy stock indicators and their relationships to commercial fishing. *Calif. Coop. Oceanic Fish. Invest. Rep.*, **21**, 54-61.

Anker-Nilssen, T. and Lorentsen, S.H. (1990) Distribution of puffins *Fratercula arctica* feeding off Røst, northern Norway, during the breeding season, in relation to chick growth, prey and oceanographical parameters. *Polar Res.*, **8**, 67-76.

Aron, W. (1962) The distribution of animals in the eastern North Pacific and its relationship to physical and chemical conditions. *J. Fish. Res. Board Can.*, **19**, 271-314.

Ashmole, N.P. (1963) The regulation of numbers of tropical oceanic birds. *Ibis*, **103**, 458-73.

Ashmole, M.J. and Ashmole, N.P. (1968) The use of food samples from seabirds in the study of seasonal variation in the surface fauna of tropical oceanic areas. *Pac. Sci.*, **22**, 1-10.

Bailey, R.S. (1986) Food consumption by seabirds in the North Sea in relation to the natural mortality of exploited fish stocks. *CM 1986/G:5*, International Council for the Exploration of the Sea, Copenhagen.

Bailey, R.S., Furness, R.W., Gauld, J.A. and Kunzlik, P.A. (1991) Recent changes in the population of the sandeel, (*Ammodytes marinus* Raitt) at Shetland in relation to estimates of seabird predation. *ICES Mar. Sci. Symp.*, **193**, 209-16.

Baird, P.H. (1990) Influence of abiotic factors and prey distribution on diet and reproductive success of three seabird species in Alaska. *Ornis Scand.*, **21**, 224-35.

Barrett, R.T. (1991) Shags (*Phalacrocorax aristotelis*) as potential samplers of juvenile saithe (*Pollachius virens* (L.)) stocks in Northern Norway. *Sarsia*, **76**, 153-6.

Barrett, R.T., Anker-Nilssen, T., Rikardsen, F. *et al.*, (1987) The food, growth and fledging success of Norwegian puffin chicks *Fratercula arctica*. *Ornis Scand.*, **18**, 73-83.

Barrett, R.T. and Furness, R.W. (1990) The prey and diving depths of seabirds on Hornøy, North Norway after a decrease in Barents Sea capelin stocks. *Ornis Scand.*, **21**, 179-86.

Barrett, R.T. and Runde, O.J. (1980) Growth and survival of nestling kittiwakes *Rissa tridactyla* in Norway. *Ornis Scand.*, **11**, 228-35.

Barrett, R.T., Røv, N., Loen, J. and Montevecchi, W.A. (1990) Diets of shags *Phalacrocorax aristotelis* and cormorants *P. Carbo* in Norway and implications for gadoid stock recruitment. *Mar. Ecol. Prog. Ser.*, **66**, 205-18.

Barrett, R.T. and Schei, P.J. (1977) Changes in the breeding distribution and numbers of cliff breeding seabirds in Sør Varanger, North Norway. *Astarte*, **10**, 29-35.

Becker, P.H. and Specht, R. (1991) Body mass fluctuation and mortality in common terns *Sterna hirundo* chicks dependent on weather and tide in the Wadden Sea. *Ardea*, **79**, 45-56.

Beddington, J.R., Beverton, R.J.H. and Lavigne, D.M. (1985) *Marine Mammals and Fisheries*. George Allen and Unwin, London.

Bergstad, O.A., Jorgensen, T. and Dragesund, O. (1987) Life history and early ecology of the gadoid resources of the Barents Sea. *Fish. Res.*, **5**, 119–61.

Berruti, A. (1983) The use of seabirds as indicators of pelagic fish stocks in the southern Benguela Current. *Proc. Symp. Birds & Man* (ed. J. Cooper), pp. 267–79.

Berruti, A. (1987) The use of cape gannets *Morus capensis* in management of the purse-seine fishery of the Western Cape. Ph.D. Thesis. University of Natal, South Africa.

Berruti, A. (1988) Distribution of predation on saury *Scomberesox saurus scombroides* in continental shelf waters off the Cape Province, South Africa. *S. Afr. J. Mar. Sci.*, **6**, 183–92.

Berruti, A. and Colclough, J. (1987) Comparison of the abundance of pilchard in cape gannet diet and commercial catches off the Western Cape South Africa. *S. Afr. J. Mar. Sci.*, **51**, 863–9.

Birkhead, T.R. and Nettleship, D.N. (1982) The adaptive significance of egg size and laying date in thick-billed murres *Uria lomvia*. *Ecology*, **63**, 300–6.

Birkhead, T.R. and Furness, R.W. (1985), Regulation of seabird populations, in *Behavioural Ecology: Ecological Consequences of Adaptive Behaviour* (eds R.M. Sibly and R. H. Smith). Blackwell Scientific Publications, Oxford, pp. 145–67.

Birkhead, T.R. and Nettleship, D.N. (1987) Ecological relationships between common murres, *Uria aalge*, and thick-billed murres, *Uria lomvia*, at the Gannet Islands, Labrador. III. Feeding ecology of young. *Can. J. Zool.*, **65**, 1638–45.

Birt, V.L., Birt, T., Goulet, D. *et al.* (1987) Ashmole's halo: evidence for prey depletion by a seabird. *Mar. Ecol. Prog. Ser.*, **40**, 205–208.

Birt-Friesen, V.L., Montevecchi, W.A., Cairns, D.K. and Macko, S.A. (1989) Activity specific metabolic rates of free-living northern gannets and other seabirds. *Ecology*, **70**, 357–67.

Blomqvist, S. and Peterz, M. (1984) Cyclones and pelagic seabird movements. *Mar. Ecol. Prog. Ser.*, **20**, 85–92.

Boekelheide, R.J. and Ainley, D.G. (1989) Age, resource availability, and breeding effort in Brandt's cormorant. *Auk*, **106**, 389–401.

Boersma, P.D. (1978) Breeding patterns of Galapagos penguins as an indicator of oceanographic conditions. *Science*, **200**, 1481–3.

Briggs, K.T., Dettman, K.F.,., Lewis, D.B. and Tyler, W.B. (1984) Phalarope feeding in relation to autumn upwelling off California, in *'Marine Birds: Their Feeding Ecology and Commercial Fishery Relationships'* (eds. D.N. Nettleship, G.A. Sanger and P.F. Springer), Canadian Wildlife Services, Ottawa, pp. 51–62.

Brown, R.G.B. (1980) Seabirds as marine animals, in *'Behaviour of Marine Animals, Vol. 4'* (eds J. Burger, B.L. Olla and W.E. Winn, Plenum Press. New York, pp. 1–39.

Brown, R.G.B., Barker, S.P. and Gaskin, D.E. (1979) Daytime surface swarming by *Meganyctiphanes norvegica* (M. Sars) (Crustacea) off Brier Island, Bay of Fundy. *Can. J. Zool.*, **57**, 2285–91.

Brown, R.G.B. and Gaskin, D.E. (1988) The pelagic ecology of the grey and red necked phalaropes *Phalaropus fulicarius* and *P. lobatus* in the Bay of Fundy, eastern Canada. *Ibis*, **130**, 234–60.

Brown, R.G.B. and Nettleship, D.N. (1984) Capelin and seabirds in the northwest Atlantic, in: *'Marine Birds: Their Feeding Ecology and Commercial Fisheries Relationships'* (eds D.N. Nettleship, G.A. Sanger and P.F. Springer), Canadian Wildlife Service, Ottawa, pp. 184–195.

Burger, A.E. (1991) Maximum diving depths and underwater foraging in alcids and penguins, in: *'Studies of High Latitude Seabirds 1: Behavioural, Energetic and Oceanographic Aspects of Seabird Feeding Ecology'* (eds W. A. Montevecchi and A.J. Gaston), *Can. Wildl. Servic. Occas. Pap.*, **68**, 9–15.

Burger, A.E. and Cooper, J. (1984) The effects of fisheries on seabirds in South Africa and Namibia, in: *'Marine Birds: Their Feeding Ecology and Commercial Fisheries Relationships'* (eds D.N. Nettleship, G.A. Sanger and P.F. Springer), Canadian Wildlife Service, Ottawa, pp. 150–160.

Burger, A.E. and Piatt, J.F. (1990) Flexible time budgets in breeding common murres: buffers against variable prey abundance. *Stud. Avian Biol.*, **14**, 71–83.

Byrd, G.V., Murphy, E.C., Kaiser, G.W. et al. (1992) Status and ecology of offshore fish-feeding alcids, murres and puffins in the North Pacific Ocean, in *'Ecology and Conservation of Marine Birds of the Temperate North Pacific'* (eds K. Vermeer, K.T. Briggs and D. Siegel Causey), Canadian Wildlife Service, Ottawa.

Cairns, D.K. (1987) Seabirds as indicators of marine food supplies. *Biol. Oceanogr.*, **5**, 261–271.

Cairns, D.K. (1989) The regulation of seabird colony size: a hinterland model. *Am. Nat.*, **134**, 141–146.

Cairns, D.K. (1992a) Bridging the gap between ornithology and fisheries biology: use of seabird data in stock assessment models. *Condor*, **94**, 811–24.

Cairns, D.K. (1992b) Population regulation of seabird colonies, in: *'Current Ornithology, Vol. 9'* (ed. D.M. Power), Plenum Press, New York, pp. 37–61.

Cairns, D.K., Bredin, K. and Montevecchi, W.A. (1987) Activity budgets and foraging ranges of common murres. *Auk*, **104**, 218–224.

Cairns, D.K., Chapdelaine, G. and Montevecchi, W.A. (1990a) Prey harvest by seabirds in the Gulf of St. Lawrence, in: *'The Gulf of St. Lawrence: Small Ocean or Big Estuary'* (ed. J.C. Therriault) *Can. Spec. Publ. Fish. Aquat. Sci.* **113**, 277–291.

Cairns, D.K., Montevecchi, W.A., Birt-Friesen, V.L. and Macko, S.A. (1990b) Energy expenditures, activity budgets, and prey harvest of breeding common murres. *Stud. Avian Biol.*, **14**, 84–92.

Cairns, D.K. and Schneider, D.C. (1990) Hot spots in cold water: feeding habitat selection by thick billed murres. *Stud. Avian Biol.*, 14, 52–60.

Clarke, M.R. (1977) Beaks, nets and numbers. *Symp. Zool. Soc. Lond.*, **38**, 89–126.

Coulson, J.C. (1968), Differences in the quality of birds nesting in the centre and on the edges of the colony. *Nature*, **217**, 478–9.

Coulson, J.C., Monaghan, P., Duncan, N. et al. (1983) Seasonal change in the herring gull in Britain – weight, moult and mortality. *Ardea*, **71**, 235–44.

Coulson, J.C. and Porter J.M. (1985) Reproductive success of the kittiwake *Rissa tridactyla*: the roles of clutch size, chick growth rates and parental quality. *Ibis.*, **127**, 450–66.

Coulson, J.C. and Thomas, C.S. (1985) Changes in the biology of the kittiwake *Rissa tridactyla*: a 31 year study of a breeding colony. *J. Anim. Ecol.*, **54**, 9–26.

Coulson, J.C. and White, E. (1958) Observations of the breeding of the kittiwake. *Bird Study*, **5**, 74–83.
Coulson, J.C. and White, E. (1961) An analysis of the factors influencing the clutch size of the kittiwake. *Proc. Zool. Soc. Lond.*, **136**, 207–17.
Crawford, R.J.M. (1981) Seasonal patterns in South Africa's Western Cape purse seine fishery. *J. Fish. Biol.*, **16**, 649–64.
Crawford, R.J.M., Cruickshank, R.A., Shelton, P.A. and Kruger, I. (1985) Partitioning of a goby resource among four avian predators and evidence for altered trophic flow in the pelagic community of an intense, perennial upwelling system. *S. Afr. J. Mar. Sci.*, **3**, 215–28.
Crawford, R.J.M. and Shelton, P.A. (1978) Pelagic fish and seabird interrelationships off the coasts of South West and South Africa. *Biol. Conserv.*, **14**, 85–109.
Crawford, R.J.M. and Shelton, P.A. (1981) Population trends for some southern African seabirds related to fish availability, in: *'Proceedings of the Symposium on Birds of the Sea and Shore.'* (ed. J. Cooper), African Seabird Group, Cape Town, pp. 15–41.
Crawford, R.J.M., Shelton, P.A., Batchelor, A.L. and Clinning, C.F. (1980) Observations on the mortality of juvenile cape cormorants *Phalacrocorax capensis* during 1975 and 1979. *Fish. Bull. S. Afr.*, **13**, 69–75.
Crawford, R.J.M., Shelton, P.A., Cooper, J. and Brooke, R.K. (1983) Distribution, population size and conservation of the cape gannet *Morus capensis*. *S. Afr. J. Mar. Sci.*, **1**, 153–74.
Croll, D.A., Gaston, A.J. and Noble, D.G. (1991) Adaptive loss of mass in thick-billed murres. *Condor*, **93**, 496–502.
Croxall, J.P. (1989) Use of indices of predator status and performance in CCAMLR fishery management strategies. *CCAMLR Papers 1989*, pp. 353–65.
Croxall, J.P. (ed.) (1991) *Seabird Status and Conservation: A Supplement*. ICBP Tech. Publ. 11, International Council for Bird Preservation, Cambridge.
Croxall, J.P., Ricketts, C. and Prince, P.A. (1984a) Impact of seabirds on marine resources, especially krill, of South Georgia waters, in *'Seabird Energetics.'* (eds G.C. Whittow and H. Rahn), Plenum, New York, pp. 285–317.
Croxall, J.P., Evans, P.G.H. and Schreiber, R.W. (eds) (1984) Status and Conservation of the World's Seabirds. *ICPB Tech Rep 2*, International Council for Bird Preservation, Cambridge.
Croxall, J.P., McCann, T.S., Prince, P.A. and Rothery, P. (1988a) Reproductive performance of seabirds and seals on south Georgia and Signy Island, South Orkney Islands, 1976–1987: Implications for Southern Ocean monitoring studies, in *'Antarctic Ocean and Resources Variability'* (ed. D. Sahrhage), Springer Verlag, Berlin, pp. 261–85.
Croxall, J.P., Davis, R.W. and O'Connell, M.J. (1988b) Diving patterns in relation to diet of gentoo and macaroni penguins at South Georgia. *Condor*, **90**, 157–67.
Croxall, J.P., Naito, U., Kato, A. *et al.* (1991) Diving patterns and performance in the antarctic blue eyed shag (*Phalacrocorax atriceps*). *J. Zool., Lond.*, **225**, 177–99.
Cushing, D.H. (1982), *'Climate and Fisheries'*, Academic Press, London.
Diamond, A.W. (1978) Feeding strategies and population size in tropical seabirds. *Am. Nat.*, **112**, 215–23.
Diamond, J.M., Karasov, W.H., Phan, D. and Carpenter, F.L. (1986) Digestive physiology is a determinant of foraging bout frequency in hummingbirds. *Nature*, **320**, 62–3.

Donmasnes, A. and Rottigen, I. (1985) Acoustic stock measurement of the Barents Sea capelin 1972–1984: a review, in *'Proceedings of the Soviet Norwegian Symposium on the Barents Sea Capelin'* (ed. H. Gjosaeter), Institute of Marine Research, Bergen, pp. 45–108.

Drent, R.H. and Daan, S. (1980) The prudent parent: energetic adjustments in avian breeding. *Ardea*, **68**, 225–52.

Duffy, D.C. (1983) Environmental uncertainty and commercial fishing: effects on Peruvian guano birds. *Biol. Conserv.*, **26**, 227–38.

Duffy, D.C., Berruti, A., Randall, R.M. and Cooper, J. (1984) Effects of the 1982–83 warm water event on the breeding of South African seabirds. *S. Afr. J. Sci.*, **80**, 65–9.

Elliot, R.D., Ryan, P.C. and Lidster, W.L. (1990) Winter diet of thick-billed murres in coastal Newfoundland waters. *Stud. Avian Biol.*, **14**, 125–38.

Erikstad, K.E. (1990) Winter diets of four seabird species in the Barents Sea after a crash in the capelin stock. *Polar Biol.*, **10**, 619–27.

Fraser, J.H. (1952) The chaetognatha and other zooplankton of the Scottish area and their value as biological indicators of hydrographical conditions. *Mar. Res. Scott.*

Furness, R.W. (1978) Energy requirements of seabird communities: a bioenergetics model. *J. Anim. Ecol.*, **47**, 39–43.

Furness, R.W. (1990) A preliminary assessment of the quantities of Shetland sandeels taken by seabirds, seals, predatory fish and the industrial fishery in 1981–83. *Ibis*, **132**, 205–17.

Furness, R.W. and Ainley, D.G. (1984) Threats to seabird populations presented by commercial fisheries. *ICPB Tech. Rept.*, **2**, 701–8.

Furness, R.W. and Barrett, R.T. (1985) The food requirements and ecological relationships of a seabird community in North Norway. *Ornis Scand.*, **16**, 305–13.

Furness, R.W. and Birkhead, T.R. (1984) Seabird colony distributions suggest competition for food supplies during the breeding season. *Nature*, **311**, 655–6.

Furness, R.W. and Cooper, J. (1982) Interactions between breeding seabird and pelagic fish populations in the southern Benguela region. *Mar. Ecol. Prog. Ser.*, **8**, 243–50.

Furness, R.W. and Monaghan, P. (1987) *'Seabird Ecology'*, Blackie, London.

Furness, R.W., Ensor, K. and Hudson, A.V. (1992) The use of fishery waste by gull populations around the British Isles. *Ardea*, **80**, 105–13.

Gales, R.P. (1988) The use of otoliths as indicators of little penguin *Eudyptula minor* diet. *Ibis*, **130**, 418–26.

Gales, R.P and Green, B. (1990) The annual energetics cycle of little penguins (*Eudyptula minor*). *Ecology*, **71**, 2297–312.

Gaston, A.J. (1985) Energy invested in reproduction by thick billed murres (*Uria lomvia*). *Auk*, **102**, 447–58.

Gaston, A.J., Chapdelaine, G. and Noble, D.G. (1983) The growth of thick billed murre chick at colonies in Hudson Strait: inter- and intra-colony variation. *Can. J. Zool.*, **61**, 2465–75.

Gaston, A.J. and Nettleship, D.N. (1981) The thick-billed murres of Prince Leopold Island – a study of the breeding biology of a colonial high Arctic seabird. *Can. Wildl. Serv. Monogr. Ser.*, **6**, Canadian Wildlife Service, Ottawa.

Gomes, M. (1991) Predictions under uncertainty: fish assemblages and food webs on the Grand Bank. Ph.D. thesis, Memorial University of Newfoundland, St. John's.

Goudie, R.I., and Ankney, D. (1988) Patterns of habitat use by sea ducks wintering in southeastern Newfoundland. *Ornis Scand.*, **19**, 249–56.

Hamer, K.C. and Furness, R.W. (1991) Age-specific breeding performance and reproductive effort in great skuas *Catharacta skua*. *J. Anim. Ecol.*, **60**, 693–704.

Hamer, K.C., Furness, R.W. and Caldow, R.W.G. (1991) The effects of changes in food availability on the breeding ecology of great skuas *Catharacta skua* in Shetland. *J. Zool., Lond.*, **223**, 75–188.

Hampton, I. Agenbag, J.J. and Cram, D.L. (1979) Feasibility of assessing the size of the South West African pilchard stock by combined aerial and acoustic survey measurements. *Fish. Bull. S. Afr.*, **11**, 10–22.

Haney, J.C. (1991) Influence of pynocline topography and water column structure on marine distributions of alcids (Aves: Alcidae) in Anadyr Straight, Northern Bering Sea, Alaska. *Mar. Biol.*, **10**, 419–35.

Harris, L. (1990) *'Independent Review of the State of the Northern Cod Stock'*, Supply and Services Canada, Ottawa.

Harris, M.P. and Wanless, S. (1990) Breeding success of British kittiwakes *Rissa tridactyla* in 1986–88: evidence for changing conditions in the northern North Sea. *J. Appl. Ecol.*, **27**, 172–87.

Harwood, J. (1983) Interactions between marine mammals and fisheries. *Adv. Appl. Biol.*, **8**, 189–214.

Harwood, J. and Croxall, J.P. (1988) The assessment of competition between seals and commercial fisheries in the North Sea and the Antarctic. *Mar. Mam. Sci.*, **4**, 13–33.

Hassell, M.P. and May, R.M. (1974) Aggregation of predators and insect parasites and its effect on stability. *J. Anim. Ecol.*, **43**, 567–94.

Hatch, S.A. (1987) Did the 1982–83 El Nino–Southern Oscillation affect seabirds in Alaska? *Wilson Bull.*, **99**, 468–74.

Hatch, S.A., Byrd, G.V., Irons, D.B. and Hunt, G.L. Jr. (in press) Status and ecology of kittiwakes in the North Pacific. *Proc. Pac. Seabird Grp. Symp. (Victoria, British Columbia)*, Canadian Wildlife Service, Ottawa.

Hatch, S.A. and Hatch, M.A. (1990b) Breeding seasons of oceanic birds in a subarctic colony. *Can. J. Zool.*, **68**, 1664–79.

Hatch, S.A. and Hatch, M.A. (1990a) Components of breeding productivity in a marine bird community: key factors and concordance. *Can. J. Zool.*, **68**, 1680–90.

Hatch, S.A. and Sanger, G.A. (1992) Puffins as predators on juvenile pollack and other forage fish in the Gulf of Alaska. *Mar. Ecol. Prog. Ser.*, **80**, 1–14.

Heinemann, D.L., Hunt, G.L. Jr. and Everson, I. (1990) The distribution of marine avian predators and their prey, *Euphausia superba*, in Bransfield Strait and southern Drake Passage. *Mar. Ecol. Prog. Ser.*, **58**, 3–16.

Heubeck, M. (1989) Breeding success of Shetland's seabirds: arctic skua, kittiwake, guillemot, razorbill and puffin, in *'Seabirds and Sandeels'* (ed. M. Heubeck), Shetland Bird Club, Lerwick, pp. 11–18.

Hewitt, R.P. and Brewer, G.D. (1983) Nearshore production of young anchovy. *Rep. Calif. Coop. Ocean. Fish. Invest.*, **24**, 235–44.

Hida, T.S. (1957) Chaetognaths and pteropods as biological indicators in the North Pacific. *US Fish and Wild. Serv. Spec. Sci. Rep. 215*, US Fish and Wildlife Service, Washington, D.C.

Hilden, M. (1988) Errors of perception in stock and recruitment studies due to wrong choices of natural mortality rate in virtual population analysis. *J. Cons. Int. Explor. Mer.*, **44**, 123–34.

Hislop, J.R.G. and Harris, M.P. (1985) Recent changes in the food of young puffins *Fratercula arctica* on the Isle of May in relation to fish stocks. *Ibis*, **127**, 234–9.

Hislop, J.R.G., Harris, M.P. and Smith, J.G.M. (1991) Variation in the calorific value and total energy content of the lesser sandeel (*Amodytes marinus*) and other fish preyed on by seabirds. *J. Zool., Lond.*, **224**, 501–17.

Hobson, K.A. (1991) Stable isotopic analysis of the trophic relationships of seabirds: preliminary investigations of alcids from coastal British Columbia, in '*Studies of High Latitude Seabirds 1: Behavioural, Energetics and Oceanographic Aspects of Seabird Feeding Ecology*' (eds W.A. Montevecchi and A.J. Gaston) *Can. Wildl. Serv. Occas. Pap.* **68**, 39–48.

Hobson, K.A. and Montevecchi, W.A. (1991) Stable isotopic determinations of tropic relationships of great auks. *Oecologia*, **87**, 528–31.

Hoenig, J.M., Restrepo, V.R. and Baird, J.W. (1991) A practical approach to risk and cost analysis of fisheries management options, with application to Northern Cod. *Intern. Council Explor. Sea*, **D**, **30**, 1–11.

Hudson, A.V. and Furness, R.W. (1988) Utilization of discarded fish by scavenging seabirds behind whitefish trawlers in Shetland. *J. Zool., Lond.*, **215**, 151–66.

Hudson, A.V. and Furness, R.W. (1989) The behaviour of seabirds foraging at fishing boats around Shetland. *Ibis*, **131**, 225–37.

Hunt, G.L. Jr. (1991) Marine ecology of seabirds in polar oceans. *Amer. Zool.*, **31**, 131–42.

Hunt, G.L. Jr. and Butler, J.L. (1980) Reproductive ecology of western gulls and Xantus' murrelets with respect to food resources in the southern California Bight. *Calif. Coop. Oceanic Fish. Invest. Rep.*, **21**, 62–7.

Hunt, G.L. Jr., Harrison, N.M. and Cooney, T. (1990) The influence of hydrographic structure and prey abundance on foraging of least aukets. *Stud. Avian Biol.*, **14**, 7–22.

Hunt, G.L. Jr, Piatt, J.F. and Erikstad, K.E. (1991) How do foraging seabirds sample their environment? *Acta XX Cong. Internat. Ornithol.*, pp. 2272–9.

Hunt, G.L. Jr, and Schneider, D.C. (1987) Scale-dependent processes in the physical and biological environment, *Seabirds: Feeding Ecology and Role in Marine Ecosystems* (ed. J.P. Croxall), Cambridge University Press, Cambridge, pp. 7–41.

Hunt, G.L. Jr., Eppley, Z. and Schneider, D.C. (1986) Reproductive performance of seabirds: the importance of population and colony size. *Auk*, **103**, 306–17.

Jouventin, P. and Weimerskirch, H. (1990) Satellite tracking of wandering albatrosses. *Nature*, **343**, 746–8.

Kampp, K. (1988) Migration and winter ranges of Brunnich's guillemots *Uria lomvia* breeding or occurring in Greenland. *Dan. Ornithol. Foren. Tidsskr.*, **82**, 117–30.

Kerr, S.R. and Ryder, R.A. (1989) Current approaches to multispecies analyses of marine fisheries. *Can. J. Fish. Aquat. Sci.*, **46**, 528–34.

Klomp, N.I. and Furness, R.W. (1992) Non-breeders as a buffer against environmental stress: declines in numbers of great skuas on Foula, Shetland, and prediction of future recruitment. *J. Appl. Ecol.*, **29**, 341–8.

Kooyman, G.L., Davis, R.W., Croxall, J.P. and Coasta, D.P. (1982) Diving depths and energy and energy requirements of king penguins. *Science*, **217**, 726–7.

Lack, D. (1966) *Population Studies of Birds*, Oxford University Press, Oxford.

Lack, D. (1967) Interrelationships in breeding adaptations as shown by marine birds. Proc XIV Intern. Ornithol. Congr., pp. 3–42.

LaCock, G.D. (1986) The Southern Oscillation, environmental anomalies, and mortality of two southern African seabirds. *Clim. Change*, **8**, 173–84.

Leggett, W.C., Frank, K.T. and Carscadden, J.E. (1984) Meteorological and hydrographic regulation of year class strength in capelin (*Mallotus villosus*). *Can. J. Fish. Aquat. Sci.*, **41**, 1193–201.

Leim, A.H. and Scott, W.B. (1966) *Fishes of the Atlantic Coast of Canada*, Fisheries Research Board of Canada, Ottawa.

Lensink, C.S. (1984) The status and conservation of Alaska seabirds, *ICBP Tech. Rep.*, **2**, 13–27.

Leveille, G.A., Fisher, H. and Feigebaum, A.S. (1961) Dietary protein and its effects on the serum proteins of the chicken. *Ann. N.Y. Acad. Sci.*, **94**, 265–71.

Lloyd, D.S. (1985) Breeding Performance of Kittiwakes and Murres in Relationship to Oceanographic and Meteorologic Conditions Across the Shelf of the Southeastern Bering Sea. M.Sc. thesis, University of Alaska, Fairbanks.

MacCall, A.D. (1979) Population estimates for the waning years of the Pacific sardine fishery. *Rep. Calif. Coop. Ocean. Fish. Invest.*, **20**, 72–82.

Martin, A.R. (1989) The diet of Atlantic puffin *Fratercula arctica* and northern gannet *Sula bassana* chicks at a Shetland colony during a period of changing prey availability. *Bird Study*, **36**, 170–80.

McEvoy, A.F. (1986) *The Fisherman's Problem*, Cambridge University Press, Cambridge.

Merriam, C.H. (1898) Life zones and crop zones of the United States. *Bull. U.S. Dept. Agric.* **10**. US Department of Agriculture, Washington, D.C.

Monaghan, P. and Metcalfe, N.B. (1986) On being the right size: natural selection and body size in the herring gull. *Evolution*, **40**, 1096–9.

Monaghan, P., Uttley, J.D. and Burns, M.D. (1991) The influences of changes in prey availability on the breeding ecology of terns. *Acta XX Cong. Internat. Orthnithol.*, pp. 2257–62.

Monaghan, P., Uttley, J.D., Burns, M.D., et al. (1989) The relationship between food supply, reproductive effort and breeding and success in Arctic terns *Sterna paradisaea*. *J. Anim. Ecol.*, **58**, 261–74.

Monaghan, P. and Zonfrillo, B. (1986) Population dynamics of seabirds in the Firth of Clyde. *Proc. R. Soc. Edinburgh*, **90B**, 363–75.

Montevecchi, W.A. (1978) Nest site selection and its survival value among laughing gulls. *Behav. Ecol. Sociobiol.*, **4**, 143–61.

Montevecchi, W.A. and Barrett, R.T. (1987) Prey selection by gannets at breeding colonies in Norway. *Ornis Scand.*, **18**, 319–22.

Montevecchi, W.A. and Berruti A. (1991) Avian bioindication of pelagic fishery conditions in the southeast and northeast Atlantic. *Acta XX Cong. Internat. Ornithol.*, pp. 2246–56.

Montevecchi, W.A. and Hufthammer, A.K. (1990) Zooarchaeological implications for prehistoric seabird distributions along the Norwegian coast. *Arctic*, **43**, 110–4.

Montevecchi, W.A. and Myers, R.A. (1992) Monitoring fluctuations in pelagic fish availability with seabirds. *Can. Atlan. Fish. Sci. Advis. Comm. Res. Doc*, **92/94**, 1–22.

Montevecchi, W.A. and Porter, J.M. (1980) Parental investments by seabirds at the breeding area with emphasis on northern gannets, in: *Behavior of*

Marine Animals. Vol. 4 (eds J. Burger, B.L. Olla and H.E. Winn), Plenum Press, New York, pp. 323–65.

Montevecchi, W.A., Birt, V.L. and Cairns, D.K. (1987a) Dietary shifts of seabirds associated with local fisheries failures. *Biol. Oceanogr.*, **5**, 153–9.

Montevecchi, W.A., Barrett, R.T., Rikardsen, F. and Strann, K. B. (1987b) The status of gannets in Norway, 1985. *Fauna Norv. Ser. C. Cinclus*, **10**, 65–72.

Montevecchi, W.A., Cairns, D.K. and Birt, V.L. (1988) Migration of post-smolt Atlantic salmon (*Salmo salar*) off northeastern Newfoundland, as inferred from tag recoveries in a seabird colony. *Can. J. Fish. Aquat. Sci.*, **45**, 568–71.

Montevecchi, W.A., Birt, V.L. and Cairns, D.K. (1992) Reproductive energetics and prey harvest by Leach's storm petrels in the northwest Atlantic. *Ecology*, **73**, 823–32.

Morrison, M.L. (1986) Bird populations as indicators of environmental change. *Curr. Ornithol.*, **3**, 429–51.

Murphy, E.C., Springer, A.M. and Roseneau, D.G. (1986) Population status of *Uria aalge* at a colony in western Alaska: results and simulations.*Ibis*, **128**, 348–63.

Murphy, E.C., Springer, A.M. and Roseneau, D.G. (1991) High annual variability in reproductive success of kittiwakes (*Rissa tridactyla* L.) at a colony in western Alaska. *J. Anim. Ecol.*, **60**, 515–34.

Nagy, K.A. (1980) CO_2 production in animals: analysis of potential errors in the doubly labeled water method. *Am. J. Physiol.*, **238**, R466–73.

Nagy, K.A. (1983), *The doubly labelled water ($^3HH^{18}O$) method: a guide to its use*, University of California at Los Angeles, Los Angeles.

Nisbet, I.C.T (1973) Courtship feeding, egg size and breeding success in common terns. *Nature*, **241**, 141–3.

Ollason, J.C. and Dunnet, G.M. (1988) Variation in breeding success, in *Reproductive Success*. (ed. T.H. Clutton Brock), University of Chicago, Chicago, pp. 263–78.

Orians, G.H. and Pearson, N.E. (1979) On the theory of central place foraging, in *Analysis of Ecological Systems* (eds D.J. Horn, R.D. Mitchell and G.R. Stairs), Ohio State University Press, Columbus, pp. 155–77.

Parsons, J. (1975) Seasonal variation in the breeding success of the herring gull: an experimental approach to prefledging success; *J. Anim. Ecol.*, **44**, 553–73.

Pearson, T.H. (1968) The feeding biology of seabird species breeding on the Farne Islands, Northumberland. *J. Anim. Ecol.*, **37**, 521–52.

Pehrsson, O. (1991) Egg size and clutch size in the mallard as related to food quality. *Can. J. Zool.*, **69**, 156–62.

Perrins, C.M. (1970) The timing of birds' breeding seasons. *Ibis*, **112**, 242–55.

Piatt, J.F. (1987) Behavioural Ecology of Common Murres and Atlantic Puffin Predation on Capelin: Implications for Population Biology. Ph.D. thesis, Memorial University of Newfoundland, St. John's.

Piatt, J.F. (1990) The aggregative response of common murres and Atlantic puffins to schools of capelin. *Stud. Avian Biol.*, **14**, 36–51.

Piatt, J.F. and Nettleship, D.N. (1985) Diving depths of four alcids. *Auk*, **102**, 293–7.

Pope, J.G. and Gray, D. (1983) An investigation of the relationship between the precision of assessment data and the precision of total allowable catches, in *Sampling Commercial Catches of Marine Fish and*

Invertebrates (eds W.G. Doubleday and D. Rivard) *Can. Spec. Pub. Fish. Aquat. Sci.*, **66**, 151–8.

Prince, P.A. and Harris, M.P. (1988) Food and feeding ecology of breeding Atlantic alcids and penguins. *Acta XIX Congressus Internationalis Ornithologici*, pp. 1195–204.

Pugesek, B.J. (1981) Increased reproductive effort with age in the California Gull (*Larus californicus*). *Science*, **212**, 822–3.

Rhymer, J.M (1988) The effect of egg size variability on thermoregulation of mallard (*Anas platyrhynchos*) offspring and its implications for survival. *Oecologia*, **75**, 20–4.

Rice, J.C. (1992) Multispecies interactions in marine ecosystems: current approaches and implications for study of seabird populations, in: *Wildlife 2001: Populations* (eds D.R. McCullough and R.H. Barrett), Elsevier, London, pp. 586–601.

Richard, J.M. (1987) The mesopelagic fish and invertebrate macrozooplankton faunas of two Newfoundland fjords with differing physical oceanography. M.Sc. thesis, Memorial University of Newfoundland, St. John's.

Ricker, W.E. (1975) Computation and interpretation of biological statistics of fish populations. *Bull. Fish. Res. Bd. Can.*, **191**.

Ricklefs, R.E. (1973) Fecundity, mortality and avian demography, in '*Breeding Biology of Birds*' (ed. D.S. Farner), National Academy of Sciences, Washington, D.C., pp. 366–434.

Ricklefs, R.E. (1990) Scaling pattern and process in marine ecosystems in '*Large Marine Ecosystems: Patterns, Processes and Yields*' (eds K. Sherman, L.M. Alexander and B.D. Gold), American Association for the Advancement of Science, Washington, D.C, pp. 169–78.

Ricklefs, R.E., Duffy, D.C. and Coulter, M. (1984) Weight gain of blue footed booby chicks: an indicator of marine resources. *Ornis Scand.*, **15**, 162–6.

Rose, G.A. and Leggett, W.C. (1990) The importance of scale to predator prey spatial correlations: an example of Atlantic fishes. *Ecology*, **71**, 33–43.

Rothschild, B.J. and Osborn, T.R. (1990) Biodynamics of the sea: preliminary observations on high dimensionality and the effects of physics on predator prey interrelationships, in *Large Marine Ecosystems: Patterns, Processes and Yields* (eds K.Sherman, L.M. Alexander and B.D. Gold), American Association for the Advancement of Science, Washington, D.C., pp. 71–81.

Russell, F.S. (1935) On the value of certain plankton animals as indicators of water movements in the English Channel and North Sea. *J. Mar. Biol. Assoc. U.K.*, **20**, 309–31.

Ryan, P.G. and Moloney, C.L. (1988) Effect of trawling on bird and seal distributions in the southern Benguela region. *Mar. Ecol. Prog. Ser.*, **45**, 1–11.

Safina, C. and Burger, J. (1988) Prey dynamics and the breeding phenology of common terns (*Sterna hirundo*). *Auk*, **105**, 720–6.

Safina, C., Burger, J., Gochfeld, M. and Wagner, R.H. (1988) Evidence for prey limitation of common and roseate tern reproduction. *Condor*, **40**, 852–9.

Schaefer, M.B. (1970) Men, birds and anchovies in the Peru Current – dynamic interaction. *Trans. Am. Fish. Soc.*, **9**, 461–7.

Schaffner, F.C. (1986) Trends in elegant tern and northern anchovy populattions in California. *Condor*, **88**, 347–54.

Schneider, D.C. (1991) The role of fluid dynamics in the ecology of marine birds. *Oceanogr. Mar. Biol. Ann. Rev.*, **29**, 487–521.

Schneider, D.C. and Duffy, D.C. (1985) Scale dependent variability in seabird abundance. *Mar. Ecol. Prog. Ser.*, **25**, 101–3.

Schneider, D.C., Duffy, D.C., McCall, A.D. and Anderson, D.W. (1988) Historical variation in guano production from the Peruvian and Benguela upwelling ecosystems. *Clim. Change*, **13**, 309–16.

Schneider, D.C. and Duffy, D.C. (in press) Seabird fisheries interactions: evolution with dimensionless rations, in *'Wildlife 2001: Populations'* (eds D.R. McCullough and R.H. Barrett), Elsevier, London, pp. 602–15.

Schneider, D.C. and Hunt, G.L. Jr. (1984) A comparison of seabird diets and foraging distribution around the Pribilof Islands, Alaska, in *'Marine Birds: Their Feeding Ecology and Commercial Fisheries Relationships'* (eds D.N. Nettleship, G.A. Sanger and P.F. Springer), Supply and Services Canada, Ottawa, pp. 86–93.

Schneider, D.C., Hunt, G.L. Jr. and Harrison, N.M (1986) Mass and energy transfer to seabirds in the southeastern Bering Sea. *Cont. Shelf Res.*, **5**, 241–57.

Schneider, D.C. and Piatt, J.F. (1986) Scale dependent correlation of seabirds with schooling fish in a coastal ecosystem. *Mar. Ecol. Prog. Ser.*, **32**, 237–46.

Schneider, D.C., Pierotti, R. and Threlfall, W. (1990) Alcid patchiness and flight direction near a colony in eastern Newfoundland. *Avian Biol.*, **14**, 23–35.

Schreiber, E.A. and Schreiber, R.W. (1989) Insight into seabird ecology from a global 'natural experiment'. *Nat. Geogr. Res.*, **5**, 64–81.

Schreiber, R.W. and Schreiber, E.A. (1984) Central Pacific seabirds and the El Nino Southern Oscillation: 1982 to 1983 retrospectives. *Science*, **225**, 713–6.

Shannon, L.V., Crawford, R.J.M and Duffy, D.C. (1984) Pelagic fisheries and warm events: a comparative study. *S. Afr. J. Sci.*, **80**, 51–60.

Shaw, D.M. and Atkinson, S.F. (1990) An introduction to the use of geographic information systems for ornithological research. *Condor*, **92**, 564–70.

Shelton, P.A., Boyd, A.J. and Armstrong, M.J. (1985) The influence of large scale environmental processes on neritic fish populations in the Benguela Current system. *Rep. Calif. Coop. Ocean. Fish. Invest.*, **26**, 72–92.

Sherman, K., Jones, C., Sullivan, L., Smith, W., Berrien, P. and Ejsymont, L. (1981) Congruent shifts in sandeel abundance in western and eastern North Atlantic ecosystems. *Nature*, **291**, 486–9.

Soutar, A. and Issacs, J.D. (1974) Abundances of pelagic fish during the 19th and 20th centuries as recorded in anaerobic sediment off the Californias. *Fish. Bull.*, **72**, 257–73.

Sparre, P. (1991) An overview of multispecies virtual population analysis. *Rapp. P. V. Reun. Cons. Int. Explor. Mer.* **190**.

Springer, A.M., Murphy, E.C., Roseneau, D.G. *et al.* (1987) The paradox of pelagic food webs in the northern Bering Sea. I. Seabird food habits. *Cont. Shelf Res.*, **7**, 895–911.

Springer, A.M., Roseneau, D.G., Lloyd, D.S. *et al.* (1986) Seabird responses to fluctuating prey availability in the eastern Bering Sea. *Mar. Ecol. Prog. Ser.*, **32**, 1–12.

Springer, A.M., Roseneau, D.G., Murphy, E.C. and Springer, M.I. (1984) Environmental controls of marine food webs: food habits of seabirds in the eastern Chukchi Sea. *Can. J. Fish. Aquat. Sci.*, **41**, 1202–15.

Steele, D.H. and Montevecchi, W.A. (1993) Leach's storm petrels prey on lower mesopelagic crustaceans: implications for crustacean and avian distributions. *Crustacea*.

Steele, J.H. (1974) *'The Structure of Marine Ecosystems'*, Harvard University Press, Cambridge, Mass.

Steele, J.H. (1979) Some problems in the management of marine resources, in *'Applied Ecology. Vol. 4.'* (ed. T.H. Cooker) Academic Press, London, pp. 103–40.

Straty, R.R. and Haight, R.E. (1979) Interactions among marine birds and commercial fish in the eastern Bering Sea, in *'Conservation of Marine Birds of Northern North America'* (eds J.C. Bartonek and D.N. Nettleship), *Wildlife Research Report II*, United States Department of the Interior, Fish and Wildlife Service, Washington, D.C. pp. 201–19.

Swatzman, G.L. and Haar, R.T. (1983) Interactions between fur seal populations and fisheries in the Bering Sea. *Fish. Bull.*, **81**, 121–32.

Taggart, C.T. and Frank, K.T. (1990) Perspectives on larval fish ecology and recruitment processes: probing the scales of relationships, in *'Large Marine Ecosystems: Patterns, Processes and Yields'* (eds K. Sherman, L. Alexander and B. Gold), American Association for Advancement of Science, Washington, D.C,. pp. 151–64.

Tasker, M.L., Furness, R.W., Harris, M.P. and Bailey, R.S. (1989) Food consumption of seabirds in the North Sea (abstract). *ICES Symp. 66. Multispecies Models to Management of Living Resources*, International Council for the Exploration of the Sea, Copenhagen.

Temple, S.A. and Wiens, J.A. (1989) Bird populations and environmental changes: can birds be bio-indicators? *Am. Birds*, **43**, 260–70.

Trivelpiece, W.Z., Ainley, D.G., Fraser, W.R. and Trivelpiece, S.G. (1990) Skua survival. *Nature*, **345**, 211.

Tucker, V.A. (1974) Energetics of natural avian flight. in *'Avian Energetics'* (ed. R.A. Paynter Jr.), Nuttall Ornithological Club, Cambridge, Mass, pp. 298–333.

Uttley, J., Monaghan, P. and White, S. (1989) Differential effects of reduced sandeel availability on two sympatrically breeding species of tern. *Ornis Scand.*, **20**, 273–7.

Uttley, J., Walton, P. and Monaghan, P. (submitted) The effects of food availability on breeding performance and adult time budgets of common guillemots. *J. Anim. Ecol.*

Vader, V., Anker Nilssen, T., Bakken, V. *et al.* (1990b) Regional and temporal differences in breeding success and population development of fish eating seabirds in Norway after collapses of herring and capelin stocks. *Trans. 19 IUGB Congr. (Trondheim)*, pp. 143–50.

Vader, W., Barrett, R.T., Erikstad, K. E. and Strann, K. B. (1990a) Differential responses of common and thick billed murres to a crash in the capelin stock in the southern Barents Sea. *Stud. Avian Biol.*, **14**, 175–80.

Vermeer, K. and DeVito, K. (1988) The importance of *Paracallisoma coecus* and myctophid fish to nesting fork tailed and Leach's storm petrels in the Queen Charlotte Islands, British Columbia. *J. Plankton Res.* **10**, 63–75.

Wanless, S., Burger, A.E. and Harris, M.P. (1991b) Diving depths of shags *Phalacrocorax aristotelis* breeding on the Isle of May. *Ibis* **133**, 37–42.

Wanless, S., Harris, M. P. and Morris, J. A. (1991a) Foraging range and feeding locations used by shags *Phalacrocorax aristotelis* during chick rearing. *Ibis*, **133**, 30–6.

Weathers, W. W., Buttemer, W.A., Hayworth, A.M. and Nagy, K.A. (1984) An evaluation of time budget estimates of daily energy expenditure in birds. *Auk*, **101**, 459–72.

Wiens, J.A. (1976) Population responses to patchy environments. *Ann. Rev. Ecol. Sys.*, **7**, 81–120.

Wiens, J.A. (1989) Spatial scaling in ecology. *Func. Ecol.*, **3**, 385–97.

Wiens, J.A. and Scott, J.M. (1975) Model estimation of energy flow in Oregon coastal seabird populations. *Condor*, **77**, 439–52.

Williams, T.D. and Croxall, J.P. (1990) Is chick fledging weight a good index of food availability in seabird populations? *Oikos*, **59**, 414–6.

Wooler, R.V., Bradley, J.S. and Croxall, J.P. (1992) Long term population studies of seabirds. *Trends in Ecology and Evolution*, **7**, 111–14.

Wynne-Edwards, V.C. (1962) *'Animal Dispersion in Relation to Social Behaviour'*, Oliver and Boyd, Edinburgh.

7
Integrated population monitoring: detecting the effects of diverse changes

J.J.D. Greenwood, S.R. Baillie, H.Q.P. Crick,
J.H. Marchant and W.J. Peach

7.1 THE SCOPE OF THIS CHAPTER

7.1.1 Why monitor populations?

An important reason for monitoring populations of birds is that their conservation is important in its own right. It is important to have a sufficient knowledge of underlying population processes to determine the probable causes of any population decline so that steps can be taken to halt or even reverse it. The subsequent success, or otherwise, of any action taken should be monitored.

Furthermore, the monitoring of populations has a wider use. As argued in Chapter 1, birds are useful indicators of the state of the environment in which they live. In other chapters, examples are given of the ways in which birds can be used to monitor specific environmental changes. But so many aspects of the environment are subject to change that we cannot hope to monitor all of them specifically. The value of monitoring the populations of a broad spectrum of birds is that it enables us to pick up the effects of environmental changes whose impact on wildlife would not otherwise be monitored. This is because birds are relatively easy to study and their ecology is well known, so

Birds as Monitors of Environmental Change. Edited by R.W. Furness and J.J.D. Greenwood.
Published in 1993 by Chapman & Hall, London. ISBN 0 412 40230 0.

that the likely causes of population change may be identified more easily than would be the case for less well-studied organisms. Furthermore, birds have a diversity of ecologies, and occupy many habitats and trophic levels, so that a broadly-based monitoring programme is likely to pick up the effects of a wide range of environmental changes.

7.1.2 Which species?

If one value of monitoring bird populations is that the use of a diversity of species covers a broad range of environmental changes, it is clearly not sensible to restrict monitoring to a few 'indicator species'. Furthermore, it is often possible to cover most of the species of a particular type (e.g. waterbirds or small passerines) with little more effort than it would take to cover just one or two of them. Indeed, it may be more satisfying for the field workers to do so and, by sustaining their enthusiasm, this would make a monitoring programme easier to manage.

7.1.3 What parameters should we monitor?

Surveillance of national or regional bird population levels is well established in North America (Collins and Wendt, 1989; Robbins, 1985; Robbins *et al.*, 1986; Robbins *et al.*, 1989; Sauer and Droege, 1990) and Europe (van Dijk, 1992; Flade and Schwarz, 1992; Hustings, 1992; Koskimies and Väisänen, 1991; Larsson, 1992; Marchant *et al.*, 1990; Tiainen, 1985), with new schemes being set up in many of the parts of Europe that have hitherto lacked them (see the various papers in Hagemeijer and Verstrael, in press) and in some other countries, such as Australia (Ambrose, in press). Such surveillance is an important component of population monitoring but in itself cannot provide an adequate monitoring programme.

The limitations of simple surveillance of population levels can be seen if we reflect on the objectives of bird population monitoring and on how these may be achieved. One objective is to provide a basis for managing the bird populations themselves. The other is to be able to use the information on the birds to monitor changes in the environment, so that the environment itself may be managed (interpreting the concept of management broadly, to include not only direct actions such as, for example, the seasonal timing of agricultural operations or the limitation of pollution but also the political processes necessary to advance such direct actions). Such management requires true monitoring, as considered in Chapter 1, rather than mere surveillance. It requires that we have some understanding of the likely causes of any deleterious changes taking place and of ways in which these changes might be reversed or halted. It requires that, in order to avoid premature or unnecessary action, we can determine thresholds for conservation action and distinguish natural from man-made causes. All this means that we need to have an understanding of the processes that lead to the population levels we observe, not just a knowledge of the levels themselves. These are the processes of reproduction, mortality, emigration and immigration. A knowledge of them, and of how they are affected by the environment and by population density, is central to building population models on which sound management can be based.

Newton (1991a) has advanced the view that 'The main contribution of research to bird population management is to identify the factors that limit population sizes. Without knowledge of the limiting factors, the management of any bird population is bound to be hit and miss.' He therefore concludes that habitat studies are crucial. We agree. The study of population processes can only be fully revealing in the context of habitat. Population monitoring thus entails recording habitat as well as measuring demographic parameters.

Newton also suggests that study of demographic parameters, though desirable, is not necessary for population management. This is going too far. We can only understand what determines the size of a population if we know not only the environmental factors that impinge upon it but also the way in which they do so. That is, we need to know how the population will react to variation in these factors and thus to understand the dynamic processes leading to its reactions.

Furthermore, it is the central contention of this book that the study of birds can be an important component of monitoring their environments. Thus, we are most interested in situations where key elements of the environment are not readily susceptible to direct surveillance. Even if we ignore the need for deep understanding of population processes, it is valuable to measure demographic parameters other than abundance alone since this may provide additional evidence about causes of population change. A simple example is that the evidence that the abundance of sedge warblers *Acrocephalus schoenobaenus* breeding in Britain is linked to rainfall in the winter quarters in the Sahel zone of West Africa was much reinforced by the finding that survival rates were correlated with the same factor (Peach *et al.*, 1991). Furthermore, other demographic parameters may sometimes be more sensitive to some environmental factors than is population size. This is because population level may be stabilized by density-dependent feedback (in relation to other environmental factors or as a result of behavioural interactions). Thus breeding populations of the tawny owl *Strix aluco* are relatively stable despite huge variations in breeding output (related to fluctuations in prey availability), because recruitment to the breeding populations is strongly density-dependent, as a result of territorial behaviour (Southern, 1970). An environmental change that depressed breeding success would not be detected by surveillance of the size of the breeding population alone. Yet such a change might have broader significance. Furthermore, it could make even the breeding population – apparently unaffected by it (in terms of abundance) – more vulnerable to changes in other environmental factors.

Another reason for studying all demographic parameters is that some of them may respond more quickly than absolute abundance to environmental changes. Croxall and Rothery (1991) point out, for example, that the long lives and delayed maturity of seabirds mean that changes in breeding productivity, even if they do lead to changes in size of the breeding population, will only do so after several years' delay. A specific case is the decline in the breeding population of the peregrine falcon *Falco peregrinus* in Britain. This was an important signal of the impact of pesticides on wildlife (Ratcliffe, 1980) but it appears to have been preceded by a decline in breeding output (Figure 7.1); the latter could have given an earlier signal had a proper monitoring programme been in place (H.Q.P. Crick quoted in Pienkowski, 1991). On a local scale, established territory-holders may remain even after substantial habitat destruction, though the population eventually declines because new birds do not take up territories falling vacant as their owners die (Temple and Wiens, 1989).

If one is monitoring birds, or other wildlife, in order to monitor environmental factors, it is important to minimize the lag between the environmental change and the wildlife response – i.e. to monitor things that change quickly in response to an environment change (Croxall *et al.*, 1988). If, in contrast, one is monitoring a species in order to be able to manage that species itself, it is more important to concentrate on factors that are responsible for setting the abundance of the population (recruitment to the breeding population in the case of the tawny owl), though without forgetting other factors.

Practical considerations may also determine which parameters are the ones that it is most important to keep under surveillance. Thus Croxall and Rothery (1991) state that it is better, in principle, to monitor adult survival rather than breeding success in K-selected species such as most seabirds (because the former is more important in determining changes in their abundances) but note that, in practice, adult survival is very difficult to measure precisely in such species.

Finally, different environmental factors will impact on different population parameters – for example, DDT on the mortality of eggs but dieldrin on that of adults (Newton, 1979). Thus, which is the most

Scope of chapter

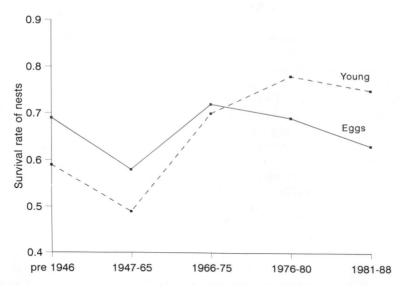

Figure 7.1 Breeding performance of peregrine falcons *Falco peregrinus*, as shown by the British Trust for Ornithology Nest Record Scheme, showing the reduction in survival of both eggs (continuous line) and young (dashed line) during 1947–65.

important parameter to monitor will depend on the circumstances. This reinforces our contention that the best procedure is to monitor all population parameters, in so far as this is feasible and cost-effective, and to do so within a context of habitat surveillance.

7.1.4 Depth of coverage

The space that we have devoted to various topics in this chapter partly reflects our own interests and priorities. But there are broader considerations. We concentrate on studies that use networks of observers working extensively because we believe that these are, for most species, more cost-effective for broad population monitoring than are intensive studies (section 7.2). We concentrate on smaller terrestrial birds partly because they are less easy to study than some others (so the methodology required is less obvious) and partly because they provide broad coverage of a variety of habitats. We concentrate on counting methods and on the analysis of counts and of data relevant to survival rates because the methods involved are not straightforward and have had much attention paid to them.

7.2 REPRESENTATIVENESS OF STUDY AREAS

It is sometimes possible to monitor the whole of a bird population, if it is very small or severely restricted in distribution. Generally, however, one can monitor only parts of a population, so that generalizations must be based on drawing inferences about the whole from what one has observed in respect of the parts. The key question is then how representative of the whole are the parts? It may be possible to study a single part of a population so carefully that one's knowledge of that part is very precise. But one has no idea of how precisely this reflects the population as a whole, since from a study of just one part one has no information on the variation between different parts. Furthermore, localized studies may give misleading results because the scale on which bird populations are determined may be greater than that of the study (Wiens, 1981, 1989).

One must therefore study more than one site. The problem is to determine how best to allocate resources: should one aim to study a few sites, each very carefully, or to study many, each in less detail? In principle, the former option is preferable if the differences between sites are small and the difficulty of making precise measurements at each site is large. In respect of the studies of bird populations in countries with significant numbers of birdwatchers, the choice is between using professional ornithologists to conduct detailed studies at a few sites or using them to organize the work of volunteers conducting less detailed studies at many sites. It is not always easy to judge which is preferable because the sort of information that can be collected by professionals and by volunteers may differ in kind. Volunteers can, for example, count, ring and weigh birds as well as (and sometimes better than) professionals but can rarely produce computer-processed assessments of patterns of post-fledging dispersal based on continuous, automated monitoring of radio-tracked birds.

The British Natural Environment Research Council (NERC) set up an Environmental Change Network 'to record, analyse and predict environmental change in the United Kingdom'. At its launch, in 1992, this consisted of eight sites, though it was hoped that others would be added. At these sites a wide range of physical, chemical, and biological variables is recorded systematically using strict, uniform protocols. The variables were selected to give a good understanding of the ecosystem processes at each site, and the uniformity of the processes of data collection enables the drawing together of information from the various sites to provide an insight into the effects of environmental change throughout the United Kingdom.

In contrast to the *modus operandi* of NERC, those responsible for bird population monitoring in Britain (and other countries) have mounted programmes that depend on large numbers of participants, usually working at large numbers of sites. They would argue that the differences between sites are so great that it is better, by dispersing the effort, to provide a nationwide monitoring programme to detect possible problems quickly, turning to intensive studies, perhaps based at only a few sites, when a particular problem has been identified. The ornithologists have been able to take that decision because of the large numbers of volunteer workers available.

So far we have been concerned with precision – the degree to which the results are repeatable by another survey using the same methods. But estimates need to be not only precise but also unbiased. Bias can only confidently be eliminated if sample units (that is the study sites in ecology) are chosen at random. None of the ornithological monitoring programmes in the world is currently based on randomly chosen sites. The main reason for this is that volunteers prefer to choose their own sites and naturally choose sites that are accessible and close to where they live. It has also been suggested that they tend to choose sites that appear likely to hold good numbers of birds, though Fuller et al. (1985) were able to show that Common Birds Census (CBC) farmland plots in south and east England were representative of lowland farms in that region, in terms of land use and cropping, so that, if there was any choice exerted by the fieldworkers who selected the plots, it must have been based on more subtle characteristics.

A potential problem with long-running surveys is that observers may discriminate against certain sorts of plots by abandoning them after having discovered that they lack birdwatching interest. It has been suggested, for example, that CBC workers may abandon plots on which bird numbers are declining particularly rapidly and that this could lead to the CBC indices underestimating the magnitude of decreases in national populations. In this case, checks have been made: the statement 'So few birds now' is rarely given as the reason for abandoning a plot, while only three of 42 observers questioned cited habitat-change (J.H. Marchant, quoted in O'Connor and Fuller, 1984) and there is little evidence that declines are more marked on plots that are abandoned than on plots that are continued (S.R. Baillie and S. Gates, in prep). Nonetheless, it is important to minimize plot turnover as far as possible and to continue to include plots in which the habitat has been grossly modified.

It is particularly important to avoid building bias into the survey design. The North American Breeding Bird Survey (BBS) is, for example, based on routes along roads. Although these roads are generally less busy than the Dutch motorways studied by Reijnen (in press), the effects on population density, breeding output, and dispersal of birds that he found were so dramatic that one wonders what bias this may introduce to the BBS results. There is the further problem that roads themselves are not distributed randomly across the landscape but tend to follow certain topographical features (Droege, 1990).

One way to reduce the importance of bias in monitoring is to use the same study sites in successive years, as is done in most surveillance programmes of bird abundance. But this will only work if the bias is constant over the years at any one site. If the BBS is affected by traffic in the same way as the areas adjacent to Dutch motorways, this will not be so: the Dutch effects were most marked in years of low population density, so year to year fluctuations in abundance (and other demographic variables) were more marked there than in the countryside generally; and, as road traffic increases with time, one might expect abundance and breeding success to decline more rapidly close to roads than elsewhere. If such potential biases have been built into the design of a surveillance programme, careful checks should be carried out to determine their likely importance.

In volunteer-based surveys, it is rarely possible to maintain absolute continuity of study sites. It is then important to check that the average characteristics of sites have not changed over the years in ways that could have produced apparent trends in the bird populations (Marchant et al., 1990). Another check on the likelihood of bias being responsible for apparent trends (or for obscuring actual trends) is to determine whether the trends are similar across different habitats and regions (Baillie, 1990; Gooch et al., 1991; Marchant et al., 1990). Paradoxically, regional (or habitat) differences in trends can be a valuable pointer to the causes of the trends (see below).

One way in which biases might, in principle, be reduced would be to stratify the distribution of sites geographically and in terms of habitat. So long as some sites were located in each stratum, differences in sampling intensity could be allowed for in the analyses. If the strata were carefully chosen, so that they accounted for a substantial proportion of differences between sites, the effect of non-random location of sites within strata would be slight. We know too little of between-habitat differences in population trends to know whether this idea would be useful in practice.

Stratification is also valuable in ensuring complete coverage of all habitats and regions. One value of this is that population changes may be more evident in some places than others. The pioneering studies of Kluijver and Tinbergen (1953) showed that tit *Parus* populations fluctuated less in deciduous than in coniferous woodlands and they suggested that the latter acted as 'buffers', occupied mainly in years of high total population. O'Connor and Fuller (1985) and Bernstein et al. (1991) have reviewed the evidence that fluctuations differ between habitats (and the possible reasons for such differences). It is clear that the sensitivity of any population monitoring programme will be enhanced by subdivision by habitat.

To return to the question of bias: in principle, it could be totally eliminated by directing observers to study-sites. In volunteer-based surveys, this may result in a loss of participants, which means that the results of the survey will be less precise. Thus in designing the survey one must strike a balance, recognizing that a slight bias in a precise estimate is often preferable to an imprecise, if unbiased, estimate. The important thing is to be able to judge when bias is likely to be so small that it does not mislead us in drawing inferences from the information we have available.

Another potential source of bias is not in the distribution of the study plots but in the type of observations being made. In the territory-mapping technique, as used in the CBC, locations of birds are plotted on maps and are then interpreted, by experienced analysts, in terms of breeding territories. This interpretation is not grossly biased but may be inaccurate for some species (O'Connor and Fuller, 1984; O'Connor and Marchant, 1981). It is important that any bias should be consistent over the years. Other techniques, such as counting the birds seen during some standardized observation period, are designed merely to provide an index of the population level. In this case, it is important that the relationship between the index and the population should not change with time. Methods that do not assure that the relationship between the inferred variable (such as population size) and the original field observations of birds is constant are inappropriate for long term monitoring. Thus the North American Christmas Bird Counts, because they are not standardized for effort or coverage, can only provide rough indications of population changes over the nine decades for which they have been run, though with care they provide useful information about some aspects of abundance and show correlations with other measures (Butcher, 1990; Butcher and McCulloch, 1990). Most list-based methods, in which abundance is indexed as the proportion of birders' lists on which a species is recorded, are inappropriate for the same reasons (*pace* Temple and Cary, 1990), though A. Cyr and J. Larivée (in prep.) have found good agreement between such a method and the BBS in Quebec and, used carefully, such lists can provide good information on geographical distribution and seasonal phenology (Cyr 1986, 1990; Fradette, 1992; Harrison and Underhill, in press; Temple and Cary, 1987a, b; Underhill *et al.*, 1992).

7.3 MONITORING DISTRIBUTION

Most of this chapter is concerned with monitoring abundance and how it changes. Before turning to this in detail, it is worth briefly considering the role of studies of distribution. In recent decades these have entered a new era, through grid-based 'atlas' surveys in many countries. Beginning with atlases of breeding birds (e.g. Sharrock, 1976) these moved on to wintering birds (e.g. Lack, 1986) and then to year-round atlases (e.g. SOVON, 1987). A second generation is now beginning (e.g. Gibbons et al., 1993), providing the opportunity to assess changes in distribution. Unfortunately, most atlases have not been based on standardized effort, so apparent changes in abundance are confounded with possible changes in recording effort. More recent atlas surveys have sometimes incorporated standard methods (e.g. Gibbons et al., 1993) or a measure of effort (e.g. Lack, 1986), so future comparisons will suffer less from this drawback.

In Britain and Ireland, many of the changes in apparent distribution have been so dramatic between the first breeding atlas of 1968–72 (Sharrock, 1976) and the second of 1988–91 (Gibbons et al., 1993) that they must reflect real changes rather than differences in coverage. Where census data are available, they show changes corresponding to those suggested by the atlases. But for many species – those too scarce to be well covered by general census schemes but not rare enough to be covered by intensive special surveys – there are no census figures. For them, atlases are an important methods of assessing long-term population changes.

The spatial coincidence between the distribution of birds and of environmental variables provides evidence of what determines the bird distributions, provided that it is assessed using the appropriate statistical techniques (Eyre et al., 1990; S. Gates, D.W. Gibbons, R.M. Fuller and D.A. Hill, in prep). Equally, changes in distributions can be related to environmental factors to test hypotheses about the causes of the changes (Gibbons and Gates, in press).

Thus repeated atlas data, though available only on a coarse time-scale, are valuable not only because they provide information on change in numbers of some species not well covered by censuses but also because they provide a wealth of spatial information. They complement annual censuses, which are more fine-grained in time, but more coarse-grained spatially.

7.4 SURVEILLANCE OF ABUNDANCE

7.4.1 General literature

Bibby et al. (1992) have recently presented a full account of methods for censusing bird populations. Further details and issues concerning

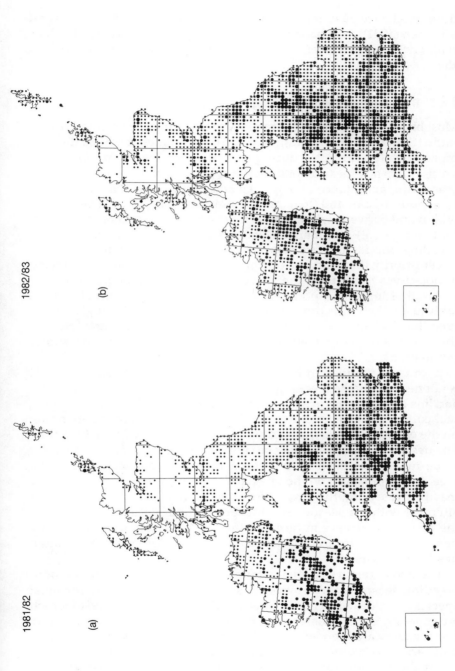

Figure 7.2 Distribution of redwings *Turdus iliacus* in Britain and Ireland in two successive winters, showing a dramatic change in local abundance. The dot sizes are proportional to abundance. (From Lack, 1986.)

the surveillance of abundance have been presented by Koskimies and Väisänen (1991), Marchant et al. (1990), Ralph and Scott (1981), Robbins et al. (1986, 1989), Sauer and Droege (1990), Taylor et al. (1985), and Verner (1985).

7.4.2 Breeding or non-breeding populations?

Most birds in most countries are monitored during the breeding season rather than the non-breeding season. This is mainly because they are more patchily distributed during winter, giving higher variances (less precision) to any measure of population level. During the non-breeding season the abundance of a species at a particular place may change from hour to hour and there may be dramatic shifts of population both within and between winters, which are reflected in differences in local (but not global) abundance (Figure 7.2) (Lack, 1986). Where non-breeding populations are surveyed, the work is therefore usually of lower priority than surveys during the breeding season or is directed at objectives other than the surveillance of overall abundance.

Another important point is that, because most breeding birds are faithful to their breeding areas and remain there throughout the breeding season, it is easier to associate demographic parameters of the population with features of the habitat than it is for winter populations.

Because a population is surveyed during the breeding season, it is not necessarily the breeding population that is counted. Point counts and line transects will usually include non-breeders. Territory mapping will come closer to covering just those birds that breed, depending on the extent to which the holding of territory is linked with breeding. The nests of some species may be conspicuous enough for direct counts to be made, but even here a count may not accurately reflect the breeding population: on the one hand, nesting attempts that fail early may be missed; on the other, birds may shift sites between breeding attempts within a season and thus be counted twice. In interpreting the results of surveys in terms of the overall dynamics of a population, it is important to be clear about what section of the population has been counted.

The above remarks notwithstanding, monitoring during the non-breeding season is preferable for some birds. This is true for most species of wildfowl (e.g., Kirby et al., 1991; Owen et al., 1986) which are generally easier to observe in the winter than when breeding, partly because they are then less secretive and partly because their populations are then more concentrated. As a result, it may be possible to count the majority of a population directly. Similar considerations apply to most shorebirds, many of which are even more difficult to count when breeding because they then inhabit such remote areas. In winter, they are readily counted at high-water roosts (Kirby et al., 1991; Prater, 1981). Winter roosts are also useful for counting some other birds.

7.4.3 Counts of migrating birds

Raptors, and other soaring birds, may concentrate in huge numbers at relatively few sites during migration. This presents an opportunity for highly cost-effective surveillance of population levels, so long as the proportion of the population moving past these special sites is great enough for changes in numbers at them to be representative of changes in total population size and so long as the counts are unbiased. Unfortunately, changes in local conditions at individual sites may cause counts to change even though total populations do not. Furthermore, numbers may vary considerably from hour to hour and from day to day, so it is important to take account both of the temporal distribution of recording effort and of weather during the analysis of the data. It is certainly true that massive long-term declines in the populations of many European raptors are matched by declines in numbers migrating through Falsterbö (Sweden) and the Bosporus (Bijleveld, 1974). Nonetheless, Fuller and Mosher (1981) and Heintzelman (1986), working in North America, where raptor counts have a high profile, believe that very great care must be taken if the counts are to be used as indicators of population levels.

The variability of counts during migration periods has led Rösner (in press) to state that these periods are not generally good ones in which to survey shorebird populations. But he points out that some species may be too inaccessible at other times of year for useful counts to be made: for these, highly standardized programmes of counting covering migration periods can be useful. (They are, of course, essential for assessing the importance of stopover sites.)

Coastal bird observatories often produce daily estimates of abundance, especially for migration periods. Though they have their advocates (e.g. Riddiford, 1983), they have generally not been used for surveillance of population levels because of their extreme dependence on weather conditions. Even away from the coast, the numbers of migrant birds seen are highly variable. Due allowance for time of day, season, and weather must therefore be made if counts of migrants are to be useful for population surveillance (Hussell, 1981), though it is not straightforward to account for the effects of weather (Darby, 1985). Hjort and Lindholm (1978) found that the autumn ringing totals at Ottenby Bird Observatory (Sweden) reflected weather conditions on the wintering grounds during the previous winter in wren *Troglodytes troglodytes* and whitethroat *Sylvia communis*, while the reversal of a decline in yellowhammers *Emberiza citrinella* coincided with a ban on the use of alkyl mercury compounds in Swedish agriculture. The crash of the whitethroat population in 1968/69 is also evident in ringing totals from Dungeness Bird Observatory in Britain (Scott *et al.*, 1976). However, these population changes were extreme and a fuller analysis by Svensson (1978) of data from three Swedish observatories showed that trapping totals generally were not well correlated with breeding population estimates and showed much more variation. Counts of visible migration were even more variable. He concluded that it was more cost-effective to survey breeding populations than to count migrants.

If migration routes or the proportion of a population migrating vary, the counts may reflect behaviour rather than population numbers. Thus Gatter and Steiof (1992), analysing counts through a mountain pass in south-west Germany, while they believed that trends in the counts of some long-distance migrants reflected real changes in numbers, considered that trends of some short-distance migrants were caused by changes in behaviour – such as more wood pigeons *Columba palumbus* overwintering in Poland and the Baltic states. Similarly, Langslow (1978) pointed out that there was evidence for changes in migration routes and wintering areas, as well as for population changes, as causes for increased counts of blackcaps *Sylvia atricapilla* at British bird observatories.

Further problems with the use of counts of migrants include not knowing for sure which populations are involved. It is also possible that changes in numbers of migrants at a particular place may be the result of changes in the habitat of surrounding areas, even when the habitat at the counting station itself is stable.

In the Netherlands, a network of stations is used for counting migrants across the country (van Gasteren, in press). The counts have not been used for studies of population trends but those from two particularly intensively worked sites have been so used (Lensink and Kwak, in press). The trends they show are correlated with known trends in the relevant breeding and wintering populations.

The numbers of migrants caught at Long Point Bird Observatory (Ontario) are correlated with levels indicated by the Breeding Bird Survey (Hussell, 1981) and standardized mist-netting programmes have often been advocated for the surveillance of migrating birds. Such projects include the MRI programme, involving three stations in Austria and Germany (Berthold *et al.*, 1986), and Operation Baltic, for which the results for three stations have been reported (Busse, in press). These two programmes have been notable for indicating decreases in a large proportion of the species covered. It is planned to continue the US Fish and Wildlife Service/Canadian Wildlife Service's Operation Recovery, a programme linking many migration banding stations, for population surveillance based on numbers of birds caught (Dawson, 1990).

The correlations between indices of migrants and of relevant breeding (or wintering) populations, where reported, are often rather low. This suggests that counts of migrants are not the best way of surveying population levels, except for species that are particularly difficult to count during the breeding season. It is particularly significant that the well-known declines of many species at the MRI (and Operation Baltic) sites are not generally reflected in trends of breeding populations in relevant countries (Marchant, 1992).

The key question yet to be answered in respect of the use of migration counts for population surveillance is how cost-effective and statistically powerful they are compared with counts of breeding (or wintering) populations. Despite several datasets being available, formal analyses aimed at answering this question have yet to be carried out.

7.4.4 Is it necessary to measure absolute abundance?

If one can measure absolute abundance precisely and accurately, it is useful to do so. Some species are sufficiently uncommon, conspicuous, and restricted in habitat for this to be possible. For example, mute swans *Cygnus olor* within individual 10 × 10 km squares in Britain can be completely censused, so a national population estimate can be made from censuses of sample squares (Greenwood *et al.*, in press). Wintering wildfowl and shorebird populations provide further examples (section 7.4.2); for some of them (e.g. godwits *Limosa* and knot *Calidris canutus* in Britain) the habitat is so restricted in terms of the manpower available that almost the whole population can be counted routinely (Moser, 1987). Species that breed in conspicuous colonies, such as some herons and most seabirds, are also readily counted, though it may be difficult to distinguish which birds are actually breeding and in some species the count is of apparently occupied nests rather than of birds (Lloyd *et al.*, 1991). Raptors also may often be counted directly within particular study areas, through intensive studies of nesting birds (section 7.6.2). For most species, however, precise and accurate population estimates are unattainable, even in small areas, unless the birds are individually marked and intensively observed.

Because sound estimates of absolute population size are generally not possible, most population surveillance depends on establishing indices of population size. In doing so, one must assume that there is a monotonic relationship between the index and true abundance. This is generally reasonable but the tacit assumption that the relationship is linear may be less easy to justify. If territorial defence activity increases as population density increases, some population indices may increase disproportionately in relation to true abundance; conversely, observers may be overwhelmed at high densities, causing indices to be too low. It is generally assumed, however, that such problems are insignificant in relation to the level of precision of the indices used.

One value of using indices, rather than striving for absolute population estimates, is that many potential biases can be removed by standardizing the methods (both in the field and during interpretation), by using the same sites and observers in successive years, and by avoiding extreme weather during counting. All of these factors are known to have substantial effects. In the discussion below, unless otherwise stated, we assume that the study design has been successful in eliminating them. Soundly-based programmes must include checks that it has.

7.4.5 Territory mapping

Territory mapping, also commonly known as spot mapping, involves the observer making a standardized set of visits to a study plot and mapping all encounters with birds, usually recording details of behaviour as well as species, sex and age. The records are analysed at the end of the breeding season, to estimate the distribution of territories.

This method is used in a number of population surveillance programmes, including the British Trust for Ornithology (BTO) Common Birds Census (CBC) and Waterways Bird Survey (WBS), for which various validation exercises have been carried out (Batten and Marchant, 1976; Fuller and Marchant, 1985; Gnielka, 1992; Marchant et al., 1990; O'Connor 1980, 1981; O'Connor and Fuller, 1984; O'Connor and Marchant, 1981; Snow, 1965; Taylor, 1965). Of all the general methods of population surveillance, territory mapping is probably the most closely linked to real breeding numbers (though the number of recorded territories is often substantially different from the real number) and is therefore likely to provide the best index, if one ignores the effort involved. Unfortunately, the effort is considerable. A typical CBC plot (20 ha in woodland, 70 ha in farmland) would require 30 h of fieldwork, 20 h for transferring field records to species maps, and 5 h for analysis.

7.4.6 Line transects

An index of numbers can be obtained if an observer walks a standard

Invertebrates (eds W.G. Doubleday and D. Rivard) *Can. Spec. Pub. Fish. Aquat. Sci.*, **66**, 151–8.

Prince, P.A. and Harris, M.P. (1988) Food and feeding ecology of breeding Atlantic alcids and penguins. *Acta XIX Congressus Internationalis Ornithologici*, pp. 1195–204.

Pugesek, B.J. (1981) Increased reproductive effort with age in the California Gull (*Larus californicus*). *Science*, **212**, 822–3.

Rhymer, J.M (1988) The effect of egg size variability on thermoregulation of mallard (*Anas platyrhynchos*) offspring and its implications for survival. *Oecologia*, **75**, 20–4.

Rice, J.C. (1992) Multispecies interactions in marine ecosystems: current approaches and implications for study of seabird populations, in: *Wildlife 2001: Populations* (eds D.R. McCullough and R.H. Barrett), Elsevier, London, pp. 586–601.

Richard, J.M. (1987) The mesopelagic fish and invertebrate macrozooplankton faunas of two Newfoundland fjords with differing physical oceanography. M.Sc. thesis, Memorial University of Newfoundland, St. John's.

Ricker, W.E. (1975) Computation and interpretation of biological statistics of fish populations. *Bull. Fish. Res. Bd. Can.*, **191**.

Ricklefs, R.E. (1973) Fecundity, mortality and avian demography, in 'Breeding Biology of Birds' (ed. D.S. Farner), National Academy of Sciences, Washington, D.C., pp. 366–434.

Ricklefs, R.E. (1990) Scaling pattern and process in marine ecosystems in 'Large Marine Ecosystems: Patterns, Processes and Yields' (eds K. Sherman, L.M. Alexander and B.D. Gold), American Association for the Advancement of Science, Washington, D.C, pp. 169–78.

Ricklefs, R.E., Duffy, D.C. and Coulter, M. (1984) Weight gain of blue footed booby chicks: an indicator of marine resources. *Ornis Scand.*, **15**, 162–6.

Rose, G.A. and Leggett, W.C. (1990) The importance of scale to predator prey spatial correlations: an example of Atlantic fishes. *Ecology*, **71**, 33–43.

Rothschild, B.J. and Osborn, T.R. (1990) Biodynamics of the sea: preliminary observations on high dimensionality and the effects of physics on predator prey interrelationships, in *Large Marine Ecosystems: Patterns, Processes and Yields* (eds K.Sherman, L.M. Alexander and B.D. Gold), American Association for the Advancement of Science, Washington, D.C., pp. 71–81.

Russell, F.S. (1935) On the value of certain plankton animals as indicators of water movements in the English Channel and North Sea. *J. Mar. Biol. Assoc. U.K.*, **20**, 309–31.

Ryan, P.G. and Moloney, C.L. (1988) Effect of trawling on bird and seal distributions in the southern Benguela region. *Mar. Ecol. Prog. Ser.*, **45**, 1–11.

Safina, C. and Burger, J. (1988) Prey dynamics and the breeding phenology of common terns (*Sterna hirundo*). *Auk*, **105**, 720–6.

Safina, C., Burger, J., Gochfeld, M. and Wagner, R.H. (1988) Evidence for prey limitation of common and roseate tern reproduction. *Condor*, **40**, 852–9.

Schaefer, M.B. (1970) Men, birds and anchovies in the Peru Current – dynamic interaction. *Trans. Am. Fish. Soc.*, **9**, 461–7.

Schaffner, F.C. (1986) Trends in elegant tern and northern anchovy populations in California. *Condor*, **88**, 347–54.

Schneider, D.C. (1991) The role of fluid dynamics in the ecology of marine birds. *Oceanogr. Mar. Biol. Ann. Rev.*, **29**, 487–521.

Schneider, D.C. and Duffy, D.C. (1985) Scale dependent variability in seabird abundance. *Mar. Ecol. Prog. Ser.*, **25**, 101–3.

Schneider, D.C., Duffy, D.C., McCall, A.D. and Anderson, D.W. (1988) Historical variation in guano production from the Peruvian and Benguela upwelling ecosystems. *Clim. Change*, **13**, 309–16.

Schneider. D.C. and Duffy, D.C. (in press) Seabird fisheries interactions: evolution with dimensionless rations, in *'Wildlife 2001: Populations'* (eds D.R. McCullough and R.H. Barrett), Elsevier, London, pp. 602–15.

Schneider, D.C. and Hunt, G.L. Jr. (1984) A comparison of seabird diets and foraging distribution around the Pribilof Islands, Alaska, in *'Marine Birds: Their Feeding Ecology and Commercial Fisheries Relationships'* (eds D.N. Nettleship, G.A. Sanger and P.F. Springer), Supply and Services Canada, Ottawa, pp. 86–93.

Schneider, D.C., Hunt, G.L. Jr. and Harrison, N.M (1986) Mass and energy transfer to seabirds in the southeastern Bering Sea. *Cont. Shelf Res.*, **5**, 241–57.

Schneider, D.C. and Piatt, J.F. (1986) Scale dependent correlation of seabirds with schooling fish in a coastal ecosystem. *Mar. Ecol. Prog. Ser.*, **32**, 237–46.

Schneider, D.C., Pierotti, R. and Threlfall, W. (1990) Alcid patchiness and flight direction near a colony in eastern Newfoundland. *Avian Biol.*, **14**, 23–35.

Schreiber, E.A. and Schreiber, R.W. (1989) Insight into seabird ecology from a global 'natural experiment'. *Nat. Geogr. Res.*, **5**, 64–81.

Schreiber, R.W. and Schreiber, E.A. (1984) Central Pacific seabirds and the El Nino Southern Oscillation: 1982 to 1983 retrospectives. *Science*, **225**, 713–6.

Shannon, L.V., Crawford, R.J.M and Duffy, D.C. (1984) Pelagic fisheries and warm events: a comparative study. *S. Afr. J. Sci.*, **80**, 51–60.

Shaw, D.M. and Atkinson, S.F. (1990) An introduction to the use of geographic information systems for ornithological research. *Condor*, **92**, 564–70.

Shelton, P.A., Boyd, A.J. and Armstrong, M.J. (1985) The influence of large scale environmental processes on neritic fish populations in the Benguela Current system. *Rep. Calif. Coop. Ocean. Fish. Invest.*, **26**, 72–92.

Sherman, K., Jones, C., Sullivan, L., Smith, W., Berrien, P. and Ejsymont, L. (1981) Congruent shifts in sandeel abundance in western and eastern North Atlantic ecosystems. *Nature*, **291**, 486–9.

Soutar, A. and Issacs, J.D. (1974) Abundances of pelagic fish during the 19th and 20th centuries as recorded in anaerobic sediment off the Californias. *Fish. Bull.*, **72**, 257–73.

Sparre, P. (1991) An overview of multispecies virtual population analysis. *Rapp. P. V. Reun. Cons. Int. Explor. Mer.* **190**.

Springer, A.M., Murphy, E.C., Roseneau, D.G. *et al.* (1987) The paradox of pelagic food webs in the northern Bering Sea. I. Seabird food habits. *Cont. Shelf Res.*, **7**, 895–911.

Springer, A.M., Roseneau, D.G., Lloyd, D.S. *et al.* (1986) Seabird responses to fluctuating prey availability in the eastern Bering Sea. *Mar. Ecol. Prog. Ser.*, **32**, 1–12.

Springer, A.M., Roseneau, D.G., Murphy, E.C. and Springer, M.I. (1984) Environmental controls of marine food webs: food habits of seabirds in the eastern Chukchi Sea. *Can. J. Fish. Aquat. Sci.*, **41**, 1202–15.

route at a steady, standard rate and counts the birds he or she sees. If the distance of each bird from the line of the route is recorded (either exactly or within certain bands such as within 25 m or beyond 25 m from the line), then it is possible to estimate population densities.

A key assumption in estimating density is that the birds do not move as a result of the presence of the observer. A second is that we know the way in which detectability declines with distance. The first is unlikely, the second untrue. Why, then, should one bother to record distances and estimate 'densities', rather than simply use total numbers as an index of abundance? The reason is that estimating 'densities' reduces problems caused by variation in detectability, such as spurious changes in numbers. Furthermore, if one combines observations from various habitats to get an overall population index (as is usual) then fluctuations or trends in the national index are disproportionately influenced by fluctuations or trends in the habitats in which birds are most detectable, unless one corrects for differences in detectability.

7.4.7 Point counts

Point counts may be considered as interrupted line transects: observers visit a standard set of points within sample plots and record what they see during a standard interval. In the North American Breeding Bird Survey (BBS), larger areas are covered than in most schemes, since the 'plots' are actually motor-routes 39.4 km long, each with 50 counting points at 805 m intervals (Droege, 1990). Observers making point counts may record distances and, as with line transects, 'densities' may then be estimated.

We know of no line-transect studies in which indices derived from densities have been compared with those derived from total counts but there is such a study for point counts (Gregory et al., in press). In this, point counts were carried out on CBC plots in four years, so that three sets of figures for year-to-year changes were available. Two sets of counts were made early and late in the season (to cover both early breeders and late arrivals) and various indices were derived from these. It was generally true that indices based on density estimation were less well correlated with those based on territory mapping than were those based merely on raw counts (Table 7.1) (Gregory et al., in press). Furthermore, each density-based index was less well correlated with other point-count indices than was the corresponding total-based index (British Trust for Ornithology, unpublished results). The reason for this paradoxical result is that the numbers observed were small, making the density estimates very unreliable.

Table 7.1 Correlations (r), across all species occurring on more than five census plots, between indices of year-on-year changes derived from point counts and those based on territory mapping (from Gregory et al., in press).

Year	Raw counts[a]	Density estimates[a]
1987–89	0.45	0.07
1989–90	0.10	0.04
1990–91	0.30	0.01

[a] Point count indices were based either on raw counts or on density estimates derived from the counts.

Other surveillance programmes are likely to involve similar numbers of observations, suggesting that line transect and point count indices should simply use total counts rather than distance-based density estimates.

7.4.8 Counting by catching

Ringers noticing declines in catches at long-term study sites may consider that these reflect general decreases in bird populations. Certainly, declines noted by Jones (1986) and Stewart (1987) fit in with the well known decreases in populations of some neotropical migrants in parts of North America (Askins et al., 1990). However, although catching birds in a standardized way has been variously used to study the numbers of migrating birds (section 7.4.3), we know of only one study of breeding populations that uses a network of bird-catching stations. This is the BTO Constant Effort Sites scheme. It involves over 100 sites, mostly in wetland, reedbeds, and scrub. These habitats are not well covered by the CBC and WBS, to which the CES thus provides a useful complement. The main use of the CES is, however, to provide estimates of productivity and survival. Population levels are indexed as the total numbers caught (Peach and Baillie, 1990, 1991, 1992; W.J. Peach, S.T. Buckland and S.R. Baillie, in prep.). At each such site, observers are expected to make catches during each of twelve 10- (or 11-) day periods through the summer. For the purposes of year-on-year comparisons, only sites covered in both years are included and, within each site, only the sample periods covered in both years are used. This ensures strict comparability. Sites not sampled in at least four of the first six sample periods and at least four of the second six are excluded.

Compared with direct observation, catching birds is labour intensive. It is very unproductive in open habitats and in mature temperate woodlands, though it is often better than direct observation in dense scrub, reedbeds and in rainforest. It is also valuable for skulking species that are rarely seen unless caught. One might imagine that catching was more readily standardized than direct observation. However, catching rates depend on the behaviour of the birds, the weather, net design and quality, and how taut the nets are. Netting is therefore probably no more highly standardized than the better methods involving direct observation.

Another apparent advantage of catching birds that are individually marked and released, is that recapture rates can provide estimates of actual population size. The key papers and reviews of the statistical methods involved are by Cormack (1968, 1979), Jolly (1965), Lebreton et al. (1992), Nichols et al. (1981), Otis et al. (1978), Pollock (1981, 1991), Pollock et al. (1990), Seber (1982, 1986) and Underhill (1990). The methods also allow survival to be estimated (see below). The estimation of population size rests on a number of assumptions and may be very sensitive to departure from them, so mark-recapture methods may give grossly incorrect estimates of population size if the probability of capture of all individuals is not equal. More refined methods have been developed to cope with such cases, but they depend on assumptions about the frequency distribution of capture probabilities (almost always unknown) and Cormack (1979) has pointed out that several alternative models may provide equally good fits to the data but give quite different population estimates. C. Vansteenwegen and B. Steck (pers. comm.) have estimated population sizes of many species on a number of sites in the French STOC programme (Vansteenwegen et al., 1990) and have compared the density estimates with those obtained on other sites by other workers who used more traditional methods. Their mark-recapture estimates are mostly much higher than those obtained by traditional methods. This may be because there are substantial numbers of non-breeders in most populations, which traditional methods overlook, or because of bias in either group of estimates.

Thus catching birds is useful for indexing abundance of some species in some habitats but we agree with Pollock et al. (1990) that methods based on observation are generally more cost-effective for this purpose.

7.4.9 Which is the best method?

This question cannot have a simple answer, since the answer depends on one's objectives, the resources available, the habitat, and the species. For present purposes, we are interested in long-term surveillance as part of a population monitoring programme, using a network of observers, in a variety of habitats, and covering as many species as possible.

Svensson (1980) pointed out some years ago that the key criteria in judging the suitability of a method were the bias involved and the precision achievable given the resources available. Unfortunately, tests of the relative merits of population surveillance techniques have often not focused on these criteria. Many papers have been written to show that the indices obtained using different methods on the same plots in the same year are correlated. While this shows that both methods are probably providing indices of true population levels, it tells us nothing, absolute or relative, about bias or precision, or about the value of the methods for surveillance over a period of years. Finding that two methods show similar trends over time suggests that either could provide valid surveillance (e.g. Flousek, 1990; Svensson, 1981), but provides no measure of bias or precision.

Bias can be measured only by direct comparison of the results using each method with those of studies that are sufficiently intensive for one to be able to believe that they reveal absolute numbers faithfully. Both territory mapping (Snow, 1965) and line transects (Hildén and Laine, 1985; Zolner, 1990) tend to underestimate true numbers. But if the degree of underestimation is constant, this will not bias estimates of changes of abundance. There appear to have been no studies aimed at assessing bias in year-on-year changes or long-term trends.

The study by Gregory et al. (in press) described in section 7.4.7 was designed to measure the relative precision of indices based on territory mapping and those based on 10 point counts distributed over the same study plots. Each point was counted for 10 minutes on each of the two visits. Averaging across species, the standard error of year-on-year changes was about twice as great for the point counts as for territory mapping (Table 7.2), so it would be necessary to have about four times as many plots to achieve the same degree of precision. This

Table 7.2 Mean standard errors, across all species occurring on more than five census plots, of indices of year-on-year changes derived from point counts and those based on territory mapping (from Gregory et al., in press)

Years	Point counts	Territory mapping
1987–89	0.26	0.16
1989–90	0.40	0.11
1990–91	0.27	0.11

would require slightly less total fieldwork time and much less analysis. Because individual plots are easier to survey by point counts than by territory mapping it is possible that more than four times as many volunteer observers could be recruited for point counts, thus increasing precision and making possible better geographical and habitat coverage than in the current CBC scheme. Systems using point counts or line transects are therefore being further explored by the BTO (Baillie and Marchant, 1992).

A comparison made in the Netherlands between territory mapping, area counts and point counts came to the opposite conclusion: point counts were so much less precise than the other methods that, though the others were more time-consuming, they were preferable (van Dijk, 1992). However, no point counts were actually made in this study: rather the point count 'data' were generated by placing circles on maps produced during territory mapping and counting the numbers of registrations falling within them. This is scarcely a valid test. Furthermore, van Dijk's study was based on measuring the between-site variances of the indices obtained within years, not on the variance of an index of year-on-year change. Since much of the between-site variance is removed in surveillance schemes by pairing plots across years, the former cannot tell us much about the latter.

Another method of exploring the relative merits of different methods is to apply them to data for which the correct answer is known – i.e. simulated data. In such studies it is essential not to oversimplify in the simulation, since the methods may be sensitive to details of the distribution of populations, the distribution of observations, and the precise way in which the programme is designed and implemented. Some of these details may be capable of being estimated from real data; if not a variety of realistic alternatives should be built into the models. Once the 'data' have been simulated, the analyses should concentrate on Svensson's key criteria – the bias and precision of measures of change in abundance.

7.5 MEASURING CHANGES IN ABUNDANCE

7.5.1 The measurement of year-on-year changes

Were there to be a surveillance programme in which random samples of the countryside were surveyed in each of two years, with observers randomized to plots, then the relative abundance of birds in the two years could properly be assessed by comparing the mean counts or densities in the two years. It would not matter what proportion of the plots was covered in both years nor whether such plots were covered by the same observer in the two years. The comparisons could be extended to any number of years. Such an approach has been applied to BBS data in order to assess annual fluctuations (Robbins et al., 1986) and to CBC data to check for drift in the standard CBC index (Moss, 1985; see below). However, in all existing surveillance programmes the samples are not random, the assignment of observers is not random, and we know that there are big differences between both sites and observers in the counts that they produce. (Furthermore, because observers usually stick to the same site in successive years, the effects of site and observer are confounded.) It is for this reason that surveillance programmes are designed to have a high proportion of sites covered in successive years (and by the same observers). This element of the design is included in the analysis of the data: to eliminate site and observer effects, in year-on-year comparisons, the analysts include only those sites counted in both years. If x is the count for a plot in the first year and y is the count for the second year then, provided the plots are representative of the region, the best unbiased estimate of the relative numbers in the region in the two years is

$$r = \Sigma y / \Sigma x$$

(summation being over all sites censused in both years). Confidence limits for this estimate can readily be calculated (W.J. Peach, S.T. Buckland and S.R. Baillie, in prep.).

Alternative approaches to annual changes have been proposed by Sauer and Geissler (1990). Both depend on fitting regressions to the data (Route Regression Analysis): in one, individual year effects are incorporated in the model as well as the long-term regression; in the other, year effects are obtained as the residuals from the underlying regression.

7.5.2 Long-term changes: chaining

In hawk migration studies the data for single sites form a sufficient sample to be worth analysing separately. Long-term trends can be tested using familiar techniques, such as linear least-squares regression or non-parametric rank trend analysis (Titus et al., 1990).

For most programmes it is necessary to combine data over several, or many sites. A simple way of doing this is to make estimates of year-on-year changes, as in section 7.5.1, and chaining these together, to provide an index that varies relative to a base year value. Thus if there is a 10% increase in abundance between year 1 and year 2 and a 20% decline between years 2 and 3, and if we take the index in year 1 to be arbitrarily 100, then the index in year 2 is 100 × 1.1 = 110 and in year 3 is 110 × 0.8 = 88.

Such chaining of successive year-on-year indices loses the information available from direct comparison of years 1 and 3, for example, – and there may be many study plots providing relevant data.

The method is also unsatisfactory in that it provides no means of calculating confidence limits for any apparent change over periods longer than one year. It would be inappropriate to assess the error variance of any such change by adding together the variances of the year-to-year values that contribute to it, since successive values are statistically non-independent (since the count for the second of three years contributes to both the year 1/year 2 and year 2/year 3 ratios).

Most importantly, errors accumulate in such chaining operations, so that the indices can show apparent trends when there are no real trends (Geissler and Noon, 1981). From simulation studies, Moss (1985) concluded that the magnitude of such random walks was probably small compared with the real changes that appear to occur in CBC data, but Greenwood (1989) concluded that they might be large enough to be misleading in some circumstances. In addition to the simulation studies, Moss used data from CBC plots to estimate mean densities of territories per unit area each year, as a basis against which to check the standard CBC indices. There was good agreement for partridge *Perdix perdix*, skylark *Alauda arvensis*, goldcrest *Regulus regulus*, and willow warbler *Phylloscopus trochilus* but not for spotted flycatcher *Muscicapa striata*. The strong autocorrelation in chained indices means that the significance of trends is better assessed using Monte Carlo methods than standard regression tests.

7.5.3 The Mountford method

The model on which this is based is that the count in a particular plot in a particular year (x_{yi}) is determined by a plot effect (a_i) and a year effect (b_y) interacting multiplicatively, with additive random error ε:

$$x_{yi} = a_i b_y + \varepsilon$$

Even if some values of x_{yi} are missing from the dataset (because not all plots are counted in every year), there is an analytical solution to fitting this model (Mountford, 1982, 1985). The b_y values provide an index of population in year y, based on comparison with all other years.

This method is less susceptible to random walk. Unfortunately, the assumption that the proportional changes in numbers are the same on all plots (the interaction of a_i and b_y is purely multiplicative) is usually demonstrably violated in data sets extending over more than about 10 years. An *ad hoc* solution to this problem has been suggested by Peach and Baillie (in press). They advocate analysing only six years' data at a time (a period over which the method usually has no difficulty); then advancing the time-frame by two years and repeating the analysis until the data are exhausted. The indices for the central pair of years in each six-year analysis are taken as the indices for those two years and the successive six-year analyses are chained together through the overlapping two-year periods. This provides an analysis covering the entire time-span of the surveillance programme without generating more than trivial random walks. Figure 7.3 shows the results of applying both this and the chaining method to some Common Birds Census data.

Mountford's original method provides confidence limits around the index values analytically. This advantage is lost in the modified method, so it will be necessary to develop Monte Carlo methods to provide confidence limits for both the year-to-year changes and long-term trends.

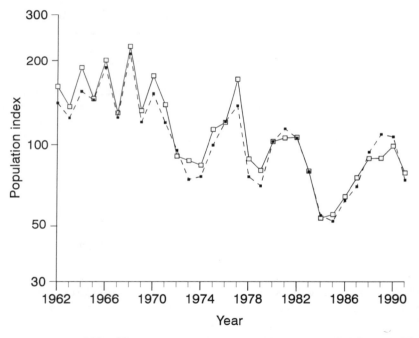

Figure 7.3 Population changes for British sedge warblers *Acrocephalus schoenobaenus* based on British Trust for Ornithology Common Birds Census data, using the chaining method (dashed line) and the extended Mountford method (continuous line). Note that the population scale is logarithmic and the index is relative to an arbitrary value of 100 for 1980.

7.5.4 Methods based on imputing

Another approach to the problems of missing observations is to impute them. Put simply, if one has an R × T array of counts for R plots and T years, missing values are filled in by comparison with other values in the same row and column. The underlying model, like that of Mountford, is that the count in a particular plot and a particular year is determined by the product of a plot effect and a year effect. These effects are estimated iteratively. To start with, informal estimates of the missing values are made. This allows column and row totals to be calculated for the table. If P is the total for a particular plot (across all years) Y the total for a particular year (across all plots), and N is the grand total (all of these including the estimates as well as the actual counts) then a revised estimate for that plot and that year is (Y × P)/N. Having produced such revised estimates for each missing value, the process is repeated iteratively, until successive estimates hardly change. The annual sums of actual counts and final estimates (over all plots) are annual indices of total population size. Confidence intervals can be obtained by Monte Carlo methods.

This method has been applied successfully to the data in the BTO Birds of Estuaries Enquiry by R.P. Prŷs-Jones, L.G. Underhill and R.J. Waters (in prep) and to three successive surveys of mute swans in Britain (Greenwood *et al.*, in press), while Moses and Robinowitz (1990) have applied it to BBS data.

Imputing methods work well when the values missing from the data matrix are sporadic. In many population surveillance programmes, however, the gaps are in blocks: thus a typical CBC plot might not have been run during the first 10 years, run for 7 years, then never run again. It is not clear how well imputing methods might work with such data.

7.5.5 Route regression analysis (RRA)

This was invented by Geissler and Noon (1981) and is the preferred method for the BBS and other North American surveys. The basic method (Robbins *et al.*, 1986) is that the count on a particular route in the yth year is given by

$$c_y = ab^y + \varepsilon$$

where a is a constant and b is the slope of the exponential trend in numbers on the site. The value of b is estimated for each site by least squares regression analysis after log transformation and the trend for an area estimated as the mean of the individual b

values for routes in that area (after weighting – section 7.5.7). Confidence limits are derived using Monte Carlo methods.

In practice, to cope with zero counts, a constant is added to the counts before log transformation. The standard back transformation (Geissler and Sauer, 1990) fails to cope with this constant and the method gives somewhat biased estimates of mean trend (Collins, 1990). For the constant usually used (0.5), the bias is towards zero – i.e. both positive and negative trends are underestimated.

Latest developments in RRA include an observer effect, which is separated from the route effect (a, in above model) (Geissler and Sauer, 1990).

A fundamental problem with this method is that it assumes that trends are uniformly exponential, which is increasingly unrealistic as longer runs of data accumulate. It provides no immediate way of examining fluctuations on time-scales shorter than the complete run of the scheme.

7.5.6 Log-linear Poisson regression

van Strien *et al.* (in press a) have recently reviewed the above methods and suggested that log-linear Poisson regression may provide a useful alternative for estimating both long-term trends and annual variations. The basic model is the same as that for RRA.

7.5.7 Weighting

In the BBS, trends for North America (or for regional subdivisions) are based on weighted averages of b values for individual routes (Geissler and Sauer, 1990; Robbins *et al.*, 1986). Three factors are used in weighting. One is the area occupied by the physiographic stratum in which the particular route is found (corrected for the number of routes in the stratum). In the absence of more detailed habitat information, this is clearly a useful weighting.

The other two weighting factors, a measure of the mean count on the route and a reciprocal measure of the error variance of the trend estimate, are of more dubious value. Collins (1990) points out that weighting by the mean may give a route undue weight if the measure of the mean is based on extrapolation beyond the range of years in which the route was actually surveyed, while weighting by the reciprocal of the variance is justifiable only if the variance represents measurement error rather than genuine variation in numbers about the route regression line.

7.5.8 Monte Carlo methods for estimating confidence intervals

Monte Carlo methods are the only ones available for generating confidence intervals for most of the methods discussed above. A general introduction to such methods is given by Manly (1991). In respect of population surveillance, they are in their infancy, though Monte Carlo confidence intervals have been developed for the BBS (Geissler and Sauer, 1990). Their calculation involves a so-called bootstrapping method. In this, one takes a random sample, with replacement, from all routes run in the region in question (the sample size being the same as the total number of routes). The RRA trend is calculated for this new sample in the usual way. The sampling and calculation are carried out 400 times. The mean or median of the 400 values of trend is taken as the best estimate of the mean trend and the standard deviation as the standard error of that estimate, confidence limits being derived therefrom using the t or z distributions. An alternative would be to arrange the 400 randomly generated values in order and take the 10th and 390th (the 2½- and the 97½-percentiles) as the lower and upper 95% confidence limits. This will always provide better limits than z and will only be outperformed by t if the number of routes is small (S.T. Buckland, pers. comm).

Collins (1990) points out that such a bootstrapping approach (or a similar 'jack-knife' method) means that one is assessing the significance of the overall pattern by looking for consistency across routes; a pattern may be highly consistent even though the trend is weak. He proposes an alternative Monte Carlo approach, which requires that the order of the counts be randomized within routes; this places more emphasis on the trends and less on consistency.

7.5.9 Evaluating the alternative methods

If one needs to judge the importance of an apparent change through using a test of statistical significance, one needs to know the power of that test. This is a measure of its ability to detect a change. Formally, it is the probability that, given that there has been a change of a certain magnitude, a significant result will be obtained. Obviously, a test with high power is more useful than one with low power. Indeed, a test of low power may be worse than useless since, by failing to give a significant result when there has actually been a substantial change, it leads one to conclude that there has been no real change. The power of a test will depend on how precisely changes are measured – i.e. on both the quality of the fieldwork and the statistical efficiency of the analysis.

General considerations of power analysis are given by Kraemer and Thiemann (1987) and by Cohen (1988), while Gerodette (1987) deals with the power of trend analyses in particular. The only such analysis applied to bird population surveillance is that of van Strien et al. (in press b) of the Dutch Breeding Bird Monitoring Programme. The results are encouraging. Such analyses should be routine for any surveillance programme, to assess its ability to detect untoward changes. Many ecological and environmental monitoring programmes have low power (Peterman and M'Gonigle, 1992). This means it is particularly important to use the best methods, to increase their power.

There is a clear need for the main methods of population trend analysis that are currently being advocated, RRA and the modified Mountford method, to be applied to the same datasets, to discover which is the more powerful. As with the assessment of the field methods, the use of simulated data will allow not just the power but also the bias of alternative analytical methods to be compared.

7.6 THE SURVEILLANCE OF REPRODUCTION

7.6.1 Nest-based studies

The reproduction of birds is easier to study than that of most other animals because it is largely tied to nests that are relatively easy to find. The number of eggs laid, the pattern of mortality and, often, its causes may be determined easily in intensive studies, through frequent visits to the nest. Since it is often possible to observe (and even catch) the parents, the reproductive activity of specific individuals may be studied (Newton, 1989). In the present context, this is valuable because it may allow age-specific fecundity rates to be determined, though this requires very intensive work. As with all intensive studies, the value of such work for population monitoring may be reduced by the locations of the studies being in habitats that are better than average. Where nest-boxes are provided (which makes the study physically easier), breeding success may be artificially increased. Conversely, very intensive study may reduce breeding success by causing parents to desert, by attracting predators, and so forth.

In a number of countries, network-based studies of breeding are conducted, the oldest and most extensive being the BTO Nest Records Scheme, started in 1939 and currently gathering details of 30 000 nests per year (British Trust for Ornithology, n.d.). Observers submit cards on which they have recorded location, habitat, and nest site details, with information on the contents of the nest on each occasion on which they visit it (with ringing details of both young and parents, if known). In recent years, standardization and the speed of processing the data have been improved by the use of carefully defined codes for much of this information. The cards are designed for easy computerization. The number of cards is over 100 per year for 51 species and surveillance reports are now provided annually for key species (Crick et al., 1992). This large dataset has allowed many studies of reproductive performance to be carried out, for example, that of lapwings *Vanellus vanellus*

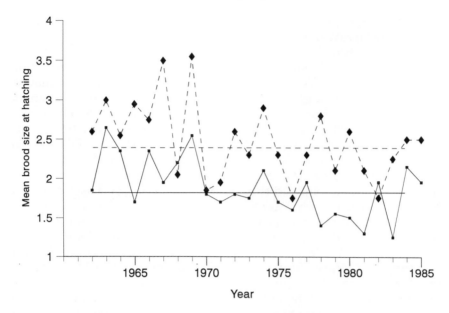

Figure 7.4 Annual mean brood size at hatching of lapwings *Vanellus vanellus* nesting in grassland (continuous line) and tillage (dashed line) in England and Wales; horizontal lines show the overall means (from Shrubb, 1990, using British Trust for Ornithology Nest Record Scheme data).

in relation to agricultural practices (Figure 7.4) (Shrubb, 1990), that of black grouse *Tetrao tetrix* in respect of factors leading to regional differences (Baines, 1991), and that of merlins *Falco columbarius* in relation to pesticide usage and habitat (Crick, 1993). The simple observation of a sustained decline in clutch-size of hen harriers *Circus cyaneus* (Crick *et al.*, 1992) has confirmed the view that this species is faring badly in Britain (Batten *et al.*, 1990).

As with intensive studies, such extensive programmes suffer from the recording activity of observers not being randomly distributed, particularly in respect of time of year, habitat and geographical location. Provided these are properly recorded, however, they can be taken into account in analyses and national estimates of breeding output can be derived using appropriate weightings (though this has yet to be implemented in practice). Whether nests covered by such surveys are more or less successful than others, because, e.g. nests easily found by birdwatchers are also those easily found by predators, is a potential bias that remains to be investigated.

Because nests are visited less frequently, the timing and causes of losses may be rather less easy to determine in extensive projects than in intensive studies, perhaps making it less easy to determine the causes of any population changes that result from changes in breeding success. Because nests may be discovered at various stages, estimation of success is not straightforward, though there is now an increasingly refined set of methods available to deal with this problem (Hensler and Nichols, 1981; Johnson, 1979; Johnson and Shaffer, 1990; Klett et al., 1986; Klett and Johnson, 1982; Mayfield, 1961, 1975; Pollock and Cornelius, 1988), to which the computer program SURVIV (White, 1983) can be applied. This method uses 'incomplete data' from nests that could not be followed throughout the season and assumes that breeding pairs fail at a constant rate during certain phases of the nesting season. A daily nest failure rate can be calculated from the number of failures divided by the total number of days' 'exposure' of the study-nests. Each nest is 'exposed' to failure from the first visit to the last and the sum of these durations for all study-nests is used to calculate the daily failure rate. Failure rates often vary over different stages of nesting and it is preferable that different rates are calculated for the incubation and nestling phases. Estimates of variance can be calculated using maximum likelihood methods and allow significance testing of the results.

Intensive studies have sometimes revealed age-specific variation in fecundity, which extensive studies are rarely able to assess (e.g. Dhondt, 1989; Newton, 1989; Partridge, 1989). They are, however, sufficiently small that changes in age structure are unlikely to have more than trivial effects on the reproductive rate of bird populations compared with comprehensive changes in fecundity brought about by environmental factors.

7.6.2 Surveillance of reproduction in raptors

Raptors, and some other birds, have characteristics that result in particular sorts of surveillance programmes, intermediate between intensive studies by professionals and the very extensive studies made by networks o volunteers on other birds. They tend to attract the attention of particular enthusiasts, who often operate collaborative projects at a local level, their studies aided by the habit of many raptors of using a relatively small number of traditional nest sites. Surveillance of both population levels and reproductive output can be conducted through studies at all the nest sites in an area. It is possible, in principle, to draw together the work of such local groups to provide national schemes for raptor monitoring (Crick et al., 1990). Fairly informal systems of this nature operate in Scotland (Dick, 1991) and Wales (Williams, in press) and a more formal one has recently been started in Germany and some neighbouring countries (Gedeon and Stubbe, 1992). Non-breeding, but territorial, pairs may be counted in some of these species if fieldwork begins early enough in the season, allowing an assessment of the extent of non-breeding. The importance of this is discussed below (section 7.6.3). In those species in which traditional nest sites are used, it may be possible to improve the precision of surveillance by making comparisons between years that take into account the individual sites, though such work is in its infancy (Geissler et al., 1990).

7.6.3 How many nesting attempts per female?

Extensive studies are limited by the lack of information they provide on the number of nesting attempts per female. Even when attempts are successful, some species, especially those resident in temperate or tropical zones, may produce several broods a year. If the first attempt fails at an early enough stage, even species that are normally single-brooded will often make a second attempt. Thus the study of output from individual nests cannot measure total reproductive output except in strictly single-brooded species.

The most dramatic variation in the number of nesting attempts occurs because some females may not breed at all in some years. Individuals may differ in their age of first breeding and the distribution of age of first breeding is then not easy to assess (Lebreton and Clobert, 1991; Lundberg and Alatalo, 1992; Newton, 1989). Even after they have bred once, some females may not breed every year (Newton, 1989). This is well known among seabirds (Croxall and Rothery, 1991) and Arctic geese (with 11–26% of adult barnacle geese *Branta leucopsis* not even establishing territories in any one year: Owen and Black, 1989) but even in passerines older birds may forgo breeding (Lundberg and Alatalo, 1992).

Two pieces of evidence have been widely cited in support of the view that there is a substantial number of apparently mature but non-breeding birds in many populations (Lebreton and Clobert, 1991; Newton, 1991b). One is that the provision of nest-boxes may rapidly increase breeding populations; however, such results have generally been observed in prime habitat, so the increases may be the result of birds moving in from sub-optimal habitats, rather than of birds that would otherwise have been unable to breed occupying the new nest-sites. The other sort of evidence is that if birds are removed from territories they are often rapidly replaced. In many cases, these results may again be explained by birds moving in from other places but this is not universally true (Newton, 1991b; Smith et al., 1991). Thus non-breeding is probably widespread but in most species we have little idea of the proportion of birds involved – and even less of how this proportion varies.

In a few species, particularly of waterfowl, it is relatively easy to count the numbers of both breeding and non-breeding birds, even in extensive, network-based studies (e.g. mute swans, Ogilvie, 1986). Even here, however, care must be taken: in ducks, for example, breeders and non-breeders may live in different places, so the whole range must be covered if one is properly to assess their relative numbers (Owen and Black, 1990).

If we are unable to measure variations in the proportion of the population breeding or in the number of breeding attempts made by those that do breed, variation in the breeding output of the population will generally be underestimated. In terms of population monitoring, if these aspects of breeding biology vary in response to environmental change, we shall be aware of it only through the consequent effects on changes in numbers and proportions of juveniles, rather than directly. It may often happen that the number of breeding attempts is greater in years when nest losses are higher (because birds replace lost clutches). In this case, we may overestimate the variation in breeding output and conclude that this is a more important component of population change than it actually is. It is possible to model the numbers of nesting attempts of multi-brooded species under the assumption that a constant proportion of those nests that fail is replaced, which will give a higher estimated number of nesting attempts in years with high nest failure rates (e.g. Baillie and Peach, 1992). However, more intensive studies are needed to provide measures of this proportion and to determine the extent to which it varies between years. We should therefore measure these aspects of breeding biology where we can and, where we cannot, we should bear in mind our ignorance when interpreting the population data.

7.6.4 Counting juvenile birds

In most species, many of the young die shortly after leaving the nest. If numbers are counted during the first summer or first winter of life, this can provide an integrated measure of initial breeding output and subsequent survival. Because survival of young birds is so poor, this may be more useful in some ways than measuring breeding success directly (especially in species that are multi-brooded or in which first-year survival is difficult to measure). Its disadvantage is that it does not allow the causes of annual fluctuations to be pinpointed so readily. The ideal, achieved in some schemes, is to survey both nesting success and juvenile numbers.

The surveillance of reproduction

The technique is widely used in wildfowl studies, especially of geese and swans, in which birds are readily counted in autumn and winter and juveniles readily distinguished in the field. Thus Nilsson (1979) was able to relate variation in the proportion of juvenile whooper swans *Cygnus cygnus* in Sweden in the winter to the weather in the

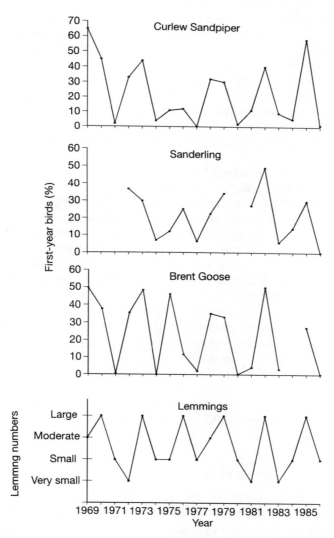

Figure 7.5 Percentage of first-year birds in wintering populations of various species that breed in the Arctic, in relation to numbers of lemmings on the breeding grounds. (From Summers and Underhill, 1987.)

previous winter and spring, which presumably determined the body condition and breeding success of females. Summers and Underhill (1987) counted the proportions of juveniles in wintering populations of brent geese *Branta bernicla* (in Europe) and waders (in Africa), finding that they fluctuated in synchrony with lemming cycles on their common breeding grounds in the Arctic (Figure 7.5). They argued that the link between the birds and the lemmings was mediated via their common predators. Ebbinge (1989) was able to use the proportion of juveniles in wintering brent geese to show the additional influence of conditions before and during spring migration on reproductive output, through their effects on body condition of females.

Similar studies have been carried out on waders, based on catching birds rather than direct observation. A project to draw together such information from the teams of volunteer ringers that operate around the coasts of Britain began in 1990. Many of the species involved breed in the Arctic, so their breeding success is not readily amenable to direct study.

Constant Effort (ringing) Sites are used not only to index changes in adult numbers (above) but also changes in the proportions of juvenile birds (Peach and Baillie, 1990, 1991, 1992). By surveying nesting birds in the same area in which they were operating a CES, du Feu and McMeeking (1991) were able to validate the technique. They showed that the numbers of juveniles captured over a 12-year period were highly correlated with the number of young produced from the nests in blackbird *Turdus merula*, blue tit *Parus caeruleus* and great tit *P. major* (though not in song thrush *Turdus philomelos*, of which few were caught, resulting in a comparatively high variance). During 1986–89, between-year changes in the percentage of juveniles were largely consistent across regions and habitats, indicating that the CES programme is likely to be useful for the surveillance of productivity at the national level (W.J. Peach, S.T. Buckland and S.R. Baillie, in prep). Formal power analyses have yet to be carried out but the 86 sites usable in the 1989/90 comparison provided data that would have shown up changes of 10% (or more) in over 60% of the species considered.

7.6.5 Indirect measure of reproductive performance

Croxall and Rothery (1991), observing that provisioning rate, chick growth rate and fledging weight may be more sensitive indicators of environmental conditions in seabirds than is the number of fledged young (Croxall *et al.*, 1988; Hunt *et al.*, 1986), suggest that these may be better measures of overall reproductive performance, because of their influence on post-fledging survival. Their use is, however, likely to be limited to intensive studies and often to be restricted by lack of detailed quantitative knowledge of their effects on survival.

7.7 THE SURVEILLANCE OF SURVIVAL

7.7.1 How survival may be measured

It is not possible to study changes in the population size of most birds during the course of a year by carrying out censuses at different seasons. One reason is that the detectability of birds may change markedly between seasons. Another is that even populations that are 'resident' on the coarse scale may redistribute themselves geographically or by habitat as the seasons progress and may be supplemented by immigrants. Thus simple counts cannot provide reliable indices of changing numbers within the year or of the causes of those changes, be they survival, reproduction or permanent immigration or emigration.

Fortunately, birds are readily marked and the study of marked individuals allows survival estimates to be made. Some information may be obtained from marks that are specific to a particular cohort of individuals or to a particular marking date but the best information is obtained when individuals are marked uniquely, so that their individual fates may be followed. The application of individually-numbered rings (bands) to birds has a long history and in many countries there are now many volunteers involved in this activity, ringing several million birds each year. These activities are concentrated in Europe and North America, where they are beginning to play an important part in population monitoring.

Ringed birds may be recaptured at (or near) the site of ringing, 'controlled' (recaptured elsewhere), or 'recovered' dead. If the rings are large enough, or are supplemented with other conspicuous marks, the birds may be resighted without having to be recaptured; resighting studies have an important role in the study of larger birds of open habitats. The information from these various forms of record can be used in the study of survival: recaptures, resightings and controls tell us that a bird was still alive on a certain date, recoveries that it was dead. Clobert and Lebreton (1991), Lebreton et al. (1992), Lebreton and North (1993), North (1990a, b), Pollock (1991), and Seber (1982, 1986) review the methods for converting such observations into estimates of survival rate.

The probabilities of recapturing, resighting, controlling, or recovering a ringed bird are less than 100% – sometimes less than 1%. This means that assumptions have to be made in order to estimate survival rates. As a result, such studies often entail statistical analyses that are not straightforward and that need to be interpreted with care.

In recent years there has been considerable development in the radio-tagging of birds (see, e.g., Pollock *et al.* (1989a, b) and several papers in Lebreton and North (1993). Radio-tags allow the fate of a bird to be exactly determined: we know either when it dies or that it has survived until the time that the tag became inactive. They thus provide information of much higher quality than rings. Currently they can only be used for specialized, intensive studies but they are becoming less costly and smaller, both of which will allow them to be more widely applied. Systems for automatic logging and processing radio-tag data are, furthermore, no longer in their infancy (Klaus and Exo, 1992). It is not fanciful to envisage that in a few decades' time even volunteer bird ringers may be routinely applying radio-tags rather than rings and that the birds' routine activities will be logged by computer-based systems, with further human intervention restricted to the detailed observations of birds that have been pinpointed by the computer. However, concerns that radio-tags may appreciably reduce survival have not yet been fully addressed. At present the routine and widespread surveillance of survival remains dependent on more traditional methods.

7.7.2 Capture-recapture methods

These involve the recapture or resighting of marked birds in a study population. Studies that use them are, therefore, basically intensive but, as we shall see below, a network of intensive studies can provide extensive coverage. Capture-recapture studies are not usually capable of discriminating losses due to death from those due to permanent emigration. For this reason, bird biologists using such methods usually concentrate on established breeders, which tend to remain in the same area in successive years; juveniles have much higher emigration rates. Even so, one must know enough about the population in question to be able to judge the extent to which any variation in apparent survival is the result of changes in actual survival or of changes in the emigration rates.

Lebreton *et al.*, (1992), Pollock (1991) and Pollock *et al.* (1990) provide reviews of capture-recapture methods in particular, additional to the more general reviews listed in section 7.7.1.

Capture-recapture models may be 'closed' or 'open'. The former assume that there are no births, deaths, immigrations, or emigrations, and so are irrelevant for present purposes. The basic 'open' model is the Jolly-Seber model. It is based on several assumptions. Three of them are that rings are not lost, that capture and release are instantaneous (rather than being so spread out that deaths, for example, may occur between the start and finish of a single capture session), and that all emigration is permanent. All three, if transgressed, can be allowed for by special modifications of the analytical methods. It is better, however, to try to ensure that they are not violated: it may be possible to design the fieldwork so that the first two are not. It is fundamental to the analysis of capture-recapture data that the probability of a bird being recaptured at a particular time depends on the probability of it being alive and the probability that, if alive, it will be captured. A further assumption of the Jolly-Seber model is that, while the probability of capture may change with time, it should be the same for all birds at any particular time. This is why it is important to have catching stations spread through the study area at a sufficiently high density. Modifications to the basic model can take account of trapping responses, such as birds tending to avoid nets or even to emigrate after trapping, or can incorporate general heterogeneity of capture probability. As discussed above (section 7.4.8), however, the exact form of such responses and heterogeneities is difficult to determine from the data. Fortunately, unlike the estimation of population size from capture-recapture data, the estimation of survival is robust in the face of heterogeneity in capture probability (Carothers, 1973, 1979).

A similar assumption is that, although survival rates may vary with time, the probability of survival at any one time should be the same for all birds. If it is not, mean survival will be overestimated (Pollock et al., 1990) – unless ringed birds have a lower survival rate than unringed, when the population survival rate will be underestimated. The basic model is that the probability of a marked bird being recaptured on the next sampling occasion depends on its probabilities of surviving to that time and of being recaptured if it does survive – that is, on survival rate and recapture rate. These rates are taken to be time-specific. They can be estimated from the pattern of recaptures.

Modifications of the basic model have been devised. Some represent special cases, involving more assumptions and thus meaning that there are fewer unknown parameters to be estimated. This increases the precision of the estimates that are made. Other modifications involve assumptions being relaxed. Some, for example, allow the probabilities of survival and recapture to vary with age (though not necessarily with time).

There is always a temptation to make one's model as general as possible, to ensure that this is realistic. But no model is more than an approximation to the truth and it is important not to build in so many parameters to the model than none of them is estimated precisely. For this reason Peach (1993), Peach et al. (1990, 1991), Pradel et al. (1990), and Pratt and Peach (1991), who were chiefly interested in changes in survival with time, all assumed that survival was constant for all age groups older than one year. (This assumption had been shown to hold, within the limits of measurement precision, in a study of reed warblers *Acrocephalus scirpaceus* by Buckland and Baillie (1987)).

Pradel et al. (1990) and Blondel et al. (1992) found that the sexes in blue tits differed in both capture and survival rates, the difference in survival rates itself varying with time, while Pratt and Peach (1991) found that capture rates differed between the sexes of willow warblers but that survival rates did not. Where one suspects that such sex differences (or, indeed, differences between other groups) may exist, it is wise to begin with the assumption that they do and only combine the data if the differences prove non-significant.

The study of Pradel et al. (1990) and Blondel et al. (1992) showed that not only were survival rates in their blue tit populations time-dependent but so were recapture rates, leaving a model with many parameters. To reduce the possibility of such a result, it is wise to standardize capture effort over time. This is done in the BTO Constant Effort Sites (CES) scheme and, as a result, capture rates at CES sites do not generally vary over time (Peach et al., 1990, 1991; Peach, 1993).

If one suspects that survival is well correlated with some environmental variable, this may be included as a covariate in the analysis. Doing so may have the effect of replacing a large number of time-related survival parameters with a few parameters that describe the relationship between survival and the environment (Clobert and Lebreton, 1985; Lebreton et al., 1992).

Another way of increasing precision is to use resighting information as well as recaptures (Clobert and Lebreton, 1985). Indeed, one can make just a single capture and then gather all survival information as resightings (Bell et al., 1993; Pollock et al., 1990). It will then be necessary to distribute the effort devoted to resighting birds evenly, to reduce heterogeneity in the probability of resightings. Even in a careful and well informed study of barnacle geese, Ebbinge and van de Voet (1990) found that individuals had different probabilities of being resighted.

An important consideration in most capture-recapture studies of birds, in which the study area tends to be small compared with the extent of the birds' movements is that transient individuals may be caught. In British Trust for Ornithology studies it is routine, for this reason, to exclude individuals that are not recaptured more than 10 days after initial capture (which are likely to be transients). By doing so, Pratt and Peach (1991) found that their estimates of willow warbler survival rose from 37% per annum to 47%.

Given that a variety of models is available, how should one choose between them? The best procedure is to begin with general models, with large numbers of parameters, and then to simplify to models with fewer parameters (and thus greater precision). The first guide in simplification should be biological intuition. The adequacy of simpler models chosen may then be tested with likelihood-ratio tests or goodness-of-fit tests. These have the advantage not only of guiding the process of simplification but of constituting formal tests of biological hypotheses. Unfortunately such tests are usually of low power in capture-recapture studies and, through applying them to a series of alternative and non-independent models, one may increase the chances of both Type I and Type II errors. A generally faster and easier approach, which avoids some of the problems inherent in making multiple tests, is to adopt the model that has the smallest value of Akaike's Information Criterion: $2(n-\ln L)$ – where n is the number of parameters in the model and L is the likelihood. Lebreton et al. (1992) illustrate its use on a number of datasets.

The fitting of various models, and testing alternatives, is nowadays relatively easy, because of the availability of a variety of powerful computer software packages (Brownie, 1987; Burnham et al., 1987; Clobert and Lebreton, 1991; Lebreton and Clobert, 1987; Lebreton et al., 1992, 1993; Pollock et al., 1990). Of these, we find the most useful to be SURGE (Lebreton and Clobert, 1987; Pradel et al., 1990). More general packages can also be used for the purpose (Cormack, 1985, 1989, 1993), but are less useful for non-statisticians than the specialist packages.

7.7.3 Networking capture-recapture studies

Capture-recapture studies have the advantage over studies that use recoveries derived from general ringing programmes that the recapture rate is so large compared with recovery rates from general ringing. Thus Peach et al. (1990), analysing data from two CES sites, found that the annual number of recaptures of reed warblers at each was greater than the total number of recoveries from the entire British and Irish ringing scheme, the same being true for sedge warblers at one of the sites. This advantage would be outweighed at a single site by the fact that it could not be properly representative of the whole population. The CES scheme, by bringing together a large number of sites, using common and standardized methodology and systems of data recording, overcomes this problem. Data arising from it are well analysed using an extended version of program SURGE, appropriate to more than one site (Clobert *et al.*, 1987; Pradel *et al.*, 1990). Program RECAPCO (Buckland, 1982) has also been used to analyse CES data but lacks the modelling flexibility of SURGE (Peach *et al.*, 1990).

The use of CES for the surveillance of survival was first presented by Buckland and Baillie (1987). Though, like Pratt and Peach (1991), they used data from just one site, they pointed out how such datasets could be combined. Subsequently, Pradel *et al.* (1990) and Peach *et al.* (1990, 1991) presented analyses based on two sites (Figure 7.6), while Peach (in press) has drawn together data from up to seven sites for each of a number of species. As mentioned above, bias from transients was reduced by allowing for the lower recapture probabilities of all birds not recaptured more than 10 days after first capture in all of the studies of Peach and his colleagues.

Not unexpectedly, recapture rates are usually site-specific. (The site differences themselves varied with time in the study of Pradel *et al.*, 1990.)

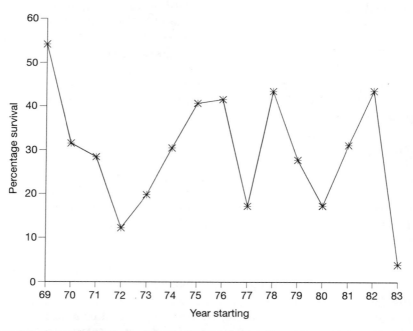

Figure 7.6 Annual survival rates of adult sedge warblers *Acrocephalus schoenobaenus*, based on capture-recapture analyses of British Trust for Ornithology data for two sites. (From Peach *et al.*, 1991.)

If survival rates do not differ between sites, a multi-site model with a single survival rate can be fitted (Pradel et al., 1990). If they do differ, a weighted mean provides a satisfactory estimate of the overall rate (Peach et al., 1990). Survival rates vary significantly over time in some of the studies: the variation depends on rainfall in the winter quarters in the sedge warbler in a way that explains fluctuations in the breeding population (Peach et al., 1991).

Formal analyses of the power of the CES programme to detect long-term changes in survival have yet to be published. Peach (in press) found, however, that even a mere 14 CES sites gave precision as good as or better than that achievable through analysis of nationwide ringing recoveries in four of the five species for which a direct comparison was possible. He considered that a change in survival rate of 10% between two 9-year periods would be detectable in the better-covered species, if only 14 sites were available. The scheme currently involves over 100 sites.

Precision could be improved if more recaptures were available per site. With this in mind, there are plans to operate some CES sites on a larger scale. This will have the added advantage of reducing the relative effect of emigration on the estimates of survival. Unfortunately, the CES scheme currently covers only a limited range of habitats (ones in which summer mist-netting is particularly productive) and, in consequence, a limited range of species. Furthermore, estimates of survival rates (even if obvious transients are omitted) are confounded with return rates. For this reason, analyses of recoveries provide a valuable cross check.

7.7.4 The use of recoveries

Survival analyses are generally based on records of birds recovered dead by members of the public and by ringers. Thus the basic data consist of a known number of birds ringed each year and the numbers of recoveries of these birds reported in each year after marking. Ideally all birds would be ringed and released on the same day but in practice the ringing period usually spans several months. Birds are usually aged as nestlings, first year birds or adults, and sex is also recorded for many species. Thus separate matrices of ringing and recovery data can be assembled for birds of different age and sex classes when ringed. General reviews of methodology for the analyses of ring recovery data are provided by North (1987–1990a, b), Clobert and Lebreton (1991), Pollock (1991) and Lebreton et al. (1993).

Such ringing and recovery data can be modelled statistically in terms of annual survival and reporting rates. The survival rate is the probability that a bird alive at the beginning of one year will be alive at the beginning of the following year while the reporting rate is the probability that a ringed bird which dies will be found and reported. These survival and reporting rates may vary with time (usually calendar year) and with age (usually number of years since hatching). The basic approach to modelling these data is concerned with such year- and age-specific variation. It can be extended to test for differences in survival between different categories of birds (e.g. sexes, birds ringed in different regions) and to fitting models which incorporate covariates of survival (weather conditions, density).

Early methods of estimating survival rates were developed by Lack (1943) and Haldane (1955) in Britain and by Hickey (1952) in North America. Some of these have since been found to be unreliable (Burnham and Anderson, 1979) but they paved the way for the development of a rigorous statistical framework for ring recovery analysis, culminating in the publication of the Brownie handbook in 1978 (second edition, Brownie et al., 1985).

The Brownie models focus on year-specific variation in survival and reporting rates, assuming that once birds become adults any age-specific variation in survival is slight. Very few individuals are likely to survive long enough to experience any effects of extreme old age, at least in the short-lived temperate species on which most ring recovery analyses have been conducted (but see Newton, 1989). It is year-specific variation in survival rates which is of greatest importance for environmental monitoring and for explaining population changes.

Brownie et al. (1985) provide a basic set of standard models incorporating time and age-specific variation in survival and reporting rates. Age-specific variation is considered only in terms of differences between first-year birds and adults, although extensions to first-year birds, subadults and adults are available if samples of all three age-classes are ringed. The approach adopted, as for mark-recapture analysis (section 7.7.2), is to seek the simplest model that provides an adequate description of the data. Most recovery datasets are 'incomplete' at the time of analysis, meaning that some of the ringed birds are still alive and remain to be recovered in future years. Partly for this reason, many of the models can only be solved iteratively using a computer. Standard computer software (programs ESTIMATE and BROWNIE) provides maximum-likelihood estimates of model parameters and their variances and covariances, likelihood-ratio tests between models, and goodness-of-fit tests to models.

Many of the key assumptions necessary for estimating survival rates from ring recoveries are similar to those needed for mark-recapture data. The properties of the Brownie models are well understood, both from experience of analysing large numbers of real datasets and through simulation studies (Brownie et al., 1985, Appendix C). They have been found generally to be robust with regard to minor violations of their assumptions.

As with any sampling programme, one must assume that the marked birds are representative of the population from which they came. One must assume that the birds were correctly aged and sexed, that the catching process does not affect survival and that the recovery information was reported accurately. Some recoveries may be reported late, by a year or more, and hence tabulated incorrectly, but small amounts of such delayed reporting cause negligible bias to the Brownie models (Anderson and Burnham, 1980). Furthermore, extensive delayed reporting would be indicated by model selection. Ringing periods of several months violate the assumptions of the models but are often necessary in practice. However, survival estimates from such programmes are largely unbiased provided that the distribution of ringing dates does not vary between years (Smith and Anderson, 1987). This is generally the case, particularly in schemes involving large numbers of volunteers where individual variations in effort will tend to cancel each other out.

One must also assume that rings are not lost. Some early survival analyses based on ring recoveries were undoubtedly affected by problems of ring loss, though this is now less of a problem in those ringing schemes that use more durable rings. Except for species with very high survival rates, time-specific survival estimates from the Brownie models only suffer slight negative bias from moderate levels of ring loss (Nelson et al., 1980). In contrast, estimates of age-specific survival from 'life table' models (below) are much more severely biased by equivalent rates of ring loss.

A further set of related assumptions is that the survival and reporting rates of all individuals are equal and independent: as with capture-recapture, breaking the sample down into subgroups prior to analysis allows one both to check for and overcome heterogeneity in so far as it depends on the the subgroup to which a bird belongs. The potential problems caused by ringed samples being made up of heterogeneous subgroups are discussed by Pollock and Raveling (1982) and Nichols et al. (1982). Such heterogeneity may give rise to underestimation of survival rates and the power of goodness-of-fit tests to detect it is low. However, it is thought that the levels of heterogeneity normally encountered are likely to cause relatively small bias to average survival estimates, at least for North American waterfowl (Nichols et al., 1982).

A wide variety of circumstances arise in which it is desirable to extend the basic Brownie models. For example, birds may not be ringed every year, giving rise to unequal time intervals between ringing occasions, or it may be desirable to incorporate covariates, such as weather factors or population density, in the models. Such extensions have been facilitated by two very flexible computer programs. MULT (Conroy and Williams, 1984; Conroy et al., 1989) adds many additional models to those provided by BROWNIE and ESTIMATE. SURVIV (White, 1983) allows the user to specify the models to be evaluated. Power is traded for ease of use and this program is only suitable for use by those who have a good understanding of modelling ring recovery data.

The main problem in applying the Brownie models to European ring recovery data is that most ringing schemes do not yet computerize their ringing data, so ringing totals categorized by age-class, season, region or other appropriate categories are not available. It would be prohibitively time-consuming to extract these numbers from handwritten records filed by ring-number. As a stop-gap, the British and Irish scheme has been gathering Age-Specific Totals for 22 passerines – i.e. numbers ringed during April–September, classed as nestlings, juveniles and adults. Analyses of the 1985–1990 data for 16 species

Table 7.3 Estimates of first-year and adult survival rates (with standard errors) for various British passerines, estimated from recoveries of ringed birds (from Baillie and McCulloch, in press)

		First-year	Adult
Swallow	*Hirundo rustica*	0.49 (0.03)	0.35 (0.05)
Pied wagtail	*Motacilla alba*	0.35 (0.02)	0.50 (0.07)
Dunnock	*Prunella modularis*	0.28 (0.04)	0.45 (0.05)
Robin	*Erithacus rubecula*	0.37 (0.03)	0.48 (0.04)
Blackbird	*Turdus merula*	0.54 (0.02)	0.67 (0.02)
Song thrush	*Turdus philomelos*	0.41 (0.05)	0.66 (0.05)
Sedge warbler	*Acrocephalus schoenobaenus*	0.24 (0.06)	0.36 (0.09)
Reed warbler	*Acrocephalus scirpaceus*	0.32 (0.06)	0.59 (0.07)
Blackcap	*Sylvia atricapilla*	0.40 (0.05)	0.41 (0.05)
Willow warbler	*Phylloscopus trochilus*	0.26 (0.04)	0.31 (0.05)
Blue tit	*Parus caeruleus*	0.33 (0.03)	0.49 (0.04)
Great tit	*Parus major*	0.27 (0.04)	0.38 (0.04)
Starling	*Sturnus vulgaris*	0.32 (0.03)	0.65 (0.03)
Chaffinch	*Fringilla coelebs*	0.55 (0.06)	0.55 (0.05)
Greenfinch	*Carduelis chloris*	0.33 (0.07)	0.49 (0.04)
Bullfinch	*Pyrrhula pyrrhula*	0.40 (0.06)	0.40 (0.05)

showed that useful estimates of survival rates could be obtained but the precision with which annual differences in survival could be assessed was not high in most species (Table 7.3; Baillie and McCullock, 1993). Greater precision might have been obtained if birds ringed as nestlings had been included (they considered only full-grown ringings) and if data from winter ringing had been available. Baillie and McCulloch also point out that most ringing schemes do not gather recapture data, since there is so much of it, but that survival estimates could be improved if they did.

It is possible to estimate survival rates without knowing the numbers ringed if one is prepared to assume that reporting rates are constant (Haldane, 1955; North and Morgan, 1979; Aebischer, 1987). There is much evidence of time- and age-specific variation in reporting rates (Anderson et al., 1981; Baillie and Green, 1987) although Baillie and McCulloch (1993) found time-specific variation in reporting rates in only three out of 16 passerines and age-specific variation in only four species. Thus species that are not hunted may be less subject to variations in reporting rates than the quarry species that have dominated ring recovery studies of survival estimation, particularly in North America. Young birds are easier targets than adults and hunting pressures vary from year to year.

Constant reporting-rate models may give rise to approximately unbiased estimates of average adult survival rates provided that there are no strong time-trends in reporting rates. Thus Dobson (1990) was able to use Haldane estimates to explore relationships between survival and life-history traits. A number of analyses based on constant recovery-rate models have produced biologically convincing results, suggesting that bias in survival estimation caused by variation in reporting rates is often small or extremely consistent. Such analyses have demonstrated relationships between survival rates and weather conditions for grey herons (North and Morgan, 1979), purple herons (Cavé, 1983) and lapwings (Peach, et al., in press). An increase in the survival rates of several species of Danish raptors following protection has been demonstrated using a simple model with constant reporting rates and constant first year and adult survival rates within two predetermined time periods (Noer, 1990). This is an important applied finding and illustrates the potential value of applying such models when more rigorous analyses are impossible.

Further analytical problems arise when only young birds are ringed (Anderson et al., 1985; Brownie et al., 1985). The main problem is that first year and adult survival rates can only be estimated from such data under the assumption that reporting rates do not vary with age. Although this may hold for some species (above), it is quite clear that for many others differences in the recovery circumstances and geographical distributions of young birds and adults lead to a strong expectation that their reporting rates will differ. The best way to overcome this problem is to ring both adults and young, in which case separate first year and adult reporting rates can be estimated using the appropriate Brownie models.

An alternative approach is to supplement the analysis with additional information. This might be an independent estimate of the survival rates of at least one age-class using mark-recapture or telemetry data (Lakhani, 1987; Freeman et al., 1992). However, there is increasing evidence that avian survival estimates based on telemetry are biased, because the radios affect survival probabilities, so studies using this method need to incorporate checks for such effects. Another approach is to incorporate census and productivity data in the model, effectively checking that the survival rates obtained from the ringing data are consistent with those expected from the observed population changes (Green et al., 1990).

The most severe problems for ring recovery analysis arise where the analyst wishes to estimate survival rates for a large number of age-classes from birds ringed as young. This is essentially an extension of the problem discussed above, the model sometimes being referred to as the 'life-table model' or the 'fully age-specific model'. A unique solution to this model can only be obtained by imposing a constraint, usually by setting two adjacent age-specific survival rates to be equal. The solution obtained may be dependent on the constraint that is used (Burnham and Anderson, 1979; Lakhani and Newton, 1983) making such analyses untrustworthy. Considerable effort has been devoted to finding solutions to this problem. Extending the model to include time-specific first-year survival rates (Morgan and Freeman, 1989; Freeman and Morgan, 1992) provides a unique solution at a cost of substantially increasing the number of parameters in the model, with consequent loss of precision. A similar, but perhaps more parsimonious, approach is to incorporate a covariate of first-year survival such as weather conditions (Rinne et al., 1990). Burnham (1990) presents some general methodology for evaluating the efficiencies of maximum likelihood estimators and the power of tests under such models. We agree with Burnham's conclusion that the main problem with these models is not that of imposing a constraint but the assumption that reporting rate does not vary with age and the unsatisfactory statistical properties of the models. Where age-specific variation in survival rates is the main focus of interest then mark-recapture methods are generally much more suitable than recovery analyses.

In summary, the Brownie models (Brownie et al., 1985) and extensions of them (White, 1983; Conroy and Williams, 1984), provide a robust and well tested methodology for using ring recoveries to study time-specific variation in survival rates. Monitoring programmes based on ring recoveries must ensure that both young birds and adults are ringed and that the ringing data are computerized. Careful analyses of existing datasets where only young birds have been ringed or where the numbers ringed are unknown may produce valuable contributions to conservation science but it will always be necessary to place caveats on such analyses. Ring recovery data are not generally suitable for detailed studies of age-specific variation in survival rates.

7.7.5 The future of survival estimation

The part that survival surveillance plays in population monitoring is poised to increase substantially. The models required for efficient data analysis are well developed, though some further work is needed, particularly a thorough investigation of their power as surveillance tools. Particular attention needs to be given to the power of ring recovery and mark-recapture analyses to detect time trends and to identify covariates of survival. Empirical studies of sample-size requirements for both approaches are needed and should lead to evaluations of their cost-effectiveness for different species. With current knowledge of survival rates it is important that at least some species should be studied using both methods.

Recovery and recapture data have common features, statistically speaking. An important development will be the combined analyses of such datasets, signalled by Buckland and Baillie (1987), Burnham (1993), and Oatley and Underhill (1993). Models need to be developed that will allow regional or national ring recovery data to be combined with data from several mark-recapture sites, such as those operated by the CES scheme. Methods are also needed for analysing recapture data arising from many sites, within each of which the number of individuals is too small for formal mark-recapture analysis. A related development is the combined analysis of controls (birds caught alive away from the place of ringing) with recoveries of birds found dead. Baillie and McCulloch (1993) have explored such data for three species. They found that controls and dead recoveries of reed warblers could be combined in a single model, resulting in more precise estimates of survival. Willow warbler controls gave a more reliable estimate of survival than dead recoveries, judged by comparison with independent data. However, controls of sand martins *Riparia riparia* could not be modelled in this way.

The most important need is to improve the gathering and collation of relevant data; national ringing schemes need to computerize general ringing as well as recoveries. The value of recording and computerizing all recapture information needs to be assessed. Schemes such as the CES need to be run in more countries, using sites that are as large as possible consistent with proper coverage. The development of networks of sites gathering recapture information in an intensive and consistent manner needs to be considered for other groups of birds such as nestbox species, seabirds and waders.

7.8 IMMIGRATION AND EMIGRATION

Of the four processes that can cause bird numbers to change, births and deaths have received far more attention than immigration and emigration. Indeed, the very existence of movement between populations is often ignored in population analysis. This is particularly unfortunate in that, in mark-recapture studies, births may be confounded with immigration and deaths with emigration because, unless all birds fledged in an area are ringed, any unringed birds caught may either have hatched locally or be immigrants; and any ringed birds that disappear may have died or they may have emigrated. In these circumstances, the independent measurement of immigration and emigration rates is difficult. This is why estimates of survival from mark-recapture studies are often lower than those from recovery analyses (Peach, in press). Interpretation of changes in apparent survival based on mark-recapture must, therefore, be particularly circumspect. The development of methods to measure movements independently of survival is recent and welcome (Hestbeck *et al.*, 1991; Nichols, 1993).

Since such methods are based on intensive mark-recapture studies, they may be of limited use in general population monitoring. The colonization of new areas by a species obviously depends on immigration and claims have been made that changes in national populations may be the result of mass immigration in some species other than those normally considered to be irruptive (Summers-Smith, 1989) but it seems unlikely that most national population changes are explicable in terms of population movement. The obvious exceptions are seabirds which may, for example, treat the whole North Sea as their homeland but this just means that, if we are to understand their population dynamics, we need to adopt a similarly broad perspective and not restrict our studies to single countries.

Even if movement of populations is judged to be unlikely as a significant cause of changes in abundance at the national scale, it would be useful to be able to measure rates of movement between populations, partly because they may respond to environmental change (and thus provide another pointer to the effects of such change), partly to ensure that they are not changing to such a degree as to mislead our studies of population dynamics, and partly to enable us to build metapopulation processes into the population models that underlie effective population monitoring. Busse (1987) and Schwarz (1993) have considered how to quantify rates of movement using ringing recoveries.

7.9 INTEGRATED POPULATION MONITORING

7.9.1 Population modelling

Bird population monitoring is undertaken in many countries. The species covered, the intensiveness of the work, the extent of the coverage, and the degree to which information from different schemes is integrated all vary considerably. The BTO Integrated Population Monitoring Programme (Figure 7.7) (Baillie, 1990, 1991; Baillie and Marchant, 1992) is the furthest developed national integrated monitoring scheme and most of our examples are drawn from it. But even that scheme is in its infancy, with much further development needed.

The basis of any integrated population monitoring scheme is population modelling. At its simplest, this can be merely the description of how annual changes in numbers are related to environmental factors and to the abundance of the population itself, using simple regression models. Since the changes in numbers are autocorrelated with the numbers themselves, there are problems with the statistical testing of density-dependence using such regressions, which have not been fully solved (Clobert and Lebreton, 1991). The models can, however, provide a useful approximation to reality, as a basis for further analysis. Thus Baillie (1990) developed an analytical model of British song thrush populations, in which 57% of the variation in the change in numbers between successive years was 'explained' by the index of numbers in the first of the two years and by relevant weather variables (number of freezing days in January and February). A plot of this model shows that it agrees reasonably with the facts (Figure 7.8), but that there are systematic deviations: there are runs of years in which the population has done better or worse than the model predicts, the poor years being concentrated in the second half of the study period. This suggests that some additional factor is causing the population to decline further than would have been expected given the weather that the birds have experienced.

An alternative to regression analysis is time-series analysis, which allows numbers at any one time to be predicted from what they have been in the past and from environmental conditions (Brillinger, 1981; Bulmer, 1975; Lebreton, 1989; Poole, 1978). Such analyses may,

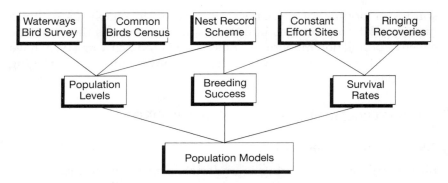

Figure 7.7 Contributions of the various schemes to population models in the British Trust for Ornithology Integrated Population Monitoring Programme.

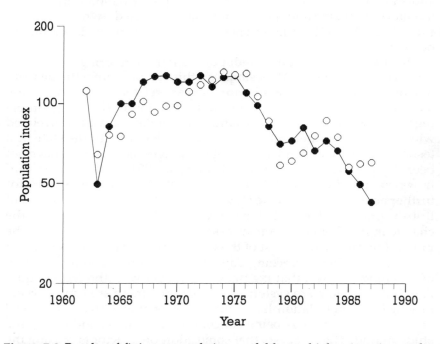

Figure 7.8 Results of fitting a population model by multiple regression to the British Trust for Ornithology Common Birds Census data for farmland song thrushes *Turdus philomelos*. The solid symbols and line show the observed indices of population size; the open circles show values predicted from a model incorporating the number of freezing days in January and February and population size in the previous year. Note that the population scale is logarithmic and the index is relative to an arbitrary value of 100 for 1966. (From Baillie, 1990.)

Integrated population monitoring 317

however, turn out not to be sufficiently powerful, given the relatively small number of years for which counts are available in most monitoring schemes.

7.9.2 Key factor analysis

Key factor analysis was developed for use on data drawn from within-generation studies of insects but has been extended to within-year studies of birds (Blank *et al.*, 1967; Krebs, 1970). It is based on the idea that the log of the ratio of numbers in two successive stages in the annual cycle is a measure of the mortality during the intervening period: these log ratios are the k factors. (Reproductive rates can be included in a complementary fashion – see below.) The sum of the k values, K, is a measure of total mortality. For such an analysis, one requires either censuses at successive stages or direct measures of productivity and survival that can take their place. Many intensive studies of bird populations can provide these. Extensive studies can also do so, particularly if all the surveys that provide the information are run by the same agency and are run with integrated analysis in mind, as is the case in Britain (Figure 7.7).

When a series of years' data is available, plots of K against each of the k values in turn reveals which of the latter make the biggest contribution to the former – i.e. which stage in the annual cycle is mostly responsible for variations in total change in numbers over the year. However, although key factor analysis is a powerful tool for detecting the life-cycle stage responsible for short-term changes in population size, it is less suitable for identifying the cause of a steady change.

The impact of environmental factors can be assessed, and modelled, via regressions of k factors on them. Since spurious correlations can easily arise if both a k value and an environmental factor are subject to long-term trends that are not causally connected, the best procedure is to model the time trends in each using auto-regressive models and then examine the correlations between the departures of the k value and the environmental factor from their respective time-series models. In this way, Peach *et al.* (1991) were able to confirm the importance of rainfall in the Sahel for overwinter survival of sedge warblers (Figure 7.9).

Plotting k values against numbers can reveal density-dependence and does so more convincingly if productivity and survival are estimated independently of numbers than if the estimates are correlated.

Baillie and Peach (1992) have applied key factor analysis to a number of Palaearctic–African migrants, using BTO data. Total K was measured from the change in numbers between successive years that the CBC

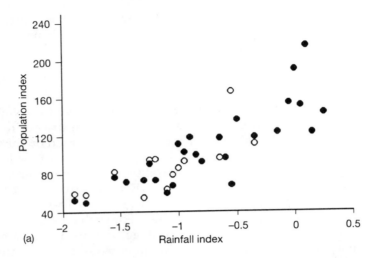

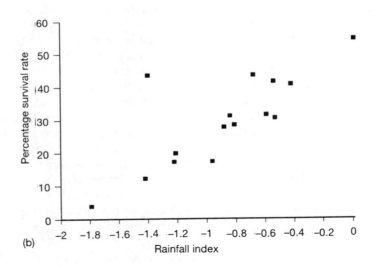

Figure 7.9 Effect of Sahelian rainfall on British sedge warbler populations: (a) indices of abundance from the Common Birds Census farmlands plots (solid symbols) and from the Waterway Birds Survey (open symbols); (b) annual survival rates of adults, from mark-recapture studies. Each point refers to a single year. (From Peach et al., 1991.)

Table 7.4 Strength of relationships between k values and total losses (K) in various European passerines, measured as coefficients of determination (r^2); three separate studies of pied flycatchers are included (from Baillie and Peach, 1992)

		Failure to lay maximum number of eggs k_1	Egg losses k_2	Nestling losses k_3	Full-grown losses k_4
Swallow	*Hirundo rustica*	0.00	0.04	0.00	0.93
Sedge warbler	*Acrocephalus schoenobaenus*	0.02	0.01	0.02	0.78
Whitethroat	*Sylvia communis*	0.00	0.02	0.05	0.78
Blackcap	*Sylvia atricapilla*	0.00	0.19	0.07	0.67
Willow warbler	*Phylloscopus trochilus*	0.02	0.07	0.06	0.67
Pied flycatcher (1)	*Ficedula hypoleuca*	0.12	0.02	0.64	0.91
Pied flycatcher (2)		0.00	0.09	0.57	0.85
Pied flycatcher (3)		0.37	0.04	0.01	0.69

Table 7.5 Density dependence in various European passerines, measured as coefficients of determination (r^2) of k factors by population density immediately before the relevant stage in the life history; three separate studies of pied flycatchers are included (from Baillie and Peach, 1992)

		Failure to lay maximum number of eggs k_1	Egg losses k_2	Nestling losses k_3	Full-grown losses k_4
Swallow	*Hirundo rustica*	0.28	0.00	0.21	0.10
Redstart	*Ph. phoenicurus*	0.06	0.09	0.25	0.53
Sedge warbler	*Acrocephalus schoenobaenus*	0.02	0.07	0.00	0.20
Whitethroat	*Sylvia communis*	0.16	0.00	0.00	0.34
Blackcap	*Sylvia atricapilla*	0.02	0.01	0.09	0.18
Willow warbler	*Phylloscopus trochilus*	0.04	0.24	0.01	0.36
Pied flycatcher (1)	*Ficedula hypoleuca*	0.01	0.19	0.14	0.02
Pied flycatcher (2)		0.03	0.00	0.11	0.24
Pied flycatcher (3)		0.11	0.00	0.17	0.55

indexes. The first k value comprised 'failure to lay the maximum number of eggs' assessed from the nest record scheme, using an adjustment to allow for very poor knowledge of how many females produce replacement clutches if their original clutch or brood is lost. Losses of eggs and of chicks, both assessed from nest records, comprised the next two k factors. The fourth such factor, losses of full-grown birds, was broken down into two: losses of adults, which could be estimated from CES data, and losses of birds in their first year of life. The last could not be directly estimated: it was obtained by subtracting the sum of the other k values from K.

Table 7.4 shows the strength of the relationships between K and the various k values: it indicates that losses of full-grown birds were the chief determinants of changes in numbers. These losses tended also to be density-dependent (Table 7.5). In the sedge warbler and the whitethroat, but not the other species, full-grown losses were correlated with Sahelian rainfall.

Such full key-factor analyses are still comparatively rare in monitoring programmes, except intensive programmes aimed at a limited set of species. More limited analyses are more common: they point the way forward to the development of monitoring that is based on full population monitoring, as well as providing pointers for practical action.

7.9.3 Incorporating the environment

If habitat underpins population performance, then we need to record the habitats in which the populations under study live. It will then be possible to relate performance at individual sites to habitat – a powerful way of identifying factors that may determine the success of a population. The recording of habitat has the added benefit of providing wider environmental monitoring than does surveillance of only the bird populations, since it provides some direct monitoring of habitat changes.

For extensive schemes, simple and objective systems for habitat description and recording are needed (though not too simple: Svensson (1992) believes that the one used in Sweden is too simple to provide satisfactory information). The system should be standardized, not only throughout each scheme but also throughout all schemes that contribute to an integrated programme. Crick (1992a) has produced such a system for BTO schemes. It employs a coding system for habitat characteristics, which promotes full standardization, increases objectivity, and aids computerization of the data. The system is based mainly on the vegetation structure of habitats, which has been shown to be an important determinant of bird community structure (e.g. MacArthur and MacArthur, 1961; May, 1982; *pace* Wiens, 1989; Willson, 1974), and in particular in Britain (Fuller, 1982). It is fully compatible with the widely used system devised by the NCC and RSNC (Nature Conservancy Council, 1990) but it has several advantages in that it embraces agricultural and man-made habitats as well as semi-natural ones, and it also includes aspects of land management and human activity.

Such habitat recording takes place at the fine scale. Environmental information is also needed on broader scales, at the regional or national level. In most countries it should be possible to obtain data on land use and land management practices, which are also likely to be important influences on bird populations. If these factors can be built into the models, we can assess their impact on wildlife and take steps to deal with any problems. Unfortunately, the surveillance of land use and other environmental information is often poor, and its integration and accessibility not much better. Furthermore, there is little standardization even between neighbouring countries, which will beome a major difficulty as we seek to develop international collaboration in bird population monitoring.

Weather data are, fortunately, widely available. They should be incorporated into any integrated population-monitoring programme, since it is a major influence on numbers of birds, as we see from the effects of freezing and snow on European birds in winter (Cawthorne and Marchant, 1980; Elkins, 1983; Greenwood and Baillie, 1991) and from the effects of Sahelian droughts on Palaearctic–African migrants (Baillie and Peach, 1992; Marchant, 1992; Peach *et al.*, 1991). Available weather records can be included in the population models, so that one can then ignore population changes caused by adverse weather and concentrate on those caused by the activities of man.

7.9.4 Identifying the causes of changes

Applied ecologists believe true population monitoring to be highly cost effective. Not only does it identify problems of declining populations but it provides evidence about the likely causes of the declines, all the difficulties of data interpretation notwithstanding. Those of a more theoretical bent sometimes question this view, on the grounds that modern population theory leads us to the conclusion that nature should be in a constant state of change, to a degree describable as chaotic (in the sense of being practically unpredictable). Even if this view were to be correct, it would remain true that the changes in a population are driven by systematic factors and can be predicted in the short-term, through the analysis of the effects of those factors. Furthermore, it is clear that the changes wrought by man to the ecosystems of many parts of the world are, in speed and magnitude, greater than those normally wrought by nature. Understanding the impact of those changes is demonstrably possible.

If a population has been through a period of rapid growth or decline, its age structure may be unstable. It may then take some time to achieve a stable age structure even in a stable environment. If fecundity and death rates are age-specific, the overall growth rate of the population will be affected by the changes in age structure. But the evidence is that age-related differences in fecundity or survival of birds are so small that the impact of age structure on population growth will be insignificant compared with the impact of the environment. Changes in age structure are unlikely therefore to obscure any relationships with the environment that we are trying to discern.

An obvious pointer to the cause of a change in a population is that some specific environmental change coincides with the population change. If both population and environment are subject to long-term change, the evidence is weak; if their short-term fluctuations are synchronous, the evidence is stronger. But an environmental factor that causes a long-term decline in population may not also cause short-term fluctuations if it has an impact that is cumulative or long-term. Geographical correlations are also useful pointers and, as with temporal correlations, the more detailed the better.

Examples of the use of such correlations are legion. The decline of many raptors in Europe and North America in the 1950s coincided with huge increases in the use of organochlorine pesticides; and those parts of Britain in which the peregrine falcon declined most were where DDT was most used (Ratcliffe, 1980). The corncrake *Crex crex* withdrew first from those parts of Britain where agricultural changes were most rapid, leading conservation ornithologists to concentrate more closely on identifying what factors in particular might be the cause of the decline (Williams *et al.*, 1991). In both of these cases, the initial pointers led in the right direction, to further field or laboratory studies that paved the way for conservation action.

As with patterns in time, spatial correlations between population changes and their causes may not be exact. If a bird flourishes more in one area than another, it may sustain a stable population in the one area in the face of a change in the environment even though that same change causes it to decline in the other less satisfactory area.

Coincidence with habitat is another pointer. Crick (1992b) investigated the possible reasons for the decline in the British population of golden plovers *Pluvialis apricaria*. Existing information suggested that heather *Calluna vulgaris* moorland was the better habitat for the species and the decline had been ascribed to (1) loss of heather moorland by afforestation or conversion to grassland and (2) the general degradation of grouse moors, as well as to (3) agricultural improvement and more intensive use of grassland and (4) more predators and disturbance (which would affect both habitats). Crick's analysis of nest record cards (Table 7.6) confirmed that golden plovers seemed to breed less well on grassland than on heather moorland, which is consistent with cause (1). He also showed that nest failure rates have remained stable on heather moors but have increased on grassland, which is consistent with cause (3) but rules out (2) and (4).

Table 7.6 Daily failure rates of British golden plover nests during incubation, according to habitat and time period (from British Trust for Ornithology Nest Record Scheme data, analysed by Crick, 1992b).

Habitat	Years		
	1943–73	1974–81	1982–89
Grassland	0.0118	0.0123	0.0328
Heather and bog	0.0075	0.0164	0.0076

Changing agricultural practices have been held to be responsible for perceived declines of population of another British plover, the lapwing. Shrubb (1990), reviewing previous evidence, found that it pointed to no clear conclusion. He showed, however, from an analysis of nest record cards, that nesting success had declined and that there were habitat differences in the extent of the decline, in ways that were consistent with changing agricultural practices and interpretable in the light of the species' breeding behaviour. His conclusions were confirmed by a study of the ways in which the numbers and distribution of lapwings varied in different parts of Britain and were related to agricultural practices (Shrubb and Lack, 1991). Peach *et al.*(in press) have analysed long-term changes in survival of full-grown birds, using ringing recoveries, and found no evidence for a decline. They also showed that the levels of breeding productivity reported in most recent studies were insufficient to balance the level of mortality they had found. It thus seems certain that the observed decline in population has been brought about by reduced breeding success, itself caused by modern agricultural practices.

In an exercise designed to monitor retrospectively the impact of increased numbers of magpies *Pica pica* on populations of smaller birds (of whose eggs and chicks they are certainly significant predators), Gooch *et al.* (1991) showed that small birds in geographical regions and habitats with particularly high magpie numbers did not suffer from generally reduced nesting success or increased population declines. In this case, the absence of either geographical or habitat patterns was taken as evidence that magpies are not having a general effect on numbers of songbirds in Britain, contrary to much prejudice. (This does not mean that individual species or particular local areas are not adversely affected by the magpie increase.)

Another type of pattern that provides a clue to causation is when several species with similar ecological requirements show similar trends, especially if other species with slightly different requirements do not show the same trends. Thus many seed-eating birds (particularly those feeding on agricultural weeds) have recently declined in numbers in Britain (Marchant et al., 1990). A preliminary analysis of ringing data indicates that mortality is highest in spring, when seeds are least available (Crick et al., 1991). Three species that had not declined in numbers (chaffinch *Fringilla coelebs*, greenfinch *Carduelis chloris*, and yellowhammer *Emberiza citrinella*) showed no decline in nest losses (from nest record cards), while two of three that had declined (linnet *Carduelis cannabina* and reed bunting *Emberiza schoeniclus*, but not corn bunting *Miliaria calandra*) showed increases in nest losses (Crick et al., 1991). Examining the trends for an entire suite of 12 seed-eating birds that use farmland, Marchant and Gregory (in press) were able to show that the decline in numbers has generally been strongest on arable farms and in parts of Britain where arable farming is at its most intense. All these results confirm the view that it is changes on arable farms, probably a reduced supply of seeds, that are responsible for most of the decreases.

Identifying which stage of the life cycle is most affected by a population decline may point to its cause. If, for example, hatching success has particularly declined, one might suspect the influence of a toxin that interfered with development. The effect of organochlorines on raptors was largely of this nature (Newton, 1979) and the pointer from field observation was confirmed by toxicological studies.

Distinguishing between effects at different stages of the life cycle requires that they all be studied. It has been suggested that less comprehensive studies can be more cost effective:

> Because the reproductive performance of birds is relatively easy to assess by observing nests, an initial investigation of reproduction should be undertaken to determine the extent to which the observed recruitment rate is falling below the potential or expected rate for the species. If such an investigation reveals that recruitment is normal, then the species' problems can unambiguously be blamed on reduced survivorship ... If, on the other hand, recruitment is found to be subnormal, there remains the possibility that survivorship as well as recruitment has been reduced. (Temple, 1986).

Green and Hirons (1991) point out, however, that we do not know precisely and generally enough what reproductive or survival rates to expect. They quote the example of Kirtland's warbler *Dendroica kirtlandii*, which was known to breed less productively than other North American warblers, mainly because of brood parasitism by brown-headed cowbirds *Molothrus ater*. A successful cowbird control programme was ineffective in reversing an earlier population decline. It may, of course, have prevented it going further, but poor breeding success was clearly not what was holding numbers down.

7.9.5 From knowledge to action

To avoid generating undue concern from studies of bird populations that indicate that some species are declining, we need to set 'thresholds'. Only if a decline exceeds the species' threshold should alarm bells be sounded (Baillie, 1990, 1991).

Some of the considerations in setting thresholds are obvious. Clearly the process will be more effective if we concern ourselves not with population declines themselves but only with those that are greater than expected given the known effects of weather or other natural forces on the populations. In the case of the sedge warbler, we are powerless to affect declining populations through action in Britain since they are driven by weather in the Sahel (Figure 7.9) (Peach *et al.*, 1991), over which we have no control (though the problems may be exacerbated by human activities in that region). But in the case of the song thrush, recent declines exceed those expected on the basis of winter weather (Baillie, 1991) and may be amenable to conservation action (Figure 7.8).

Even if we do not know what causes them, we must recognize that many fluctuations in populations are entirely natural, so that we should not be concerned with those that fall within the normal range for a species. Only long-term surveillance can define that range.

Other aspects of threshold-setting are less easy to determine. For example, is a 20% decline in a species over 20 years more or less important than a 20% decline in one year following 19 years of stability?

Another problem relates to statistical significance. In conventional approaches to science, one would find a population decline of 7% that had 95% confidence limits of 4–10% more important (because significant) than a decline of 20% with limits ranging from 10% increase to 50% decrease. Indeed, in conventional significance testing, because the second change was not significant one would conclude that it was zero. But, if we take the 95% confidence limits as our guide, the second population might well have declined by 50% whereas the first is unlikely to have declined by more than 10%. Which population deserves our greater attention? The answer is not obvious.

Even if we can be sure that a population has declined we shall probably be uncertain of the cause. Most ecologists, having been brought up as scientists, prefer to deal in certainties and are reluctant to make decisions based on inadequate data. But if conservation is to be effective, this may be essential. What is important is that decisions should be rational and based on the best available information. Formal decision theory may help (e.g. Lindley, 1971).

Of course, one possible line of action is to do more research, to provide a better foundation for decisions about practical conservation action. But, although scientists are always keen to do more research and politicians always keen to put off difficult decisions, one should only opt to do further research after careful consideration of its value relative to the likely value of direct conservation action. Otherwise there will be unnecessary delays; an average of 10 years between detection of a decline and effective conservation action has been claimed by Green and Hirons (1991). The cost of the research plus the cost of not taking immediate conservation action (there will be such a cost if that action had been the correct action to take) must be balanced against the cost of taking action should that action be wrong.

Just as further research may be considered a form of action, so all direct conservation action must be considered an opportunity for research. That is, its consequences must be monitored. All too often this is forgotten. Yet 'If population biologists had monitored a fraction of the conservation actions that have been attempted, we might now be in a position to predict the outcome of similar actions for other threatened species' (Rands, 1991). Thus Green and Hirons (1991) state that a model of the declining stone curlew *Burhinus oedicnemus* population in eastern England suggests that action to boost the number of birds fledging successfully could halt the decline, at a cost of £10 000–£20 000 per annum. This is a small expense compared with the cost of further intensive research aimed at refining the model in order to be sure that the action will work. It is obvious that the action should be implemented and that its success (or otherwise) be monitored. The action itself thus becomes an experiment, a test of the model, and a significant tool in extending our knowledge. It may be impossible to arrange that such action should be organized according to the best principles of experimental design, employing replicates and controls, but it cannot be emphasized too strongly that the inferences one can draw from monitoring its results will be much more certain (so less likely to be misleading) if it is so organized. Skalski and Robson (1992) consider the design and interpretation of environmental assessment studies, mainly in relation to the use of mark-recapture studies to measure population size, but covering important general principles.

In contrast, studies of Palaearctic–African migrants indicate the need for further research to discover what, if any, action is required. Work in central Europe, based on constant-effort ringing of migrant passerines from a wide area of northern Europe, suggested widespread population declines (Berthold et al., 1986). The problem was investigated further, through a comparative analysis of trends in breeding populations in five countries and found to be less serious than earlier thought, with approximately equal numbers of increases and decreases (Marchant, 1992). Integrated analyses of BTO data, combined with a review of published studies of populations, indicated that sedge warblers and whitethroats are limited by winter resources (in Africa), while willow warblers, pied flycatchers *Ficedula hypoleuca*, and perhaps blackcaps are limited on the breeding grounds (Baillie and Peach, 1992). Thus what initially seemed to be a general problem, which might have required a general solution, appears to be a set of species-specific problems.

Should further research appear necessary, what form should it take? Monitoring provides pointers through correlations. But correlation does not prove causation, particularly in ecology, where the complexity of the systems means that many factors may be changing at once – any subset of which could be the 'cause' of the change in any other subset. Following Karl Popper, there has been much emphasis on the role of experiment in testing scientific hypothesis. This is justified: experiments (if properly conducted) identify causes clearly. But the Popperian approach is not the full story of science, as Thomas Kuhn pointed out. Indeed, some areas of science cannot be investigated by experiment. Even in those that can, the role of observation of unmanipulated nature must not be forgotten, not only in formulating hypotheses in the first place but in actually testing those hypotheses in an experiment-like manner. That is to say, one predicts what one would expect to observe were one's hypothesis true – e.g. that the seed-eaters affected most by the declining availability of seeds on arable farms should have declined the most on such farms and in regions where arable farming is most intense. In the last example, the prediction was broadly correct. This increases the likelihood that the hypothesis is correct. Had it not been, our faith in the explanation should have been weakened. Of course, some uncertainty will always remain: it would (in principle) persist even after the most careful manipulative experiment.

Thus our monitoring programmes will provide us with pointers to factors underlying changes in our wildlife and further research will strengthen those indications. But eventually, despite remaining uncertainty, conservation action must be undertaken, otherwise the whole monitoring programme and the associated research are pointless. And when the action is taken, its effects must be monitored. These principles apply whether the objective of monitoring birds is to manage the birds themselves or whether it is to monitor the environment in order to manage it for broader purposes than the conservation of birds alone.

REFERENCES

Aebischer, N.J. (1987) Estimating time-specific survival rates when the reporting rate is constant. *Acta Ornithologica*, **23(1)**, 35–40.

Ambrose, S. (in press) The RAOU/BP Australian Bird Count, in *Bird Numbers 1992. Distribution, Monitoring and Ecological Aspects* (eds W. Hagemeijer and T. Verstrael). Proceedings of 12th International Conference of IBCC and EOAC, SOVON, Beek-Ubbergen.

Anderson, D.R. and Burnham, K.P. (1980) Effect of delayed reporting of band recoveries on survival estimates. *Journal of Field Ornithology*, **51**, 244–7.

Anderson, D.R., Wywialowski, A.P. and Burnham, K.P (1981), Tests of the assumptions underlying life table methods for estimating parameters from cohort data. *Ecology*, **62**, 1121–24.

Anderson, D.R., Burnham, K.P. and White, G.C. (1985) Problems in estimating age-specific survival rates from recovery data of birds ringed as young. *Journal of Animal Ecology*, **54**, 89–98.

Askins, R.A., Lynch, J.F. and Greenberg, R. (1990) Population declines in migratory birds in eastern North America. *Current Ornithology*, **7**, 1–57.

Baillie, S.R. (1990) Integrated population monitoring of breeding birds in Britain and Ireland. *Ibis*, **132**, 151–66.

Baillie, S.R. (1991) Monitoring terrestrial breeding bird populations, in *Monitoring for Conservation and Ecology* (ed. F.B. Goldsmith), Chapman & Hall, London, pp. 112–32.

Baillie, S.R. and Green, R.E. (1987) The importance of variation in recovery rates when estimating survival rates from ringing recoveries. *Acta Ornithologica*, **23**, 41–60.

Baillie, S.R. and Marchant, J.H. (1992) The use of breeding bird censuses to monitor common birds in Britain and Ireland – current practice and future prospects. *Die Vogelwelt*, **113**, 172–82.

Baillie, S.R. and McCulloch, M.N. (1993), Modelling the survival rate of passerines ringed during the breeding season from national ringing and recovery data, in *Marked Individuals in the Study of Bird Population* (eds J.-D. Lebreton and P.M. North), Birkhauser Verlag, Basel, pp. 123–39.

Baillie, S.R. and Peach, W.J. (1992) Population limitation in Palaearctic–African migrant passerines. *Ibis*, **134** (suppl. 1), 120–32.

Baines, D. (1991) Factors contributing to local and regional variation in black grouse breeding success in northern Britain. *Ornis Scandinavica*, **22**, 264–9.

Batten, L.A. and Marchant, J.H. (1976) Bird population changes for the years 1973–74. *Bird Study*, **23**, 11–20.

Batten, L.A., Bibby, C.J., Clement, P., et al. (eds) (1990) *Red Data Birds in Britain*, T. & A.D. Poyser, London.

Bell, M.C., Fox, A.D., Owen, M. et al. (1993) Estimation of survival rates in two arctic-nesting goose species, in *Marked Individuals in the Study of Bird Population* (eds J.-D. Lebreton and P.M. North) Birkhauser Verlag, Basel, pp. 141–55.

Bernstein, C., Krebs, J.R. and Kacelnik, A. (1991) Distribution of birds amongst habitats: theory and relevance to conservation, in *Bird Population Studies. Relevance to Conservation and Management* (eds C.M. Perrins, J.-D. Lebreton and G.J.M. Hirons), Oxford University Press, Oxford, pp. 317–45.

Berthold, P., Fliege, G., Querner, U. and Winkler. H. (1986) Die Bestandsentwicklung von Kleinvögeln in Mitteleuropa: Analyse von Fangzahlen. *J. Orn.*, **127**, 397–437.

Bibby, C.J., Burgess, N.D. and Hill, D.A. (1992) *Bird Census Techniques*, Academic Press, London.

Bijleveld, M. (ed.) (1974) *Birds of Prey in Europe.*, Macmillan, London.

Blank, T.H., Southwood, T.R.E. and Cross, D.J. (1967) The ecology of the partridge. 1. Outline of the population processes with particular reference to chick mortality and nest density. *Journal of Animal Ecology*, **36**, 549–56.

Blondel, J., Pradel, R. and Lebreton, J.-D. (1992) Low fecundity insular blue tits do not survive better as adults than high fecundity mainland ones. *Journal of Animal Ecology*, **64**, 205–13.

Brillinger, D.R. (1981) Some aspects of modern population mathematics. *Canadian Journal of Statistics*, **9**, 173–94.

British Trust for Ornithology (n.d.) *The Nest Record Scheme*. British Trust for Ornithology, Tring.

Brownie, C. (1987) Recent models for mark-recapture and mark-resighting data. *Biometrics*, **43**, 1017–19.

Brownie, C., Anderson, D.R., Burnham, K.P. and Robson, D.S. (1985) *Statistical inference from band recovery data – a handbook*, 2nd edn, U.S. Fish Wildl. Serv., Resource Pub. 156., U.S. Fish and Wildlife Service, Washington, D.C.

Buckland, S.T. (1982) A mark-recapture survival analysis. *Journal of Animal Ecology*, **51**, 833–47.

Buckland, S.T. and Baillie, S.R. (1987) Estimating bird survival rates from organized mist-netting programmes. *Acta Ornithologica*, **23(1)** 89–100.

Bulmer, M.G. (1975) The statistical analysis of density-dependence. *Biometrics*, **31**, 901–11.

Burnham, K.P. (1990) Survival analysis of recovery data from birds ringed as young: efficiency of analyses when numbers ringed are not known. *Ring*, **13**, 115–32.

Burnham, K.P. (1993) A theory for combined analysis of ring recovery and recapture data, in *Marked Individuals in the Study of Bird population* (eds J.-D. Lebreton and P.M. North). Birkhauser Verlag, Basel, pp. 192–213.

Burnham, K.P. and Anderson, D.R. (1979) The composite dynamic method as evidence for age-specific waterfowl mortality. *Journal of Wildlife Management*, **43**, 356–66.

Burnham, K.P., Anderson, D.R., White, G.C. et al. (1987) Design and analysis methods for fish survival experiments based on release-recapture. *American Fisheries Society Monographs*, **5**.

Busse, P. (1987) Interpretation of recovery patterns – contradictory points of view. *Acta Ornithologica*, **23(1)**, 115–19.

Busse, P. (in press) The bird migrants population trends 1961–1990 at the southern Baltic Coast, in *Bird Numbers 1992. Distribution, Monitoring and Ecological Aspects* (eds W. Hagemeijer and T. Vestrael). Proceedings of 12th International Conference of IBCC and EOAC, SOVON, Beek-Ubbergen.

Butcher, G.S. (1990) Audubon Christmas Bird Counts, in *Survey designs and statistical methods for the estimation of avian population trends* (eds J.R. Sauer and S. Droege), *U.S. Fish Wildl. Serv., Biol. Rep.*, **90(1)**, 5–13.

Butcher, G.S. and McCulloch, C.E. (1990) Influence of observer effort on the number of individual birds recorded on Christmas Bird Counts, in *Survey designs and statistical methods for the estimation of avian population trends* (eds J.R. Sauer and S. Droege) *U.S. Fish Wildl. Serv., Biol. Rep.*, **90(1)**, 120–29.

Carothers, A.D. (1973) The effects of unequal catchability on Jolly-Seber estimates. *Biometrics*, **29**, 79–100.

Carothers, A.D. (1979) Quantifying unequal catchability and its effect on survival rates in an actual population. *Journal of Animal Ecology*, **48**, 863–9.

Cavé, A.J. (1983) Purple Heron survival and drought in tropical West Africa. *Ardea*, **71**, 217–24.

Cawthorne, R.A. and Marchant, J.H. (1980) The effects of the 1978/79 winter on British bird populations. *Bird Study*, **27**, 163–72.

Clobert, J. and Lebreton, J.-D. (1985) Dependence de facteurs de milieu dans les estimations de taux de survie par capture-recapture. *Biometrics*, **41**, 1031–7.

Clobert, J. and Lebreton, J.-D. (1991) Estimation of demographic parameters in bird populations, in *Bird Population Studies. Relevance to Conservation and Management* (eds C.M. Perrins, J.-D. Lebreton and G.J.M. Hirons), Oxford University Press, Oxford, pp. 75–104.

Clobert, J., Lebreton, J.-D. and Allainé, D. (1987) A general approach to survival rate estimation by recaptures or resightings of marked birds. *Ardea*, **75**, 133–42.

Cohen, J. (1988) *Statistical Power Analysis for the Behavioural Sciences*, 2nd edn, Lawrence Erlbaum Associates, Hillsdale, N.J.

Collins, B.T. (1990) Using rerandomizing tests in route-regression analysis of avian population trends, in *Survey designs and statistical methods for the estimation of avian population trends.* (eds J.R. Sauer and S. Droege) *U.S. Fish Wildl. Serv., Biol. Rep.* **90(1)**, 63–70.

Collins, B.T. and Wendt, J.S. (1989) *The Breeding Bird Survey in Canada 1966–83: analysis of trends in breeding bird populations*, Technical Report Series **75**, Canadian Wildlife Service, Ottawa.

Conroy, M.J., Hines, J.E. and Williams, B.K. (1989) *Procedures for the analysis of band recovery data and user instructions for program MULT*, *U.S. Fish Wildl. Serv., Resource Pub.* **175**, U.S. Fish and Wildlife Service, Washington, D.C.

Conroy, M.J. and Williams, B.K. (1984) A general methodology for maximum likelihood inference from band-recovery data. *Biometrics*, **40**, 739–48.

Cormack, R.M. (1968), The statistics of capture-recapture methods. *Oceanography and Marine Biology Annual Reviews*, **6**, 455–506.

Cormack, R.M. (1979) Models for capture-recapture, in *Sampling Biological Populations. Statistical Ecology Series, Vol. 5.* (eds R.M Cormack, G.P. Patil and D.S. Robson), International Co-operative Publishing House, Fairland, Md, pp. 217–55.

Cormack, R.M. (1985) Examples of the use of GLIM to analyse capture-recapture studies, in *Statistics in Ornithology* (eds B.J.T. Morgan and P.M. North), Springer-Verlag, Berlin, pp. 243–73.
Cormack, R.M. (1989) Log-linear methods for capture-recapture. *Biometrics*, **45**, 395–413.
Cormack, R.M. (1993), The flexibility of GLIM analyses of multiple recapture or resighting data, in *Marked Individuals in the Study of Bird Population* (eds J.-D. Lebreton and P.M. North), Birkhauser Verlag, Basel, pp. 39–49.
Crick, H.Q.P. (1992a) A bird-habitat coding system for use in Britain and Ireland incorporating aspects of land management and human activity. *Bird Study*, **39**, 1–12.
Crick, H.Q.P. (1992b) Trends in the breeding performance of Golden Plover in Britain, *BTO Research Report*, **76**, British Trust for Ornithology, Thetford.
Crick, H.Q.P. (1993) Trends in breeding success of Merlins (*Falco columbarius*) in Britain from 1937–89, in *Biology and Conservation of Small Falcons Conference*, Hawk & Owl Trust, London (eds M.K. Nicholls and R. Clarke), pp. 30–38.
Crick, H.Q.P., Baillie, S.R. and Percival, S.M. (1990) A review of raptor population monitoring, *BTO Research Report*, **49**, British Trust for Ornithology, Thetford.
Crick, H.Q.P., Donald, P.F and Greenwood, J.J.D. (1991) Population processes in some British seed-eating birds. *BTO Research Report*, **80**, British Trust for Ornithology, Thetford.
Crick, H.Q.P., Dudley, C., Glue, D. and Turner, J. (1992) Breeding birds in 1990. *BTO News*, **179**, 8–9.
Croxall, J.P. and Rothery, P. (1991) Population regulation of seabirds: implications of their demography for conservation, in *Bird Population Studies. Relevance to Conservation and Management* (eds C.M. Perrins, J.-D. Lebreton and G.J.M. Hirons), Oxford University Press, Oxford, pp. 272–96.
Croxall, J.P., McCann, T.S., Prince, P.A. and Rothery, P. (1988) Reproductive performance of seabirds and seals at South Georgia and Signy Island, South Orkney Island, 1976–1987: implications for Southern Ocean monitoring studies, in *Antarctic Ocean and Resources Variability* (ed. D. Sahrhager), Springer-Verlag, Berlin, pp. 261–85.
Cyr, A. (1986) Un atlas saisonnaire des oiseaux du Québec. *Le Kakawi*, **8**, 43–59.
Cyr, A. (1990) A seasonal atlas of the birds of Quebec. *Picoides*, **4**, 9–10.
Darby, K.V. (1985) Migration counts and local weather at British bird observatories – an examination by linear discriminant analysis, in *Statistics in Ornithology* (eds B.J.T. Morgan and P.M. North), Springer-Verlag, Berlin. pp. 37–64.
Dawson, D.K. (1990) Migration banding data: a source of information on bird population numbers?, in *Survey designs and statistical methods for the estimation of avian population trends* (eds J.R. Sauer and S. Droege), *U.S. Fish Wildl. Serv., Biol. Rep.*, **90(1)**, 37–40.
Dhondt, A.A. (1989) The effect of old age on the reproduction of Great and Blue Tit. *Ibis*, **191**, 268–80.
Dick, D. (1991) Scottish monitoring of breeding raptors in 1989, in *Britain's Birds in 1989–90* (eds D.A. Stroud and D. Glue), British Trust for Ornithology/Nature Conservancy Council, Thetford.
Dobson, A. (1990) Survival rates and their relationship to life-history traits in some common British birds. *Current Ornithology*, **7**, 115–46.

Droege, S. (1990) The North American Breeding Bird Survey, in *Survey designs and statistical methods for the estimation of avian population trends* (eds J.R. Sauer and S. Droege), *U.S. Fish Wildl. Serv., Biol. Rep.*, **90(1)**, 1–4.
du Feu, C. and McMeeking, J. (1991) Does Constant Effort Netting estimate juvenile abundance? *Ringing & Migration*, **12**, 118–23.
Ebbinge, B.S. (1989) A multifactorial explanation for variation in breeding performance of Brent Geese *Branta bernicla*. *Ibis*, **131**, 196–204.
Ebbinge, B.S. and van de Voet, H. (1990) Have all Barnacle Geese (*Branta leucopsis*) the same probability of being sighted? Implications for estimation of survival rate. *Ring*, **13(1–2)**, 37–43.
Elkins, N. (1983) *Weather and Bird Behaviour*, T. & A.D. Poyser, Calton.
Eyre, M.D., Foster, G.N. and Foster, A.P. (1990) Factors affecting the distribution of water beetle species assemblages in drains in eastern England. *Journal of Applied Entomology*, **109**, 207–25.
Flade, M. and Schwarz, J. (1992) Stand und erste Ergebnisse des DDA-Monitorprogramms. *Die Vogelwelt*, **113**, 210–22.
Flousek, J. (1990) Do point counts and line transects provide comparable results of population tendencies?, in *Bird Census and Atlas Studies. Proceedings of the XI International Conference on Bird Census and Atlas Work* (eds K. Stastný and V. Bejek), Institute of Applied Ecology and Ecotechnology, Agricultural University, Prague, pp. 63–8.
Fradette, P. (1992) *Les oiseaux des Îsles-de-la-Madeleine: populations et sites d'observation*, Attention Frag'îles, Cap-aux-Meules, Québec.
Freeman, S.N. and Morgan, B.J.T. (1992) A modelling strategy for recovery data from birds ringed as nestlings. *Biometrics*, **48**, 217–36.
Freeman, S.N., Morgan, B.J.T. and Catchpole, E.A. (1992) On the augmentation of ring recovery data with field information. *Journal of Animal Ecology*, **61**, 649–57.
Fuller, M.R. and Mosher, J.A. (1981) Methods of detecting and counting raptors: a review, in *Estimating Numbers of Terrestrial Birds* (eds C.J. Ralph and J.M. Scott), *Studies in Avian Biology*, **6**, 235–246.
Fuller, R.J. (1982) *Bird Habitats in Britain*, T. & A.D. Poyser, Calton.
Fuller, R.J. and Marchant, J.H. (1985) Species-specific problems of cluster analysis in British mapping censuses, in *Bird Census and Atlas Studies. Proceedings of the VIII International Conference on Bird Census and Atlas Work* (eds K. Taylor, R.J. Fuller and P.C. Lack), British Trust for Ornithology, Tring, pp 83–6.
Fuller, R.J., Marchant, J.H. and Morgan, R.A. (1985) How representative of agricultural practice in Britain are Common Birds Census farmland plots? *Bird Study*, **32**, 56–70.
Gatter, W. and Steiof, K. (1992) Ermittlung von Bestandstrends durch Zugbeobachtungen. *Die Vogelwelt*, **113**, 240–55.
Gedeon, K. and Stubbe, M. (1992) Monitoring Greifvögel und Eulen – Beispiel für integriertes Populationsmonitoring. *Die Vogelwelt*, **113**, 255–62.
Geissler, P.H. and Noon, B.R. (1981) Estimates of avian population trends from the North American Breeding Bird Survey, in *Estimating Numbers of Terrestrial Birds* (eds C.J. Ralph and J.M. Scott), *Studies in Avian Biology*, **6**, 42–51.
Geissler, P.H. and Sauer, J.R. (1990) Topics in route-regression analysis, in *Survey designs and statistical methods for the estimation of avian population trends* (eds J.R. Sauer and S. Droege), *U.S. Fish Wild. Serv., Biol. Rep.*, **90(1)**, 54–7.

Geissler, P.H., Fuller, M.R. and McAllister, L.S. (1990) Trend analyses for raptor nesting productivity: an example with peregrine falcon data, in *Survey designs and statistical methods for the estimation of avian population trends* (J.R. Sauer and S. Droege), *U.S. Fish Wild. Serv., Biol. Rep.* **90**(1), 139–43.

Gerodette, T. (1987) A power analysis for detecting trends. *Ecology*, **68**, 1364–72.

Gibbons, D.W. and Gates, S. (in press) Hypothesis testing with ornithological atlas data: two case-studies, in *Bird Numbers 1992. Distribution, Monitoring and Ecological Aspects* (eds W. Hagemeijer and T. Verstrael), Proceedings of 12th International Conference of IBCC and EOAC, SOVON, Beek-Ubbergen.

Gibbons, D.W., Reid, J.B. and Chapman, R.A. (1993) *The New Atlas of Breeding Birds in Britain and Ireland: 1988–91*, Academic Press/Poyser.

Gnielka, R. (1992) Möglichkeiten und Grenzen der Revierkartierungsmethode. *Die Vogelwelt*, **113**, 231–40.

Gooch, S.,, Baillie, S.R. and Birkhead, T.R. (1991) Magpie *Pica pica* L. and songbird populations. Retrospective investigation of trends in population density and breeding success. *Journal of Applied Ecology*, **28**, 1068–86.

Green, R.E., Baillie, S.R. and Avery, M.I. (1990) Can ringing recoveries help to explain the population dynamics of British terns? *Ring*, **13**, 133–7.

Green, R.E. and Hirons, G.J.M. (1991) The relevance of population studies to conservation of threatened birds, in *Bird Population Studies. Relevance to Conservation and Management* (eds C.M. Perrins, J.-D. Lebreton and G.J.M. Hirons), Oxford University Press, Oxford, pp. 594–633.

Greenwood, J.J.D. (1989) Bird population densities. *Nature*, **338**, 627–8.

Greenwood, J.J.D. and Baillie, S.R. (1991) Effects of density-dependence and weather on population changes of English passerines using a non-experimental paradigm. *Ibis*, **133** (suppl. 1), 121–33.

Greenwood, J.J.D., Delany, S. and Kirby, J. (in press) Estimating the size of the British Mute Swan population, in *Bird Numbers 1992. Distribution, Monitoring and Ecological Aspects* (eds W. Hagemeijer and T. Verstrael), Proceedings of 12th International Conference of IBCC and EOAC, SOVON, Beek-Ubbergen.

Gregory, R.D., Marchant, J.H., Baillie, S.R. and Greenwood, J.J.D. (in press) A comparison of population changes among British breeding birds using territory mapping and point count data, in *Bird Numbers 1992. Distribution, Monitoring and Ecological Aspects* (eds W. Hagemeijer and T. Verstrael). Proceedings of 12th International Conference of IBCC and EOAC, SOVON, Beek-Ubbergen.

Hagemeijer, W. and Verstrael, T. (eds) (in press) *Bird Numbers 1992. Distribution, Monitoring and Ecological Aspects*. Proceedings of 12th International Conference of IBCC and EOAC, SOVON, Beek-Ubbergen.

Haldane, J.B.S. (1955) The calculation of mortality rates from ringing data. *Acta XI Congressus Internationalis Ornithologici*, **11**, 454–8.

Harrison, J. and Underhill, L. (in press) The Southern African Bird Atlas Project: methods and progress, in *Bird Numbers 1992. Distribution, Monitoring and Ecological Aspects* (eds W. Hagemeijer and T. Verstrael), Proceedings of 12th International Conference of IBCC and EOAC, SOVON, Beek-Ubbergen.

Heintzelman, D.S. (1986) *The Migrations of Hawks*, Indiana University Press, Bloomington.

Hensler, G.L. and Nichols, J.D. (1981) The Mayfield method of estimating nesting success: a model, estimators, and simulation results. *Wilson Bulletin*, **93**, 42–53.

Hestbeck, J.B. , Nichols, J.D. and Malecki, R.A. (1991) Estimates of movement and site fidelity using mark-resight data of wintering Canada Geese. *Ecology*, **72**, 523–33.

Hickey, J.J. (1952), *Survival studies of banded birds, U.S. Fish Wildl. Serv. Spec. Sci. Rep. Wildl.* **15**, U.S. Fish and Wildlife Service, Washington, D.C.

Hildén, O. and Laine, L.J. (1985) Accuracy of single line transects in Finnish woodland habitat, in *Bird Census and Atlas Studies*, Proceedings of the VIII International Conference on Bird Census and Atlas Work (eds K. Taylor, R.J. Fuller and P.C. Lack), British Trust for Ornithology, Tring, pp. 111–16.

Hjort, C. and Lindholm, C.-G. (1978) Annual bird ringing totals and population fluctuations, *Oikos*, **30**, 387–92.

Hunt, G.L., Eppley, Z.A. and Schneider, D.C. (1986) Reproductive performance of seabirds: the importance of population and colony size. *The Auk*, **103**, 306–17.

Hussell, D.J.T. (1981) The use of migration counts for monitoring bird population levels, in *Estimating Numbers of Terrestrial Birds* (eds C.J. Ralph and J.M. Scott), *Studies in Avian Biology*, **6**, 92–102.

Hustings, F. (1992) European monitoring studies on breeding birds: an update. *Bird Census News*, **5(2)**, 1–56.

Johnson, D.H. (1979) Estimating nest success: the Mayfield method and an alternative. *The Auk*, **96**, 651–61.

Johnson, D.H. and Shaffer, T.L. (1990) Estimating nest success: when Mayfield wins. *The Auk*, **107**, 595–600.

Jolly, G.M. (1965) Explicit estimates from capture-recapture data with both death and immigration – stochastic model. *Biometrika*, **52**, 225–47.

Jones, E.T. (1986) The passerine decline. *N. Am. Bird Bander*, **11**, 74–5.

Kirby, J.S., Ferns, J.R., Waters, R.J. and Prŷs-Jones, R.P. (1991) *Wildfowl and Wader Counts 1990–91*, Wildfowl & Wetlands Trust, Slimbridge.

Klaus, V. and Exo, M. (1992) Methoden zur Aufnahme von Raum-Zeit-Budgets bei Vögeln, dargestellt am Beispiel des Austernfischers (*Haematopus ostralegus*). *Die Vogelwarte*, **36**, 311–25.

Klett, A.T. and Johnson, D.H. (1982) Variability in nest survival rates and implications to nesting studies. *The Auk*, **99**, 77–87.

Klett, A.T., Duebbert, H.F., Faanes, C.A. and Higgins, K.F. (1986) *Techniques for studying nest success of ducks in upland habitats in the prairie pothole region. US Fish Wildl. Serv., Resource Publ.* **158**, U.S. Fish and Wildlife Service, Washington, D.C.

Kluijver, H.N. and Tinbergen, L. (1953) Territoriality and the regulation of density in titmice. *Archives Néerlandaises de Zoologie*, **10**, 265–89.

Koskimies, P. and Väisänen, R.A. (1991) *Monitoring Bird Populations. A Manual of Methods Applied in Finland*, Zoological Museum, Finnish Museum of Natural History, Helsinki.

Kraemer, H.C. and Thiemann, S. (1987) *How Many Subjects? Statistical Power Analysis in Research*, Sage Publications, London.

Krebs, J.R. (1970) Regulation of numbers in the Great Tit (Aves: Passeriformes). *Journal of Zoology*, **162**, 317–33.

Lack, D. (1943) The age of the Blackbird. *British Birds*, **36**, 166–75.

Lack, P. (1986) *The Atlas of Wintering Birds in Britain and Ireland*, T. & A.D. Poyser, Calton.

Lakhani, K. (1987) Efficient estimation of age-specific survival rates from ring recovery data, of birds ringed as young, augmented by further field information. *Journal of Animal Ecology*, **56**, 969–87.

Lakhani, K. and Newton, I. (1983) Estimating age-specific bird survival rates from ring recovery – can it be done? *Journal of Animal Ecology*, **52**, 83–91.

Langslow, D.R. (1978) Recent increases of Blackcaps at bird observatories. *British Birds*, **71**, 345–54.

Larsson, T. (ed.) (1992) *Bird Monitoring Programmes in the Nordic Countries 1991. Report from a Working Group under the Nordic Council of Ministers*, Swedish Environmental Protection Agency, Solna.

Lebreton, J.-D. (1989) Statistical methodology for the study of animal populations. *Bulletin International Statistical Institute*, **53**, 267–82.

Lebreton, J.-D., Reboulet, A.-M. and Banco, G. (1993), An overview of software for terrestrial vertebrate population dynamics, in *Marked Individuals in the Study of Bird Population* (eds J.-D. Lebreton and P.M. North), Birkhauser Verlag, Basel, pp. 357–72.

Lebreton, J.-D., Burnham, K.P., Clobert, J. and Anderson, D.R. (1992) Modelling survival and testing biological hypotheses using marked animals: a unified approach with case studies. *Ecological Monographs*, **62**, 67–118.

Lebreton, J.-D. and Clobert, J. (1987) *User's manual for program Surge, version 2.0*, CEPE/CNRS, Montpellier.

Lebreton, J.-D. and Clobert, J. (1991) Bird population dynamics, management, and conservation: the role of mathematical modelling, in *Bird Population Studies. Relevance to Conservation and Management* (eds C.M. Perrins, J.-D. Lebreton and G.J.M. Hirons), Oxford University Press, Oxford, pp. 105–25.

Lebreton, J.-D. and North, P.M. (eds) (1993), *Marked Individuals in the Study of Bird Population*. Birkhauser Verlag, Basel.

Lensink, R. and Kwak, R. (in press) The number of birds in broad front migration in autumn over The Netherlands compared to results of monitoring schemes of breeding birds in northern Europe and wintering birds in western Europe, in *Bird Numbers 1992. Distribution, Monitoring and Ecological Aspects* (eds W. Hagemeijer and T. Verstrael). Proceedings of 12th International Conference of IBCC and EOAC, SOVON, Beek-Ubbergen.

Lindley, D. (1971) *Making Decisions*, John Wiley & Sons Ltd., London.

Lloyd, C., Tasker, M.L. and Partridge, K. (1991) *The Status of Seabirds in Britain and Ireland*, T. & A.D. Poyser, London.

Lundberg, A. and Alatalo, R.V. (1992) *The Pied Flycatcher*, T. & A.D. Poyser, London.

MacArthur, R.H. and MacArthur, J.W. (1961) On bird species diversity. *Ecology*, **42**, 594–8.

Manly, B.F.J. (1991) *Randomization and Monte Carlo Methods in Biology*, Chapman & Hall, London.

Marchant, J.H. (1992) Recent trends in breeding populations of some common trans-Saharan migrant birds in northern Europe. *Ibis*, **134** (suppl. 1), 113–19.

Marchant, J.H. and Gregory, R.D. (in press) Recent population changes among seed-eating passerines in the United Kingdom, in *Bird Numbers 1992. Distribution, Monitoring and Ecological Aspects* (eds W. Hagemeijer and T. Verstrael) Proceedings of 12th International Conference of IBCC and EOAC, SOVON, Beek-Ubbergen.

Marchant, J.H., Hudson, R., Carter, S.P. and Whittington, P. (1990) *Population Trends in British Breeding Birds*, British Trust for Ornithology, Tring.

May, P.G. (1982) Secondary succession and breeding bird community structure: patterns of resource utilization. *Oecologia*, **55**, 208–16.
Mayfield, H. (1961) Nesting success calculated from exposure. *Wilson Bulletin*, **73**, 255–61.
Mayfield, H. (1975) Suggestions for calculating nest success. *Wilson Bulletin*, **87**, 456–66.
Morgan, B.J.T. and Freeman, S.N. (1989) A model with first-year variation for ring-recovery data. *Biometrics*, **45**, 1087–101.
Moser, M.E. (1987) A revision of population estimates for waders (Charadrii) wintering on the coastline of Britain. *Biological Conservation*, **39**, 153–64.
Moses, L.E. and Robinowitz, D. (1990) Estimating (relative) species abundance from route counts of the breeding bird survey, in *Survey designs and statistical methods for the estimation of avian population trends* (eds J.R. Sauer and S. Droege), *U.S. Fish Wildl. Serv., Biol. Rep.* **90**(1), 71–9.
Moss, D. (1985) Some statistical checks on the BTO Common Birds Census index – 20 years on, in *Bird Census and Atlas Studies*, Proceedings of the VIII International Conference on Bird Census and Atlas Work (eds K. Taylor, R.J. Fuller and P.C. Lack) British Trust for Ornithology, Tring, pp. 175–9.
Mountford, M.D. (1982) Estimation of population fluctuations with application to the Common Bird Census. *Applied Statistics*, **31**, 135–43.
Mountford, M.D. (1985) An index of population change with application to the Common Bird Census, in *Statistics in Ornithology* (eds B.J.T. Morgan and P.M. North), Springer-Verlag, Berlin, pp. 121–32.
Nature Conservancy Council (1990) *Handbook for Phase 1 Habitat Survey, Field Manual*, Nature Conservancy Council, Peterborough.
Nelson, L.J., Anderson, D.R. and Burnham, K.P. (1980) The effect of band loss on estimates of annual survival. *Journal of Field Ornithology*, **51**, 30–8.
Newton, I. (1979) *Population Ecology of Raptors*, T. & A.D. Poyser, Berkhamsted.
Newton, I. (ed.) (1989) *Lifetime Reproduction in Birds*, Blackwell Scientific Publications, Oxford.
Newton, I. (1991a), Concluding remarks, in *Bird Population Studies. Relevance to Conservation and Management* (eds C.M. Perrins, J.-D. Lebreton and G.J.M. Hirons), Oxford University Press, Oxford. pp. 637–54.
Newton, I. (1991b), Population limitations in birds of prey: a comparative approach, in *Bird Population Studies. Relevance to Conservation and Management* (eds C.M. Perrins, J.-D. Lebreton and G.J.M. Hirons), Oxford University Press, Oxford, pp. 3–21.
Nichols, J.D. Brownie, C., Hines, J.E. *et al.* (1993). The estimation of exchanges among populations or subpopulations, in *Marked Individuals in the Study of Bird Population* (eds. J.-D. Lebreton and P.M. North), Birkhauser Verlag, Basel, pp. 265–79.
Nichols, J.D., Noon, B.R., Stokes, S.L. and Hines, J.E. (1981) Remarks on the use of mark-recapture methodology in estimating avian population size, in *Estimating Numbers of Terrestrial Birds* (eds C.J. Ralph and J.M. Scott), *Studies in Avian Biology*, **6**, pp. 121–36.
Nichols, J.D., Stokes, S.L., Hines, J.E. and Conroy, M.J. (1982) Additional comments on the assumptions of homogeneous survival rates in modern bird banding estimation models. *Journal of Wildlife Management*, **46**, 953–62.

Nilsson, L. (1979) Variation in the production of young swans wintering in Sweden. *Wildfowl*, **30**, 129–34.

Noer, H. (1990) Estimation of survival rates of Danish birds of prey before and after protection. *Ring*, **13(1–2)**, 75–86.

North, P.M. (ed.) (1987) Ringing recovery analytical methods. Proceedings of the EURING technical conference and meeting of the Mathematical Ecology Group of the Biometric Society (British Region) and British Ecological Society. *Acta Ornithologica*, **23(1)**, 1–175.

North, P.M. (ed.) (1990a) The statistical investigation of avian population dynamics using data from ringing recoveries and live recaptures of marked birds. Proceedings of the EURING technical conference and meeting of the Mathematical Ecology Group of the Biometric Society (British Region) and British Ecological Society. *Ring*, **13(1–2)**, 1–314.

North, P.M. (1990b) Analysis of avian ring recovery and live recapture data: where have we come from Wageningen to Sempach? *Ring*, **13(1–2)**, 11–22.

North, P.M. and Morgan, B.J.T. (1979) Modelling Heron survival using weather data. *Biometrics*, **35**, 667–81.

Oatley, T.B. and Underhill, L.G. (1993) Merging recoveries and recaptures to estimate survival probabilities, in *Marked Individuals in the Study of Bird Population* (eds J.-D. Lebreton and P.M. North), Birkhauser Verlag, Basel, pp. 77–90.

O'Connor, R.J. (1980) The effects of census data on the results of intensive Common Birds Census surveys. *Bird Study*, **27**, 126–36.

O'Connor, R.J. (1981) The influence of observer and analyst efficiency in mapping method censuses, in *Estimating Numbers of Terrestrial Birds* (eds C.J. Ralph and J.M. Scott), *Studies in Avian Biology*, **6**, 372–6.

O'Connor, R.J. and Fuller, R.J. (1984) A re-evaluation of the aims and methods of the Common Birds Census. *BTO Research Report*, **15**, British Trust for Ornithology, Tring.

O'Connor, R.J. and Fuller, R.J. (1985) Bird population responses to habitat, in *Bird Census and Atlas Studies*. Proceedings of the VIII International Conference on Bird Census and Atlas Work (eds K. Taylor, R.J. Fuller and P.C. Lack), British Trust for Ornithology, Tring, pp. 197–211.

O'Connor, R.J. and Marchant, J.H. (1981) A field validation of some Common Birds Census techniques, *BTO Research Report*, **4**, British Trust for Ornithology, Tring.

Ogilvie, M. (1986) The Mute Swan *Cygnus olor* in Britain 1983. *Bird Study*, **33**, 121–37.

Otis, D.L., Burnham, K.P., White, G.C. and Anderson, D.R. (1978) Statistical inference from capture data on closed animal populations. *Wildlife Monographs*, **62**, 1–135.

Owen, M. and Black, J.M. (1989) The Barnacle Goose, in *Lifetime Reproduction in Birds* (ed. I. Newton), Blackwell Scientific Publications, Oxford, pp. 349–62.

Owen, M. and Black, J.M. (1990) *Waterfowl Ecology*, Blackie, Glasgow.

Owen, M., Atkinson-Willes, G.L. and Salmon, D.G. (1986) *Wildfowl in Great Britain*, 2nd Edn, Cambridge University Press, Cambridge.

Partridge, L. (1989) Lifetime reproductive success and life-history evolution, in *Lifetime Reproduction in Birds* (ed. I. Newton), Blackwell Scientific Publications, Oxford, pp. 421–40.

Peach, W.J. (1993), Combining mark-recapture data sets for small passerines, in *Marked Individuals in the Study of Bird Population* (eds J.-D. Lebreton and P.M. North), Birkhauser Verlag, Basel, pp. 107–22.

Peach, W. and Baillie, S. (1990) Population changes on Constant Effort sites 1988–1989. *BTO News*, **167**, 6–7.
Peach, W. and Baillie, S. (1991) Population changes on Constant Effort sites 1989–1990. *BTO News*, **173**, 12–14.
Peach, W. and Baillie, S. (1992) Population changes on Constant Effort sites 1990–1991. *BTO News*, **179**, 12–13.
Peach, W.J. and Baillie, S.R. (in press) Implementation of the Mountford indexing method for the Common Birds Census, in *Bird Numbers 1992. Distribution, Monitoring and Ecological Aspects* (eds W. Hagemeijer and T. Verstrael), Proceedings of 12th International Conference of IBCC and EOAC, SOVON, Beek-Ubbergen.
Peach, W.J., Baillie, S.R. and Underhill, L.G. (1991) Survival of British Sedge Warblers *Acrocephalus schoenobaenus* in relation to west African rainfall. *Ibis*, **133**, 300–305.
Peach, W.J., Buckland, S.T. and Baillie, S.R. (1990) Estimating survival rates using mark-recapture data from multiple ringing sites. *Ring*, **13(1–2)**, 87–102.
Peach, W.J., Thompson, P.S. and Coulson, J.C. (in press) Survival of British Lapwings: an analysis of British ringing recoveries. *Journal of Animal Ecology*.
Peterman, R.M. and M'Gonigle, M. (1992) Statistical power analysis and the precautionary principle. *Marine Pollution Bulletin*, **24**, 231–4.
Pienkowski, M.W. (1991) Long-term ornithological studies and conservation. *Ibis*, **133**(suppl. 1), 62–75.
Pollock, K.H. (1981) Capture-recapture models: a review of current methods, assumptions, and experimental design, in *Estimating Numbers of Terrestrial Birds* (eds C.J. Ralph and J.M. Scott), *Studies in Avian Biology*, **6**, 426–35.
Pollock, K.H. (1991) Modelling capture, recapture, and removal statistics for estimation of demographic parameters for fish and wildlife populations: past, present, and future. *Journal of the American Statistical Association*, **86**, 225–38.
Pollock, K.H. and Cornelius, W.L. (1988) A distribution-free nest survival model. *Biometrics*, **44**, 397–404.
Pollock, K.H. and Raveling, D.G. (1982) Assumptions of modern band-recovery models, with emphasis on heterogeneous survival rates. *Journal of Wildlife Management*, **46**, 88–98.
Pollock, K.H., Bunck, C.M. and Curtis, P. D. (1989a) Survival analysis in telemetry studies: the staggered entry design. *Journal of Wildlife Management*, **53**, 7–15.
Pollock, K.H., Winterstein, S.R. and Conroy, M.J. (1989b) Estimation and analysis of survival distributions for radio-tagged animals. *Biometrics*, **45**, 99–109.
Pollock, K.H., Nichols, J.D., Brownie, C. and Hines, J.E. (1990) Statistical inference for capture-recapture experiments. *Wildlife Monographs*, **107**, 1–97.
Poole, R.W. (1978) The statistical prediction of population fluctuations. *Annual Review of Ecology and Systematics*, **9**, 427–48.
Pradel, R., Clobert, J. and Lebreton, J.-D. (1990) Recent developments for the analysis of capture-recapture multiple data sets. An example concerning two Blue Tit populations. *Ring*, **13(1–2)**, 193–204.
Prater, A.J. (1981) *Estuary Birds of Britain and Ireland*, T. & A.D. Poyser, Calton.

Pratt, A. and Peach, W. (1991) Site tenacity and annual survival of a Willow Warbler *Phylloscopus trochilus* population in Southern England. *Ringing & Migration*, **12**, 128–34.

Ralph, C.J. and Scott, J.M. (eds) (1981) Estimating Numbers of Terrestrial Birds. *Studies in Avian Biology*, **6**, Cooper Ornithological Society, Lawrence, Kansas.

Rands, M.R.W. (1991) Conserving threatened birds: an overview of the species and the threats, in *Bird Population Studies. Relevance to Conservation and Monitoring*, (eds C.M. Perrins, J.-D. Lebreton and G.J.M. Hirons), Oxford University Press, Oxford, pp. 581–93.

Ratcliffe, D. (1980) *The Peregrine Falcon*, T. & A.D. Poyser, Calton.

Reijnen, R. (in press) The impact of car traffic on the distribution and numbers of breeding birds, in *Bird Numbers 1992. Distribution, Monitoring and Ecological Aspects* (eds W. Hagemeijer and T. Verstrael), Proceedings of 12th International Conference of IBCC and EOAC, SOVON, Beek-Ubbergen.

Riddiford, N. (1983) Recent declines of Grasshoper Warblers *Locustella naevia* at British bird observatories. *Bird Study*, **30**, 143–8.

Rinne, J., Lokki, H. and Saurola, P. (1990) Survival estimates of nestling recoveries: forbidden fruits of ringing. *Ring*, **13**, 255–68.

Robbins, C.S. (1985) Summary of bird censusing and atlasing in North America, in *Bird Census and Atlas Studies*. Proceedings of the VIII International Conference on Bird Census and Atlas Work (eds K. Taylor, R.J. Fuller and P.C. Lack), British Trust for Ornithology, Tring, pp. 15–24.

Robbins, C.S., Bystrak, D. and Geissler, P.H. (1986) *The Breeding Bird Survey: Its First Fifteen Years, 1965–1979. U.S. Fish Wildl. Serv., Resource Publ.* **157**, U.S. Fish and Wildlife Survey, Washington, D.C.

Robbins, C.S., Droege, S. and Sauer, J.R. (1989) Monitoring bird populations with Breeding Bird Survey and atlas data. *Ann. Zool. Fennici*, **26**, 297–304.

Rösner, H.-U. (in press) Spring-tide counts in the Wadden Sea: is trend-monitoring of migratory wetland birds possible?, in *Bird Numbers 1992. Distribution, Monitoring and Ecological Aspects* (eds W. Hagemeijer and T. Verstrael), Proceedings of 12th International Conference of IBCC and EOAC, SOVON, Beek-Ubbergen.

Sauer, J.R. and Droege, S. (eds) (1990) *Survey designs and statistical methods for the estimation of avian population trends*. U.S. Fish Wild. Serv., Biol. Rep., **90(1)**, 1–166.

Sauer, J.R. and Geissler, P.H. (1990) Estimation of annual indices from roadside surveys, in *Survey designs and statistical methods for the estimation of avian population trends* (eds J.R. Sauer and S. Droege), *U.S. Fish Wild. Serv., Biol. Rep.*, **90(1)**, 58–62.

Schwarz C.J. (1993) Estimating migration rates using tag-recovery data, in *Marked Individuals in the Study of Bird Population* (eds J.-D. Lebreton and P.M. North), Birkhauser Verlag, Basel, pp. 255–66.

Scott, B., Cawkell, H. and Riddiford, N. (1976) Dungeness, in *Bird Observatories in Britain and Ireland* (ed. R. Durman), T. & A.D. Poyser, Berkhamsted, pp. 94–114.

Seber, G.A.F. (1982) *The Estimation of Animal Abundance and Related Parameters*, 2nd edn, Griffin, London.

Seber, G.A.F. (1986) A review of estimating animal abundance. *Biometrics*, **42**, 267–92.

Sharrock, J.T.R. (1976) *The Atlas of Breeding Birds in Britain and Ireland*, T.& A.D. Poyser, Calton.

Shrubb, M. (1990) Effects of agricultural change on nesting Lapwings *Vanellus vanellus* in England and Wales. *Bird Study*, **37**, 115–27.

Shrubb, M. and Lack, P.C. (1991) The numbers and distribution of Lapwings *V. vanellus* nesting in England and Wales in 1987. *Bird Study*, **38**, 20–37.

Skalski, J.R. and Robson, D.S. (1992) *Techniques for Wildlife Investigations*, Academic Press, London.

Smith, D.R. and Anderson, D.R. (1987) Effects of lengthy ringing periods on estimators of annual survival. *Acta Ornithologica*, **23**, 69–76.

Smith, J.N.M., Arcese, P. and Hochachka, W.M. (1991) Social behaviour and population regulation in insular bird populations: implications for conservation, in *Bird Population Studies. Relevance to Conservation and Management* (eds C.M. Perrins, J.-D. Lebreton and G.J.M. Hirons), Oxford University Press, Oxford, pp. 148–67.

Snow, D.W. (1965) The relationship between census results and the breeding population of birds on farmland. *Bird Study*, **12**, 287–304.

Southern, H.N. (1970) The natural control of a population of Tawny Owls (*Strix aluco*). *J. Zool., Lond.*, **162**, 197–285.

SOVON (1987) *Atlas van de Nederlandse Vogels*, SOVON, Arnhem.

Stewart, P.A. (1987) Decline in numbers of wood warblers in spring and autumn migrations through Ohio. *N. Am. Bird Bander*, **12**, 58–60.

Summers, R.W. and Underhill, L.G. (1987) Factors related to the breeding production of Brent Geese *Branta b. bernicla* and waders (*Charadrii*) on the Taimyr Peninsula. *Bird Study*, **34**, 161–71.

Summers-Smith, J.D. (1989) A history of the status of the tree sparrow *Passer montanus* in the British Isles. *Bird Study*, **36**, 23–31.

Svensson, S.E. (1978) Efficiency of two methods for monitoring bird population levels: breeding bird censuses contra counts of migrating birds, *Oikos*, **30**, 373–86.

Svensson, S. (1980) Comparison of recent bird census methods, in *Bird Census Work and Nature Conservation*. Proceedings VI International Conference on Bird Census Work, IV Meeting European Ornithological Atlas Committee (ed. H. Oelke), DDA, Lengede, pp. 13–22.

Svensson, S. (1981) Do transect counts monitor abundance trends in the same way as territory mapping in study plots?, in *Estimating Numbers of Terrestrial Birds* (eds C.J. Ralph and J.M. Scott), *Studies in Avian Biology*, **6**, 209–14.

Svensson, S. (1992) Experiences with the Swedish bird monitoring programme. *Die Vogelwelt*, **113**, 182–96.

Taylor, S.M. (1965) The Common Birds Census – some statistical aspects. *Bird Study*, **12**, 268–86.

Taylor, K., Fuller, R.J. and Lack, P.C. (eds) (1985) *Bird Census and Atlas Studies*, Proceedings of the VIII International Conference on Bird Census and Atlas Work, British Trust for Ornithology, Tring.

Temple, S.A. (1986) The problem of avian extinctions. *Current Ornithology*, **3**, 453–85.

Temple, S.A. and Cary, J.R. (1987a) *Wisconsin Birds: A Seasonal and Geographical Guide*, University of Wisconsin Press, Madison.

Temple, S.A. and Cary, J.R. (1987b) Climatic effects on year-to-year variations in migration phenology: a WSO research report. *Passenger Pigeon*, **49**, 70–5.

Temple, S.A. and Cary, J.R. (1990) Using checklist records to reveal trends in bird populations, in *Survey designs and statistical methods for estimation of avian population trends* (eds J.R. Sauer and S. Droege), *U.S. Fish Wildl. Serv., Biol. Rep.*, **90(1)**, 98–104.

Temple, S.A. and Wiens, J.A. (1989) Bird populations and environmental changes: can birds be bio-indicators? *American Birds*, **43**, 260–70.

Tiainen, J. (1985) Monitoring bird populations in Finland. *Ornis Fennica*, **62**, 80–9.

Titus, K., Fuller, M.R. and Jacobs, D. (1990) Detecting trends in hawk migration count data, in *Survey designs and statistical methods for the estimation of avian population trends* (eds J.R. Sauer and S. Droege), *U.S. Fish Wildl. Serv., Biol. Rep.*, **90(1)**, 105–13.

Underhill, L.G. (1990) Bayesian estimation of the size of closed populations. *Ring*, **13(1–2)**, 235–44.

Underhill, L.G., Prŷs-Jones, R.P., Harrison, J.A. and Martinez, P. (1992) Seasonal patterns of occurrence of Palaearctic migrants in southern Africa using atlas data. *Ibis*, **134** (suppl. 1), 99–108.

van Dijk, A.J. (1992) The breeding bird monitoring programme of SOVON in The Netherlands. *Die Vogelwelt*, **113**, 197–209.

van Gasteren, H. (in press) Regional differences in visible bird migration during autumn 1981–90 in The Netherlands, in *Bird Numbers 1992. Distribution, Monitoring and Ecological Aspects* (eds W. Hagemeijer and T. Verstrael). Proceedings of 12th International Conference of IBCC and EOAC, SOVON, Beek-Ubbergen.

van Strien, A.J., Meijer, R., ter Braak, C.J.F. and Verstrael, T.J. (in press a) Analysis of monitoring data with many missing values: which method?, in *Bird Numbers 1992. Distribution, Monitoring and Ecological Aspects* (eds W. Hagemeijer and T. Verstrael). Proceedings of 12th International Conference of IBCC and EOAC, SOVON, Beek-Ubbergen.

van Strien, A., Hagemeijer, W. and Verstrael, T. (in press b) Estimating the probability of detecting trends in breeding birds: often overlooked but necessary, in *Bird Numbers 1992. Distribution, Monitoring and Ecological Aspects* (eds W. Hagemeijer and T. Verstrael). Proceedings of 12th International Conference of IBCC and EOAC, SOVON, Beek-Ubbergen.

Vansteenwegen, C., Hémery, G. and Pasquet, E. (1990) Une réflexion sur le programme français que suivi temporel du niveau d'abondances des populations d'oiseaux terrestre communs (S.T.O.C.). *Alauda*, **58**, 36–44.

Verner, J. (1985) Assessment of counting techniques. *Current Ornithology*, **2**, 247–302.

White, G.C. (1983) Numerical estimation of survival rates from band-recovery and biotelemetry data. *Journal of Wildlife Management*, **47**, 716–28.

Wiens, J.A. (1981) Scale problems in avian censusing, in *Estimating Numbers of Terrestrial Birds* (eds C.J. Ralph and J.M. Scott), *Studies in Avian Biology* **6**, 513–21.

Wiens, J.A. (1989) *The Ecology of Bird Communities*, Cambridge University Press, Cambridge.

Williams, G., Stowe, T. and Newton, A. (1991) Action for Corncrakes. *RSPB Conservation Review*, **5**, 47–53.

Williams, I. (in press) Raptor monitors in Wales 1991, in *Britain's Birds in 1991–92* (ed. S. Carter), British Trust for Ornithology/Joint Nature Conservation Committee, Thetford, pp. 116–19.

Willson, M.F. (1974) Avian community organization and habitat structure. *Ecology*, **55**, 1017–29.

Zolner, J. (1990) The problem of accuracy of some methods of counting songbirds (Passeriformes), in *Bird Census and Atlas Studies*. Proceedings of the XI International Conference on Bird Census and Atlas Work (eds K. Stastný and V. Bejcek), Institute of Applied Ecology and Ecotechnology, Agricultural University, Prague, pp. 97–8.

Index

All references in *italics* are to tables and those in **bold** type represent figures.

Abundance
  absolute abundance 281–2
  techniques for measuring 276–94
  changes in 287–94
Abundance indices
  bird populations 281–2
  precision of 286–7
  fish stock 223, 242
Acanthis flavirostris 66
Accipiter spp. 74
A. gentilis 115, **116**
A. nisus 59, 60, 193
Acidification 75–6, 127–8, 199–209
  effect on dippers *Cinclus cinclus* 200–7
  effect on duck populations 127–8
  effect on eggshell formation 127
Acridotheres tristis 69
Acrocephalus schoenobaenus 34, 48, 269, **290**, 305, **306**, 307, 317, **318**, 320, 325, 326
A. scirpaceus 304, 305, 313
Actinide elements 149–52
  as alpha emitters 149
  routes of uptake 149
  see also Plutonium
Aerial photography 9
Aerosols, effect on the ozone layer 53–4
Aethia pusilla 247

Afforestation 61–2, 65–6, 322–3
  and acidification 75
Agrostis stolonifera 66
Air pollution 129–31
  effect on insect abundance 129
Air temperature, *see* Mean annual temperature
Aix sponsa 165
Alauda arvensis 56, 65, 289
Albatross, wandering 32, 108
Albizia falcateria 62
Alca impennis 3
A. torda 252
Alcedo atthis 99, 189, 190
Alcids, effect of overfishing on 233
Aldrin, *see* Pesticides
Alpha radiation 148
Aluminium 127, 200
Aminolevulinic acid dehydratase 119
Ammodytes spp. 220, 233, 245
A. hexapterus 63
A. marinus 111, 114
Anas penelope 66
A. platyrhynchos 95, 113, 145, 160–1, 167
A. rubripes 95
Anchovies 220, 233
Anser spp. 98
A. anser 3
Anthus pratensis 65

A. trivalis 59
Apteryx australis 61
Apus apus 31
Aquila audax 70
A. chrysaetos 66, 108
Ardea cinerea 99, 103, 189, 190, 311
A. purpurea 34, 311
A. ralloides 34
Atlas surveys 276
Auk, great 3
Auklet, least 247
Averrhoa bilimbi 70
Aythya affinis **163**
A. collaris **163**

Basal metabolic rate (BMR) 221–2
Beached-bird surveys, to monitor oil pollution 128–9
Beta radiation 148
Betula sp. 45
Bioaccumulation 87
Bioamplification 87
Bioenergetics models *221*, 221–3
Biotic indices 15–16, 198
　Biological Monitoring Working Party (BMWP) Score System 15–16
　　reliability of 16
　Chandler's Score System 15, 207
　Trent Biotic Index 15
Birch 45
Bird banding, *see* Bird-ringing
Bird behaviour
　in response to prey density 229
　as weather predictor 3, 31–3
　see also Foraging methods
Bird community, *see* Community composition
Bird counts
　to assess organic nutrients 125
　to assess prey fluctuations 234
Bird observatories, role in monitoring 279–80
Bird ringing
　to monitor caesium levels 168
　to monitor populations 284–5, 301–13
　ring loss 309
　ringing dates 309
　see also Constant Effort Sites Scheme
Birds of Estuaries Enquiry 291
Bittern 66
Blackcap 280, 327
Blackbird 65, 300
Bluebird, eastern 69
Body burden
　of caesium 156–60
　in coots 162–6
　differences between juveniles and adults 156–9
　seasonal changes **163**
Body condition
　effect of acidification on 201, 202
　variation with prey abundance 234–5
Boiga irregularis 69
Bombycilla garrulus 30
Bonasa umbellus 115
Booby, blue-footed 231, 238
Boreogadus saida 220
Botaurus stellaris 66
Brambling 47
Branta bernicla 300
B. leucopsis 297
Breeding Bird Survey 280
Breeding population
　stability of 269–70
　suitability for monitoring 278–9
Breeding success **271**
　effect of farming on 322–3
　effect of foraging on 62–3, 228, 248
　effect of overfishing on 231–4, *232*
　effect of prey availability on 238–9
　monitoring of 294–301
　in relation to acidification 200–4
British Trust for Ornithology (BTO) *see* Long-term studies; Monitoring programmes in Britain
Brood parasitism 325
Brood size
　effect of acidification on 201, 202
　effect of DDT on 91

Brownie models 308–13
Bucephala albeola **163**
B. clangula 209
Bufflehead **163**
Bunting
  corn 57, 324
  reed 324
Burhinus oedicnemus 65, 326
Buteo buteo 74
Buzzard 74

Cadmium 103, 115–18
  atmospheric deposition of 115, 118
  levels in eggs 117
  levels in feathers 115, **116**, 118
Caesium 146, *147*, 148–9
  in birds 126, 156–60
    effect of diet on 159
    within colony variation 157
    from Chernobyl 166–8
    concentration *150*, **156**, *163*
    concentration in muscle 149, *150*
    half-life *147*, 165
    ratio of caesium 134:137 154, 156
    for establishing source 167
    from Sellafield 126
  uptake of 151–2, 168
    by American coots *Fulica americana* **161**, 161–6
Calcium
  effect of pH on 203
  in streams **184**, 185
Calidris alpina 8, 124
C. canutus 281
Calluna vulgaris 322
Calonectris diomedea 114
Capelin 63, 220, 233, 247
Capercaillie *30*, 58
Capture-recapture methods 302–7, **306**, 314
Carbon dioxide level 48–53, **49**
Carduelis cannabina 324
C. chloris 324
C. flammea 59
C. spinus 58
Cat 69, 70
Catch per unit effort (CPUE) 218–20, 223, 244

Catching
  of birds for monitoring 168, 284–5
  sex bias in rate of 304
Catchment area 183
Catharacta skua 13, 103, **104**, 109, 240
Censusing methods 23–6
Cepphus grylle 108, 109, 252
Certhia familiaris 59
CFCs, effect on the ozone layer 53–4
Chaffinch 47, 65, 324
Chaining of data 288–9
Chen caerulescens caerulescens 113
Chernobyl 144, 166–8
  atmospheric transport of radiation from 174
Chick feeding 241, 244
Chick growth
  effect of acidification on 201, 202
  effect of prey availability on 238–9
Chicken, as sentinel species *150*, *151*, 151–2
Chlidonias niger **163**
Chough 57
Ciconia ciconia 34
Cinclus cinclus 7, 76, 127, 190–5, 200–9
Circus cyaneus 295
Climatic change
  birds' ability to respond to 43–6
  in Finland 46–7
  monitoring of 22, 33–4
  natural variation in 46–8
  time-scale of 45
  warming 46–8
  see also Global warming
Climatic models 50–2
Clupea harengus 26, 63, 111, 220, 246
Clutch size
  effect of acidification 201, **202**, 202
  variation with prey availability 235
Clyde Estuary, sewage pollution in 124–5

Cod 220
  arctic 220
Colinus virginianus 169, 172
Cololabis saira 247
Colony attendance 32, 228
Columba palumbus 56, 280
Common Birds Census (BTO) 17, 23, 34, 273, 282, 288
Community composition
  effects of afforestation on 64–6
  effects of deforestation on 58–60
  effects of drainage on 66
  heathland 65
  moorland 65–6
  scrubland 65
  woodland 58–62, 65
Constant Effort Sites Scheme (BTO) 24, 284, 300, 304, 327
Coot, American 161–6, 168–9
Coppicing 59
Coracopsis nigra barklyi 70
Cormorant 126, 187
  double-crested 230
Corncrake 322
Corophium volutator 124–5
Corvus brachyrhyncos 159
C. corax 66
C. frugilegus 31, 56
C. ossifragus 159
Coturnix coturnix 112
Cowbird, brown-headed 325
Cranes 3
Creeper, brown 61
Crex crex 322
Crossbill 58
Crow 159
Curlew 124
  stone 65, 326
Cyanocitta cristata 172
Cygnus spp. 66
C. columbianus bewickii 173
C. cygnus 299
C. olor 74, 118, 195, 281, 291, 298

Data analysis 287–94, 305–13
  computer software 305–6, 308–10
DDE, see DDT
DDT 72–3, 89–94, **97**, 270
  effects on birds of prey 16–17, 89–93, **91**, 322
  effects on egg-shell thickness 91, **91**
  feminization due to 92
  in freshwater birds 189
  illicit use of 197
  legal use of 197
  levels in eggs **96**, 97, **97**
Deer, white-tailed 170
Deforestation 58–60, 63–4
Delichon urbica 129
Demographic parameters
  importance of monitoring 269–71
  speed of response of 270
Dendroica kirtlandii 33–4, 52, 325
Dieldrin, see Pesticides
Diet
  effect on caesium body burden 159
  mercury in 111
  of seabirds 220
    correlation with fisheries catches 245–6
  stable isotope ratios in 111
Diomedea exulans 32, 108
Dipper 7, 76, 127, 190–5, 200–9
Distribution
  of bird species 8–9, **277**
    correlation with environmental variables 276
    effect of drought on 47–8
    effect of global warming on 52–3
    techniques for monitoring 276
Diver 209
  black-throated 209
  great northern 31
  red-throated 31
Diversity
  of bird species 9
    in hedgerows 55
    in woodland 60–2
Doubly labelled water (DLW), to measure energy turnover 221–2
Drainage of land 66
Dromaius novaehollandiae 9
Drought 34, 46–8
Dryocopus martius 30

Index

Duck
  black 95
  ring-necked **163**
  ruddy **163**
  wood 165
Dunlin 8, 124
Dunnock 55
Dutch Breeding Bird Monitoring Programme 293

Eagle
  bald 17, 71, 95
  golden 66, 108
  wedge-tailed 70
EC50 14
Eggs
  cadmium in 117
  effect of acidification 201, *202*
  heavy metals in 105, 111–14
    see also individual metals
  influence of prey available on 235
  lead in 120
  mercury in 111–4
    effect of diet on 113
    effect of laying sequence on 112–3
  PCBs in 190–2, **191**, **192**
  pesticides in 94–5, *95*, **96**, 97, 188, 193–5
  selenium in 111–3
  within clutch variation 193
Egg-shell thickness **90**
  and acidification 76, 127
  effect of pesticides on 17, **19**, **91**
  as a monitor for DDT 20
El Nino 62, 231
Elaphe obsoleta 165
Emberiza citrinella 55, 280, 324
E. schoeniclus 324
Emigration 314–5
Emu 9
Energy budget 229
Energy expenditure 32, 222, 229
Engraulis capensis 233
E. mordax 220
Environmental change, time-scale 43–4
Environmental Change Network 272

Environmental stress, indices of 13–14
Enzyme measurements, as pollution indicators 188
Etheostoma sp. 171
Euphausia superba 12, 220

Falco columbarius 59, 60, 295
F. peregrinus 89–91, 270, **271**, 322
F. rusticolus 74
F. tinnunculus 57
Falcon, peregrine 89–91, 270, **271**, 322
Farming 54–7, 66, 322–3
Fat reserves
  organochlorines in 93–4
  PCBs in 101
  in starvation 93–4
Feathers
  cadmium in 115, **116**
  lead in 119
  mercury in 106, 107–11, **110**
  for monitoring heavy metals 106
Fecundity 296
Felis catus 69, 70
Feminization, as a result of DDT 92
Ficedula hypoleuca 327
Field metabolic rates (FMR) 222
Fieldfare 3, 55
Finschia novaeseelandiae 61
Fir, Douglas 61
Fish
  abundance assessment 223–5
  age distributions 243
  in seabird diets 220
Fisheries 4, 62–3, 218–20
  effect on seabirds 231–4
  seabirds as indicators of 244–50, **244**, **245**
Flapping 32, **33**
Fledging, effect of prey availability on 238–9
Florida caerulea 157
Flycatcher
  pied 327
  spotted 289
Foraging effort 244
  effect on mortality 240
  flexibility of 239–44

348 Index

Foraging methods
  effect on prey availability 62–3
  flexibility of 239–40
  influence on response to
    oceanographic variation 226
  monitoring of 229
  of seabirds 227, 247–50, *248*, **249**
  variation in reproductive success
    with 231
Foraging range
  of seabirds 222
    breeding birds 240
    non-breeding birds 240–1
Forest, *see* Woodland
Forest Health, monitoring of 22
Forestry 57–62
  effect on water quality 183
Fox, red 70
Fratercula arctica 12, 109, **110**, 233,
    246, 247, 252
F. cirrhata 62, 245
Fringilla coelebs 47, 324
F. montifringilla 47
Fulica americana 161–6, 168–9
Fulmar 25, 32, **33**, 75, 250
Fulmarus glacialis 25, 32, **33**, 75, 250

Gadus morhua 220
Gallinago chlorops **163**
G. gallinago 56
Gallinule, common **163**
Gallirallus australis 69
Gamma radiation 148
Gannet **19**, 95, 111, 112, 114, 121,
    220, 243, 247, 252
  cape 233, 241, 243
Gavia spp. 209
G. arctica 209
G. immer 31
G. stellata 31
Geese
  barnacle 297
  brent 300
  grey 98
  greylag 3
  lesser snow 113
Geographic Information Systems
    9–11
Global arming 34, 45, 48–53
Glyceria maxima 66

Goat 70
Godwit 281
  bar-tailed 103
  black-tailed 66
Goldcrest 58, 65, 289
Goldeneye 209
Goosander 209
Goshawk 115, **116**
Grebe 187
  great-crested 189, 190
  horned (Slavonian) **163**
  pied-billed **163**
Greenfinch 324
Greenhouse effect, *see* Global
    warming
Greenhouse gases **49**
  see also Carbon dioxide level;
    Global warming
Grosbeak, pine 30
Grouse
  black 295
  ruffed 115
Grousemoor 322–3
Grus grus 3
Guano birds 233
Guillemot 9, **97**, 108, 109, 114, 229,
    238, 239, 244, 247, 251–2
  black 108, 109, 252
Gull 187, 250
  Audouin's 114
  black-headed 126
  glaucous-winged 62
  herring 95, 109, 112, 113, 114, 188
  laughing 117
  red-billed 103
  yellow-legged herring 112
Gyrfalcon 74

Habitat alteration 63–8
Habitat assessment 8–9, 29–31
Habitat characteristics 320–1
Habitat edge 66–8
Habitat fragmentation 63–8
Habitat heterogeneity 59
Habitat 'islands' 66–8
  genetic effects 67
Haematopus ostralegus 103, 124
Half-life of radioactive
    contaminants 147, 149
Haliaeetus leucophalus 17, 71, 95
Harrier, hen 295

Heather 322
Heathland habitat 64, 65
Heavy metals 22, 73–4, 102–20
　in blood 105
　in chicks 106–7
　effect of age on levels of 103–4, **104**
　in eggs 105
　in feathers 106
　in freshwater birds 195–6
　relative levels in males and females 103–4
　sampling problems 103–7
　tissue distribution 104
　see also under individual metals
Hedgerows 55
HEOD
　in birds of prey 91
　in freshwater birds 189
Heptachlor, see Pesticides
Heron 187, 281
　grey 99, 103, 189, 190, 311
　little blue 157
　night 34
　purple 34, 311
　squacco 34
Herring 26, 63, 111, 220, 246
Hirundo rustica 31, 34
Hydrocarbons 22

Illex illecebrosus 243
Immigration 314–15
Imputing 291
Indicator species 186, 198–9, 208–9, **208**
　bird communities as 209
　plankton as 218
　selection of 15, 17
　see also Sentinel species
Insect abundance 129
Insecticides, see Organochlorines
Integrated population monitoring 315–28
Integrated Population Monitoring Programme (BTO) 24, 315–16
Introduced species 68–70
　effects on New Zealand avifauna 69
　as food items 70
　as pests 69–70

Iodine *147*, 171
Island biogeography theory 67

Jay
　blue 172
　Siberian 30
Jolly-Seber model 302–3
Juvenile birds, counting of 298–300

Kestrel 57
Key factor analysis 317–20
　for european passerines *319*
kittiwake 25, 26, 62–3, 231, 240, 247, 250, 251
Kingbird, eastern 157
Kingfisher 99, 187, 189, 190
Kiwi, brown 61
Knot 281
Krill 12, 220

Lakes
　calcium levels in 185
　pollution of 182–5
Land management 321
　effect on bird populations 55–7, 58–62
　see also Farming; Forestry
Landsat 8–9
　accuracy of 9
Lapwing 56, 124, 294, **295**, 311, 323
Larch, japanese (hybrid) 58, 60
Larix kaempferi × *eurolepsis* 58
Larus spp. 250
L. argentatus 95, 109, 112, 113, 114, 188
L. atricilla 117
L. audouini 114
L. glaucescens 62
L. michahellis 112
L. novaehollandiae scopulinus 103
L. ridibundus 126
Laying date, effect of acidification 201, 202
LC50 13–14, 73
Lead 43, 73, 118–20, 195
　atmospheric 119
　from angling weights and shotguns 74
　in eggs 120

in feathers 119
ingestion of 118
Limosa spp. 281
L. lapponica 103
L. limosa 66
Lindane 197
Line transect censusing 17, 31, 282–3, 286
Linnet 324
Logging 61
Long-term studies
  BTO Common Bird Census (CBC) 17
  BTO Constant Effort Sites Scheme 24
  BTO Nest Record Scheme 24
  BTO Ringing Scheme 24
  BTO Waterways Bird Survey 23
  Continuous Plankton Recorder Survey 2
  line transect censusing 17
  migrant trapping 17
  precipitation data 2
  Seabird Colonies Register 23
Loon, common 31
Lophodytes cucullatus 209
Loxia curvirostra 58
Lutra lutra 195

Mackerel 111, 220, 243
Magpie 115, **118**, 323
Mallard 95, 113, 145, 160–1, 167
Mallotus villosus 63, 220, 233, 247
Mangifera indica 70
Marking of individuals 301–13
Martin
  house 129
  sand 313
Mean annual temperature 46, 50, 63
Meleagris gallapavo 169
Mercury 73, 107–11
  age effect 103–4, **104**
  in agriculture 280
  atmospheric deposition of 108
  in birds 73–4, 108–11, 189
  in diet 111
  in eggs 111–15
    effect of diet on 13
    effect of laying sequence on 112–13
  in feathers 106, 107–11, **110**
  geographical variation 114
  historical trends 108–11
  methylmercury 73, 103, 108
  relative levels in males and females 103–4
Merganser
  common 209
  hooded 209
Mergus merganser 209
Merlin 59, 60, 295
Migratory birds
  counting 279–81
  effect of drought on 48
  as transporters of radionuclides 160–6
  transequatorial migrants 34–5
miliaria calandra 57, 324
Mist-netting 26, 280, 285
Models
  bioenergetic *221*, 221–3
  climatic 50–2
  population 315–17
Mohoua albicilla 61
Moluthrus ater 325
Monitor selection 12
  for radionuclide contamination 169–72
Monitoring
  abundance 276–94
  bias in 273–5, 286
  choice of parameters 268–71
  choice of study areas 272–5
  definition 5–7
  distribution 276
  juvenile birds 298–300
  non-biological 11–12
    continuous v. periodic sampling 11
  reproduction 294–301
  survival 301–14
Monitoring programmes
  in Asia 22
  in Austria 280
  in Baltic 280
  in Britain 17, 23, 24, *130*, 273, 282, 284, 291, 294, 304, 315–7
  in Canada *130*
  factor-monitoring 6

in Finland 17
in Germany 17, 23, *130*, 280
National Status and Trends Program 22
in The Netherlands *130*, 293
in Norway *130*
reliability of 20–1
sensitivity of 20–1
in Sweden *130*
in Switzerland 23
target-monitoring 6, 14
in USA 17, 22, 73, *130*, 273, 275, 281, 283, 291
World Climate Impact Programme 22
see also under individual programmes
Monte Carlo methods
for estimating confidence limits 292–3
Mortality
of fish
estimates 223, 246
in population models 317–20
of seabirds
due to oil 128–9
effect of foraging effort on 240
Motacilla cinerea 190–2, 204
Moult cycle, effect on mercury levels of 107–8
Mountford method 289, **290**
Murrelets, ancient 9
Muscicapa striata 289
Muscle, caesium levels in 149, *150*, 170
Museum collections 28, 91, 106
contamination of 109
Mussel 22
Mynah, indian 69
Mysis relicta 51

National Rivers Authority 210
Nereis diversicolor 124, 125
Nest-based studies for monitoring reproduction 294–6
Nest boxes 297
Nest failure 296, 298, 323, *323*
Nest Record Scheme (BTO) 24, 294, 323

Nesting attempts 297–8
Nitrogen oxides
and acidification 75
effect on the ozone layer 53–4
North American Breeding Bird Survey (BBS) 273, 283, 288, 291
North American Christmas Bird Counts 275
Numenius arquata 124
Nuthatch 59
Nutrition
effect on body condition 234–5
effect on clutch size 235
effect on egg production 235
effect on reproductive success 238–9
Nycticorax nycticorax 34

Oak 58
Oceanites oceanicus 75
Oceanographic activity, effect on seabirds 230–1
Odocoileus virginianus 170
Oenanthe oenanthe 65
Oil
degradation of 71
impact on birds 71
pollution 70–2, 128–9
Operation Baltic 280
Operation Recovery 280
Organic nutrients, as pollutants 122–5, 207
Organochlorines 87, 89–98
effect on birds of prey 16–17, 322
in eggs 188, 193–5, *193*
legal use of 197–8
transport of 97
see also DDE, DDT, PCB
Organophosphates 98–9
Oryctolagus cuniculus 70
Osprey 17, 76, 187
Otter 195
Ousel, ring 66
Overfishing 4, 231
effect on seabirds 227, 231–3, 232
see also Fisheries

Owl
 barn 69, 100
 tawny 269, 270
Oyster 22
Oystercatcher 103, 124
Ozone layer 53–4
 Antarctic 53–4
 Europe 54
 North America 54

Palaeomonetes sp. 171
Pandion haliaetus 17, 76, 187
Parrot, Seychelles black 70
Partridge 74, 99, 289
Parus spp. 274
P. ater 58
P. caeruleus 300, 304
P. cinctus 30, 47
P. cristatus 47, 58, 59
P. major 300
PCBs, see Polychlorinated biphenyls
Pelecanus occidentalis 220
Pelican 187, 233, 234
 brown 220
Penguin 220, 247
 Adelie 52
Perdix perdix 74, 99, 289
Perisoreus infaustus 30
Pesticides 89–100, 109
 aldrin 72, 89
 DDE, see DDE
 DDT, see DDT
 dieldrin 89, 92–3, 97, 197, 270
 heptachlor 72, 89
 measurement of levels in birds 93–4, 99
 measurement of levels in eggs 94–5, 188, 193–5
 organochlorines, see Organochlorines
 see also DDE; DDT
 organophosphates 72, 98–9
 rodenticides 100
 see also under individual compounds
Petroica australias 61
P. macrocephala 61
Phalocrocorax aristotelis 97, 121, 245

P. auritus 230
P. carbo 126
Phasianus colchicus 73
Pheasant 73
Phoenicurus phoenicurus 30
Phylloscopus trochilus 65, 289, 304, 305, 313, 327
Pica pica 115, **118**, 323
Picea abies 47, 58
P. engelmannii 45
P. sitchensis 58
Pigeon 74, 119
 wood 280
Pilchard 233, 243
Pine, 60
 corsican 59
 jack 34, 52
 lodgepole 58
 Monterey 61
 scot's 58
Pinicola enucleator 30
Pinus banksiana 34, 52
P. caribaea 62
P. contorta 58
P. nigra 59
P. radiata 61
P. sylvestris 58
Pipit
 meadow 65
 tree 59
Plankton, as indicator species 218
Plantations 59–60, 65
Plastic pollution 74–5, 120–2
 effect on birds 75
 entanglement in 120–1
 ingestion of 121–2
Ploughing, timing of 56
Plover, golden 322, 323, *323*
Plutonium *147*, 148, 149–52
 as an alpha emitter 149
 in birds 126
 concentration *150, 151*
 uptake of 149, 151–2
Pluvialis apricaria 322, 323, *323*
Podiceps auritus **163**
P. cristatus 189, 190
Podilymbus podiceps **163**
Point counts 283–4, 286–7
Pollachius virens 245
Pollock, walleye 245

Pollution
  air 129–31
  definition 183
  heavy metals 102–20
  in lakes 182–5
  oil 70–2
  organic nutrients 122–5
  PCBs 100–2
  pesticides 89–100
  plastic 120–2
  in rivers 182–5
  sewage 124–5
  see also under individual pollutants
Pollution stress 13–15
  indices of 14–15
Polychlorinated biphenyls **97**, 100–2
  atmospheric 198
  in bird tissues 101, 189
  effect on birds of prey 17
  in eggs 190–2, **191**, *192*
  reduction in use of 101, 196
  toxicity of 101, 196–7
Population regulation 229–30
Population size
  of birds
    effect of acidification 127–8, 200–1
    effect of DDT on 92
    effect of dieldrin on 92
    effects of nutrients on 124–5, 207
    effect of overfishing on 231–4, *232*
    fluctuations in 217
    heathland v. plantation 65
    in woodland 60–2
  of fish 223–4
Prairies 64
Precipitation, see Rainfall
Predator-prey interactions 224–9
Prey abundance, assessment of 223–5, 234
Prey availability 62–3, 246
  effect of acidification on 202–3
  seabirds as monitors of 218, 220–3, 225–50
Prey distribution 222–3, 246

Prey selection, generalist v. specialist 228
Prey switching 63, 228
Prunella modularis 55
Pseudotsuga menziesii 61
Puffin 251
  atlantic 12, 109, **110**, 233, 246, 247
  tufted 62, 245
Puffinus puffinus 109
Pygoscelis adeliae 52
Pyrrhocorax pyrrhocorax 57

Quail 112
  bobwhite 169, 172
Quercus robur 58
Q. petrae 58

Rabbit 70
Radio-tagging 302
Radiocaesium, see Caesium
Radionuclides 125–7
  characteristics of 146–52, *147*
  in freshwater birds 195–6
  monitor selection 169–72
  transport by birds 160–6
    at Chernobyl 167
  see also under individual elements
Rail, water 66
Rainfall
  correlation with bird productivity 18–19, 34
  variation in 46–8
Rallus aquaticus 66
Raptors
  counting of 279
  monitoring of reproduction in 296–7
  see also under individual species
Rat 69
Rat snakes 165
Rattus spp. 69
Raven 66
Razorbill 252
Recapture rate 285, 306
Recovery data 307–13, *310*
  see also Bird ringing
Recruitment indices 245–6
Red tides 97
Redpoll 59

Redshank 56, 124
Redstart *30*
Redwing 55, **277**
Reedgrass 66
Regulus regulus 58, 65, 289
Remote sensing 7–9
Reporting rates, of bird recoveries 308, 311–12
Reproduction
  effect on pesticides on 17
  techniques for monitoring 294–301
Reproductive success 228, **271**
  effect of farming on 323
  effect of overfishing 231–4, *232*
  effect of prey availability on 238–9
  effect of sea temperaature 231
  monitoring of 294–301
  variation with foraging method 231, *248*
Resighting information 304
Ringing Schemes
  BTO 24
  see also Constant Effort Sites Scheme
Riparia riparia 313
Rissa tridactyla 25, 26, 62–3, 231, 240, 247, 250, 251
Rivers
  calcium levels in **184**, 286
  pollution of 182–5
Robin, New Zealand 61
Rodenticides 100
Rook 31, 56
Route regression analysis (RRA) 291–2

Sahel 34, 46–8, 269, **318**
Saithe 245
Salmo salar 246
S. trutta 204, **205**, 206
Salmon, atlantic 246
Salvelinus namaycush 51
Sandeel 63, 111, 114, 220, 233, 245
Sardine 220
Sardinops ocellata 233, 243
Satellite imagery 7–9
Saury
  atlantic 247

  pacific 247
Savannah River Site 154–60
Saxicola rubetra 65
S. torquata 65
Scaup, lesser **163**
Scientific Committee on the Problems of the Environment 21
Scomber scomberesox 247
S. scombrus 111, 220, 243
Sea level, implications of global warming for 51
Sea temperature, effect on seabirds 230–1
Seabird Colonies Register 23
Seabird foraging 227
  comparison of 247–50, *248*, **249**
Seabirds
  as fisheries indicators 244–50
  monitoring of 225–9, 281
  prey availability 229–34
  responses to prey availability 234–44
  role in food web 220–3
  see also under individual species
Selenium, in eggs 111–13
Sellafield 126
Sentinel species
  chicken as *150*, *151*, 151–2
  for monitoring pollutants 88–9
  for monitoring radionuclides 168–9
  see also Indicator species
Sequential population analysis 223–4
Sewage 124–5
Shag 97, 121, 245
Shearwater
  Cory's 114
  Manx 109
Sheep 66, 70, **194**
Sheep dip 193
  dieldrin in 89
Shorebirds, *see* Waders
Shrimp 171
  opossum 51
Sialis sialia 69
Siskin 58
Site fidelity, by coots 164
Sitta europaea 59

Skua, great 13, 103, **104**, 109, 240
Skylark 56, 65, 289
Snipe 56
Soil
  erosion 54–5
  implications of global warming for 51
Sowing, timing of 56
Sparrowhawk 59, 60, 193
Species distribution, *see* Distribution
Species diversity, *see* Diversity
Spot mapping, *see* Territory mapping
Sprat 114, 246
Sprattus sprattus 114, 246
Spruce 47, 60
  Engelmann 45
  Norway 58
  sitka 58
Squid 247
  short-finned 243
Stable isotope ratios 111
Starling, european 69, 95
Sterna spp. 246, 250, 251
S. *elegans* 234
S. *forsteri* 101
S. *hirundo* 95, 114, 120, **163**
S. *maxima* 112
S. *paradisea* 234, 245
Stonechat 65
Stork, white 34
Storm petrel 250
  Wilson's 75
Strix aluco 269, 270
Strontium 147, 148
  as a calcium analogue 146, 148
  concentration *150*, *151*
  uptake of 151–2
Stubble burning 56
Study site selection 272–5
Sturnus vulgaris 69, 95
Sula bassana **19**, 95, 111, 112, 114, 121, 220, 243, 247, 251
S. *capensis* 233, 241, 243
S. *nebouxii* 231, 238
Sulphur dioxide, and acidification 75
Surrogate species 169
  chickens as *150*, *151*, 151–2

Survival
  effect of foraging on 62–3
  estimation of 285, 301–14, *310*
  effect of immigration and emigration on 314–5
  passerines *310*
  of juveniles 298–300
  monitoring of 301–14
  sex differences in 304
Swallow 31, 34
Swan 66
  Bewick's 173
  mute 74, 118, 195, 281, 291, 298
  whooper 299
Swamp habitat 64
Swift 31
Sylvia atricapilla 280, 327
S. *communis* 17, 34, 55, 65, 280, 320, 327
S. *undata* 64
Synthliboramphus antiquus 9

Teak 62
Tectonia grandis 62
Telemetry 312
Tern 3–4, 187, 240, 246, 250, 251
  arctic 234
  black **163**
  common 95, **163**, 114, 120
  elegant 234
  Forster's 101
  royal 112
Territory mapping 26, 275, 278, 282, 286–7
Territory requirements 60
Territory size
  in dippers
    effect of acidification on 200–1
Tetra-ethyl lead 43
Tetrao tetrix 295
T. *urogallus* 30, 58
Theragra chalcogramma 245
Thrush
  mistle 30
  song 167, 300, 315, **316**, 325
Tit 274
  blue 300, 304
  crested 47, 58, 59
  great 300

pied 61
siberian *30*, 47
Treecreeper 59
Tringa totanus 56, 124
Trogolodytes trogolodytes **23**, 65, 280
Trout
  brown 204, **205**, 206
  lake 51
Tuna 3–4, 247
Turdus iliacus 55, **277**
T. merula 65, 300
T. philomelos 167, 300, 315, **316**, 325
T. pilaris 3, 55
T. torquatus 66
T. viscivorous 30
Turkey 169
Twite 66
Tyrannus tyrannus 157
Tyto alba 69, 100

Uria aalge 9, 108, 109, 114, 229, 238, 239, 244, 247, 251–2

Vanellus vanellus 56, 124, 294, **295**, 311, 323
Vegetation distribution 320–1
Vegetation mapping 8–9
  accuracy of 9
Virtual population analysis (VPA) 246
Volunteer-based surveys 273–5, 287
Vulpes vulpes 70

Wadden Sea, pollution in 124
Waders
  effect of sewage on 124–5
  timing of monitoring 278–9, 281, 300
  see also under individual species
Wagtail, grey 190–2, 204
Warbler
  Dartford 64
  reed 304, 305, 313
  sedge 34, 48, 269, **290**, 305, **306**, 307, 317, **318**, 320, 325, 327
  Kirtland's 33–4, 52, 325
  willow 65, 289, 304, 305, 313, 327
Warfarin 100
Water quality 15–16, 182–5, 187–96, 198–210
  standards 209
Water temperature 51
Waterways Bird Survey (BTO) 23, 282
Waxwing *30*
Weather data 321
Weather forecasting 3, 31–3
Weighting 292
Weka 69
Wheatear 65
Whinchat 65
Whitehead 61
Whitethroat 17, 34, 55, 65, 280, 320, 327
Wigeon 66
Wildfowl
  timing of monitoring 278, 281
  see also under individual species
Wind speed 32–3
Woodland
  in Australia 60–1
  bird communities in 58–62
  in Britain 58–60
  composition 58, 61–2
  distribution 58, 59, 61
  habitat heterogeneity 59
  in New Zealand 61
Woodpecker
  black *30*
  three-toed *30*
Woodpigeon 56
Wren 23, 280

X-radiation 148

Yellowhammer 55, 280, 324